Solar Energy Application in Buildings

CONTRIBUTORS

MEHDI N. BAHADORI

ERTUGRUL BILGEN

W. W. S. CHARTERS

ALBERT G. H. DIETZ

WILLIAM J. D. ESCHER

H. GEHRKE

M. C. GUPTA

D. HARIHARAN

M. IQBAL

J. KEABLE

KEN-ICHI KIMURA

T. A. LAWAND

M. A. S. MALIK

JACQUES MICHEL

J. K. PAGE

B. SAULNIER

A. A. M. SAYIGH

J. STEPHENSON

ROBERT K. SWARTMAN

T. NEJAT VEZIROGLU

J. RICHARD WILLIAMS

Solar Energy Application in Buildings

Edited by *A. A. M. SAYIGH*
MECHANICAL ENGINEERING DEPARTMENT
COLLEGE OF ENGINEERING
UNIVERSITY OF RIYADH
RIYADH, SAUDI ARABIA

ACADEMIC PRESS New York San Francisco London 1979
A Subsidiary of Harcourt Brace Jovanovich, Publishers

ACADEMIC PRESS, INC.
111 Fifth Avenue, New York, New York 10003

United Kingdom Edition published by
ACADEMIC PRESS, INC. (LONDON) LTD.
24/28 Oval Road, London NW1 7DX

Library of Congress Cataloging in Publication Data

Sayigh, A A M
Solar energy application in buildings.

Includes bibliographical references.
1. Solar energy. 2. Solar heading. 3. Solar air conditioning. I. Title.
TJ810.S29 Date 697'.78 78–67882
ISBN 0–12–620860–3

PRINTED IN THE UNITED STATES OF AMERICA

79 80 81 82 9 8 7 6 5 4 3 2 1

Contents

7 Passive Cooling of Buildings

A. A. M. SAYIGH

8 Solar Cooling for Buildings

ROBERT K. SWARTMAN

9 Natural Cooling in Hot Arid Regions

MEHDI N. BAHADORI

10 Hydronic Solar Heating and Cooling in Georgia

J. RICHARD WILLIAMS

11 Some Solar-Heated Buildings in Canada

T. A. LAWAND AND B. SAULNIER

12 Solar House Heating with Heat Pipe Collectors

H. GEHRKE

13 Solar Houses in Japan

KEN-ICHI KIMURA

14 A Low-Energy House in New Zealand

J. STEPHENSON

List of Contributors

Numbers in parentheses indicate the pages on which the authors' contributions begin.

MEHDI N. BAHADORI (81, 195), Department of Mechanical Engineering, School of Engineering, Pahlavi University, Shiraz, Iran

ERTUGRUL BILGEN (389), École Polytechnique, Université de Montréal, Montreal, Quebec, Canada

W. W. S. CHARTERS (139), Department of Mechanical Engineering, University of Melbourne, Melbourne, Australia

ALBERT G. H. DIETZ (17), Department of Architecture, Massachusetts Institute of Technology, Cambridge, Massachusetts 02139

WILLIAM J. D. ESCHER (105), Escher Technology Associates, St. Johns, Michigan 48879

H. GEHRKE (279), Dornier System GmbH, Postfach 1360, D-7990 Friedrichshafen, Federal Republic of Germany

M. C. GUPTA (377), Solar Energy Division, Department of Mechanical Engineering, Indian Institute of Technology, Madras, India

D. HARIHARAN (377), Building Technology Division, Department of Civil Engineering, Indian Institute of Technology, Madras, India

M. IQBAL (371), Department of Mechanical Engineering, The University of British Columbia, Vancouver, British Columbia, Canada

J. KEABLE (327), Helix Multiprofessional Services, Mortimer Hill, Mortimer, Reading, England

KEN-ICHI KIMURA (287), Department of Architecture, Waseda University, 3-4-1 Okubo, Shinjuku, Tokyo 160, Japan

T. A. LAWAND (253), Brace Research Institue, MacDonald College, McGill University, Montreal, Quebec, Canada

M. A. S. MALIK* (345), University of Delaware, Newark, Delaware 19711

*Permanent address: Engineering Division, Kuwait Institute for Scientific Research, Kuwait, Kuwait.

JACQUES MICHEL (389), Paris, France

J. K. PAGE (41), Department of Building Science, University of Sheffield, Sheffield, United Kingdom

B. SAULNIER (253), Brace Research Institue, MacDonald College, McGill University, Montreal, Quebec, Canada

A. A. M. SAYIGH (1, 81, 147), Mechanical Engineering Department, College of Engineering, University of Riyadh, Riyadh, Saudi Arabia

J. STEPHENSON (317), Department of Mechanical Engineering, University of Auckland, Auckland, New Zealand

ROBERT K. SWARTMAN (171), Faculty of Engineering Science, The University of Western Ontario, London, Ontario, Canada

T. NEJAT VEZIROGLU (105), Clean Energy Research Institute, University of Miami, Coral Gables, Florida 33124

J. RICHARD WILLIAMS (227), College of Engineering, Georgia Institute of Technology, Atlanta, Georgia 30332

Preface

The energy potential of the Sun has been exploited in some way for heating, and of course as a source of light, since the dawn of civilization. Remains of dwellings from early settlements indicate an awareness of this potential in attempts to adapt to climatic conditions. In the recent technological era, materials and nonrenewable energy sources have been used lavishly as if supplies of both were unlimited. Present day building systems reflect this type of usage. This assumption is faulty, and events have led to an increasing interest in alternative energy sources. Solar energy is one such abundant, renewable resource.

Scientists and engineers are looking more closely at how nontechnological societies used solar energy and, in particular, at their use of solar energy in buildings. This book sets out to document successful use of the Sun's energy in various cultures, continents, and climates. The various chapters also analyze current plans to take advantage of the Sun's bounty.

The book consists of 19 chapters. Chapters 1–6 are devoted to the fundamentals: climate, storage of solar energy, and material properties. The concept of passive heating and cooling in buildings is discussed in Chapters 7–10. The remaining chapters deal with various applications of solar energy in old and modern buildings in the United States, Iran, Canada, Germany, Japan, New Zealand, Great Britain, India, and France. The book contains over 300 references.

The editor extends his thanks and appreciation to all of his colleagues, the Dean of the College of Engineering, and the Vice Rector for Research and Post Graduate Studies, University of Riyadh, Saudi Arabia, for their valuable comments and help in producing this book.

Nomenclature

a — shorter dimension of rectangular duct (ft)

A — area of duct cross section = ab (ft^2)

A_a — aperture area (m^2)

b — longer duct dimension (ft); collector tilt from horizontal

c_c — fraction of sky covered by cloud (dimensionless)

C_p — specific heat at constant pressure of dry air (Btu (lb °F)$^{-1}$)

$C_{p,a}$ — specific heat at constant pressure of moist air (Btu (lb °F)$^{-1}$)

d_h — hydraulic diameter of rectangular duct = $2ab/(a + b)$ (ft)

D_i, D_0 — diffuse radiation falling on a given surface with i tilt to horizontal and zero tilt, respectively

E_i, E_0 — total or global radiation at i and zero to horizontal

f — coefficient of friction, rough duct

f_0 — coefficient of friction, smooth duct

g — acceleration due to gravity (ft/s^2)

G_{bh} — direct solar beam irradiance on a horizontal surface (W/m^2)

G^*_{bn} — direct solar beam irradiance on a surface normal to the beam after transmission through a perfectly clear atmosphere with a stated content of precipitable water vapor, corrected to mean solar distance (W/m^2)

G_{bn} — direct solar beam irradiance on a surface normal to the solar beam (W/m^2)

G_{dh} — diffuse irradiance on a horizontal surface (W/m^2)

G_h — global irradiance on a horizontal surface (W/m^2)

G_{sc} — solar constant, or extraterrestrial direct beam solar irradiance on a surface normal to the beam at mean solar distance (W/m^2)

h — convective heat transfer coefficient, $m_a C_{p,a} \Delta t_B/(t_p - t_B)$ (Btu (h ft^2 °F)$^{-1}$); hour angle

h_{cw} — surface conductance for heat transfer from a surface due to forced convection by wind ($W/m^2/°C$)

$\bar{H}_{dh}$ — monthly mean daily diffuse irradiation on a horizontal surface ($MJ/m^2/day$)

H_h — global irradiation on a horizontal surface in a stated period (MJ/m^2/stated period)

$\bar{H}_h$ — monthly mean daily global irradiation on a horizontal surface ($MJ/m^2/day$)

H_i — direct solar radiation at slope i

H_n — normal component of solar radiation

H_0	direct component of solar radiation on a horizontal surface
$\bar{H}_{0h}$	monthly mean daily irradiation on a horizontal surface outside the earth's atmosphere (MJ/m²/day)
i	arbitrary angle of incidence
k	thermal conductivity of air (Btu (h ft °F)$^{-1}$)
L	duct length (ft); latitude
L_e, L_0, L_z	intake length (ft); radiation of sky at angle 0 and the zenith (W/m²/sr)
m	optical air mass (dimensionless)
m_a	rate of flow of dry air (lb/h)
Nu, Nu_{rough}	used interchangeably for Nusselt number of rough duct = hd_h/k
$\text{Nu}_{\text{smooth}}$	Nusselt number of smooth duct
p	pressure of air at distance x (feet of air); azimuth of a plane
P_h	atmospheric pressure at station height h (mb)
Pr	Prandtl number $C_p\mu/k$)
P_{sl}	standard atmospheric pressure at sea level (mb)
Re	Reynolds number = $[m_a(1 + w)/A](d_h/\mu)$
R_i	reflected radiation
t, t_a	temperature (°F), screen air temperature (°C)
T_a	air temperature (°K)
t_B	bulk air temperature (°F)
t_{in}	air temperature at inlet (°F)
t_{out}	air temperature at outlet (°F)
t_p, $\bar{T}_p$	plate temperature (°F); average plate temperature
U_L	heat loss coefficient (W/°C)
v	wind velocity over surface (m/s)
x	axial distance from entrance (ft)
z	zenith angle
α_1	angular height
α	wavelength exponent; solar altitude; absorptivity
α_g	solar albedo
β	Ångström coefficient
δ	declination angle of the sun
ϕ_{th}	incoming long-wave thermal flux on a horizontal surface (W/m²)
λ	wavelength (μm)
η	efficiency of collector = $\dfrac{m_a C_{p,a}(t_{out} - t_{in})}{(\text{incident radiation}) \times (\text{collector area})}$
μ	dynamic viscosity of air (lb (ft h)$^{-1}$)
τ	transmissivity
τ_a	Monteith and Unsworth turbidity coefficient (dimensionless)

1

The Solar Radiation Spectrum and Its Utilization

A. A. M. SAYIGH

MECHANICAL ENGINEERING DEPARTMENT
COLLEGE OF ENGINEERING
UNIVERSITY OF RIYADH
RIYADH, SAUDI ARABIA

1.1 INTRODUCTION

The earth receives solar energy at a rate of 5.4×10^{24} J/yr. This is equivalent to about 30,000 times the currently used sources of energy. Harnessing this power requires knowledge of the nature of solar insolation, the factors that influence its intensity, and the tools that utilize such energy. For example, in building applications the total solar radiation (global radiation) is the source of energy. The intensity of this source depends on whether the radiation is direct or diffuse, or both. As for the tools that utilize the sun's energy, the most efficient are flat plate collectors. For solar concentrators, direct solar radiation is the energy source. This chapter will not deal with the storage problem. The storage of solar energy is explained in Chapter 2.

1.2 THE SOLAR CONSTANT AND THE EXTRATERRESTRIAL SOLAR SPECTRUM

The most important parameter in the survey of solar radiation measurement is the energy received outside the earth's atmosphere. This parameter has a degree of constancy when compared to the energy received on the ground. Therefore it is called the *solar constant*. It is defined as the energy received from the sun per unit area exposed normally

ISBN 0-12-620860-3

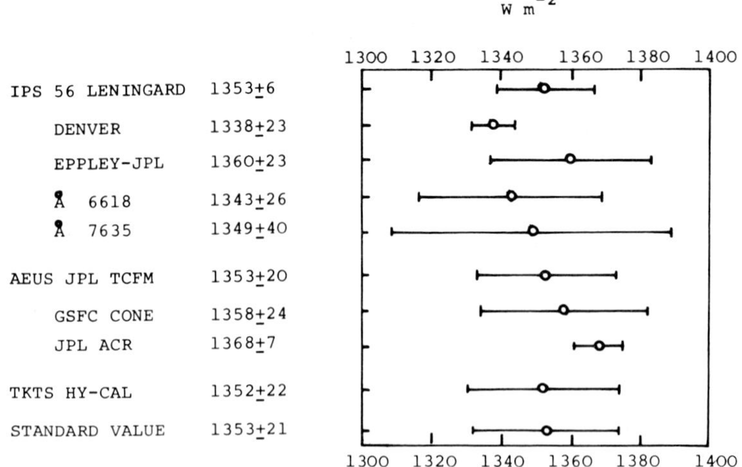

FIG. 1 Values of the solar constant derived from high-altitude measurements (units W/m^2).

to the sun's rays at the average sun–earth distance in the absence of the earth's atmosphere. The 1970 ISES Congress in Melbourne accepted the value of 1353 W/m^2 for the solar constant, with an error of about ± 1.5%. This value is an average value for nine long series of measurements, all made from high-altitude platforms (Convair 990, balloons, X-15 aircraft, and Mariner Mars probe) during the period 1967–1970. Figure 1 shows these measurements (Thekaekara, 1975). A more recent suggestion states

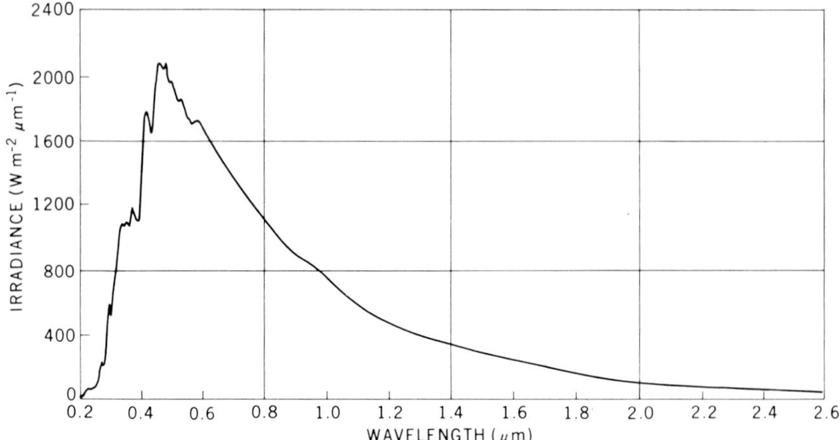

FIG. 2 Solar spectrum irradiance, standard curve, solar constant 1353 W/m^2.

TABLE 1

Recent Values of the Solar Constant

Author	Platform	Measurement date	Solar constant reference scale (SCRS) (W/m^2)	Original value (W/m^2)
Kondratyev and Nikolsky (1973)	Balloon	1962/1967	1376 + 18	1356
Drummond *et al.* (1968)	Aircraft: B-578	1966/1967	1387 ± 17	1359
Thekaekara *et al.* (1969)	Aircraft: CV 990	1967	1377 ± 40	1349
		1967	1372 ± 24	1364
		1967	1375 + 30	1343
Hickey (1976)	Aircraft: X-15	1967	1385 ± 14	1361
Drummond (1973)	Aircraft: CV 900	1967/1968	1387 ± 19	1359
Murcray (1969)	Balloon	1967/1968	1373 ± 12	1339
Kendall (1973)	Aircraft: CV 900	1968	1373 ± 14	1370
Plamondon (1969)	Mariner VI and VII	1969	1362 ± 18	1352
Willson and Stallkamp (1971)	Balloon	1969	1369 ± 11	1366
Hikey (1976)	Nimbus 6	1975	1369 ± 14	1390
	Aerobee rocket	1976	1367 ± 7	—

the value of the solar constant to be between 1368 and 1377 W/m^2 as deduced from Table 1 (Frohlich, 1976). The extraterrestrial solar spectrum is shown in Fig. 2. The spectrum consists of the ultraviolet region, 0.115–0.455 μm [the percentage of energy collected in this region is 15.891% (Sayigh, 1977a)], the visible region, 0.455–0.750 μm (the percentage of energy collected in this region is 35.800%), and the infrared region, 0.750–1000 μm (the percentage of energy collected in this region is 48.301%). It is worth mentioning here that the percentage of energy in the infrared region, 0.750–4.000 μm, is 47.367%.

1.3 TOTAL AND SPECTRAL SOLAR IRRADIANCE AT GROUND LEVEL

The variability of solar energy incident on a collector surface on the ground is considerably greater than that of the extraterrestrial solar energy. On a clear sunny day the energy increases from zero at sunrise to a maximum at solar noon, and decreases to zero at sunset. At any moment clouds may intercept the rays of the sun and decrease the energy to a low value because of the diffuse radiation. Figures 3 and 4 show such varia-

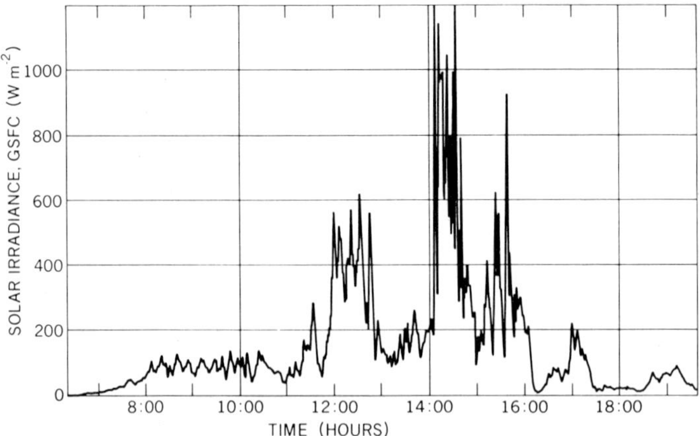

FIG. 3 Global irradiance due to the sun and sky on a horizontal surface, measured at GSFC on May 13, 1971. Total energy received during the day, 175 cal/cm² (732 J/cm²).

tions for May 13 and 14, 1971. The two figures are based on measurements made on two successive days as part of the GSFC international comparison of working standard pyranometers. The instrument was an Eppley pyranometer, model 2, mounted on a rooftop. The readings were taken every 4 s. On May 13 it was overcast during most of the morning. A short interval of sunshine in the afternoon was followed by a heavy cloudburst at 18:00 hour. The next day was one of relatively clear sunshine with a few passing clouds. The high value of solar irradiance was near 1200

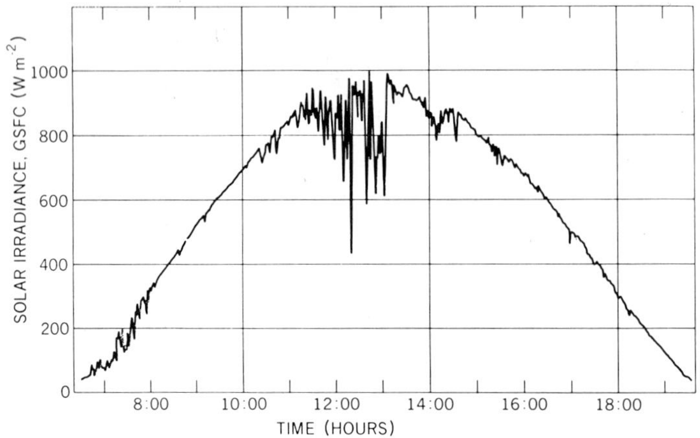

FIG. 4 Global irradiance due to the sun and sky on a horizontal surface, measured at GSFC on May 14, 1971.

W/m^2 after 14:00 hour on May 13, and this value is 30% higher than on a clear day for the given solar elevation; this is obviously due to reflection from the clouds. Such abnormally high values are of short duration.

Attenuation of solar energy depends on the Rayleigh attenuation coefficient C_1 and the ozone attenuation coefficient C_2. Both factors are based on the data developed by Elterman (1968). They are valid for US standard atmosphere (see Sayigh, 1977a). For atmospheric turbidity, Ångström has developed the equation

$$C_3 = \beta/\lambda^{\alpha} \tag{1}$$

where β is Ångström coefficient, α the wavelength exponent, and λ the wavelength (μm). This equation permits a greater flexibility in choosing α and β parameters corresponding to different levels of atmospheric pollution. In the infrared a fourth parameter is required to account for the molecular absorption bands (Gates and Harrop, 1963).

It is now important to define a common term, *air mass:* Air mass is the ratio of the path length of the radiation through the atmosphere at any given angle to the sea level path straight through the atmosphere (vertically). Air mass zero refers to the absence of atmospheric attenuation at 1 astronomical unit from the sun.

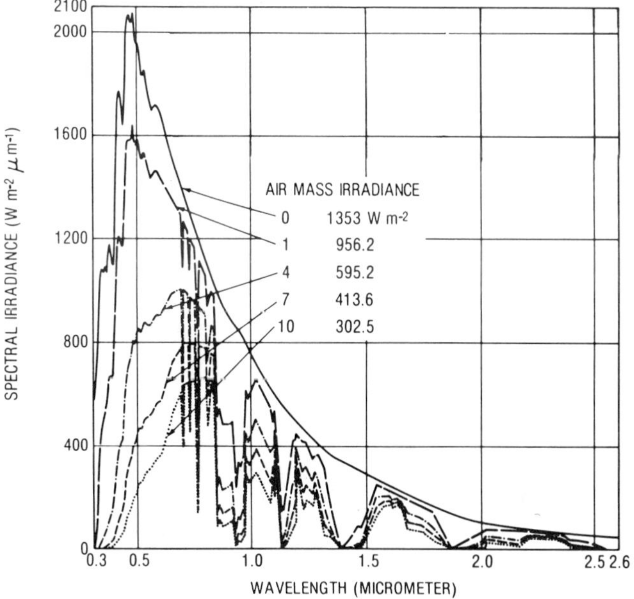

FIG. 5 Solar irradiance for different air mass values, US standard atmosphere; H_2O = 20 mm, O_3 = 3.4 mm, $\alpha = 1.3$, $\beta = 0.02$.

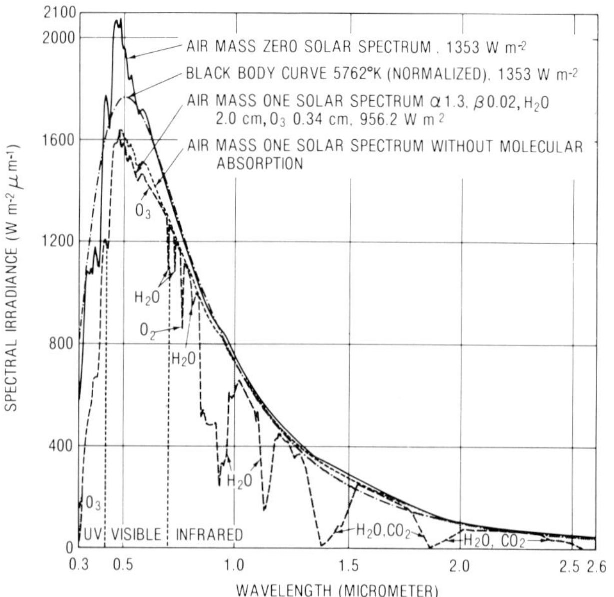

FIG. 6 Four curves related to solar spectral irradiance.

For simplicity it can be assumed that the zenith angle is small, and therefore the air mass m can be expressed as

$$m = \sec Z \tag{2}$$

where Z is the zenith angle. For large values of Z, account has to be taken of the curvature of the solar ray due to refraction in increasingly denser atmospheric layers. Figure 5 shows the effect of various air masses on solar irradiance for US standard atmospheric turbidity; that is, H_2O = 20 mm, O_3 = 3.4 mm, $\alpha = 1.3$, and $\beta = 0.02$. The various air masses are air mass 0 for no atmosphere, air mass 1 corresponding to $Z = 0$, air mass 4 corresponding to $Z = 75.5°$, air mass 7 corresponding to $Z = 81.7°$, and air mass 10 corresponding to $Z = 84.3°$. Figure 6 shows four different curves related to solar spectrum irradiance.

The investigation of solar radiation on inclined surfaces and the predictions based on the horizontal component of the radiation, which is well established in most parts of the world, are of great importance in utilizing solar radiation effectively.

1.4 DIRECT SOLAR RADIATION

The problem that deals with direct solar radiation is well explained by various authors (Kondratyev, 1969; Zarem and Erway, 1963; Heywood

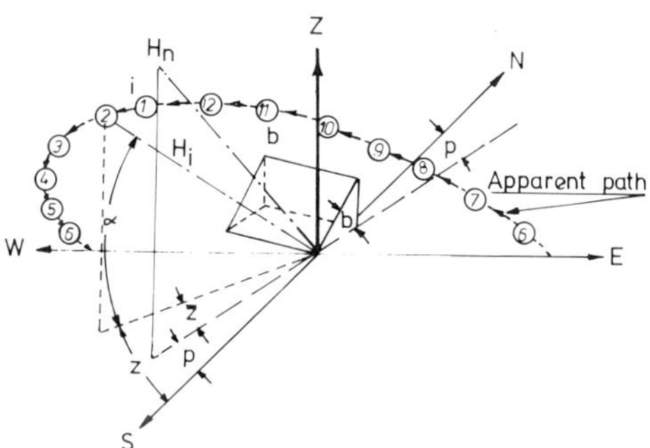

FIG. 7 Angle diagram for finding solar radiation on sloped surfaces.

1965). Using Fig. 7, the direct solar radiation flux H_i to an arbitrarily oriented inclined surface, where the inclination angle of the surface is b, can be expressed by the solar radiation component normal to the surface H_n using the expression (Sayigh, 1977c)

$$H_i = H_n \cos i \tag{3}$$

where

$$\cos i = \cos b \sin \alpha + \sin b \cos \alpha \cos \overline{Z} \tag{4}$$

where α is the solar altitude, $\overline{Z} = Z - p$, where p is the azimuth of the plane, and Z is the sun's azimuth from the south. α and Z can be determined from the equations

$$\sin \alpha = \sin L \sin \delta + \cos L \cos \delta \cos h \tag{5}$$

$$\cot Z = (\sin L - \cos L \tan \delta)/\sin h \tag{6}$$

$$\sin Z = (\cos \delta \sin h)/\cos \alpha \tag{7}$$

where L is the northern latitude of the place, δ the declination of the sun, and h the hour angle (zero at noon, positive after noon, and negative before noon). Therefore, using the previous equations, Eq. (3) can be written as

$$H_i/H_n = \sin b \cos \delta \cos(Z - p) + \cos b \sin \delta \tag{8}$$

For a plane on any north–south slope, in other words, truly facing south,

$$\cos i = \cos(L - b) \cos \delta \cos h + \sin(L - b) \sin \delta \tag{9}$$

and if $h = 0$, which occurs at noon, then

$$\cos i = H_i/H_n = \cos L \cos b \cos \delta + \sin L \sin b \cos \delta + \sin L \cos b \sin \delta - \sin b \cos L \sin \delta \tag{10}$$

TABLE 2

Declination Angle and Various Solar Radiation Values Falling on Different Sloped Surfaces in Riyadh (g · cal/cm²/day)

	Jan	Feb	Mar	Apr	May	June	Jul	Aug	Sept	Oct	Nov	Dec
	δ: −21	−13.5	−3.5	10.5	21	23	21.5	14.5	4	−9	−18	−23
E_0	389	468	536	575	622	669	672	642	619	544	429	388
E_n	560	598	610	594	624	669.5	673	653	663	656	587	580
E_{15}	480	548	593	594	613	652	659.5	651	660	597	518	486
E_{30}	538	591	610	572	561	591	602	615	655	654	572	551
E_{45}	560	594	585	512	472	490	504	539	606	644	587	579
E_{60}	560	598	575	469	402	411	428	482	576	649	587	580
E_{75}	493	481	420	293	204	196	213	281	390	495	498	516
E_{90}	403	372	291	149	44	24	41.5	120	238	367	401	430

For example, in the town of Riyadh, Saudi Arabia, $L = 25°$ N, and for a collector placed horizontally at noon during the middle of the month,

$$H_0/H_n = \sin L \sin \delta + \cos L \cos \delta \tag{11}$$

Table 2 shows the values of δ and the various components of solar radiation at various angles. The angles were zero, normal, 15, 30, 45, 60, and 90°, respectively. Figures 8 and 9 show these results in graph form (Sayigh, 1977c). Figure 10 shows a nomogram for calculating the length of the day that includes a curve for the declination angle (Sayigh, 1977d).

Equation (8) is the general expression of the dependence of incoming

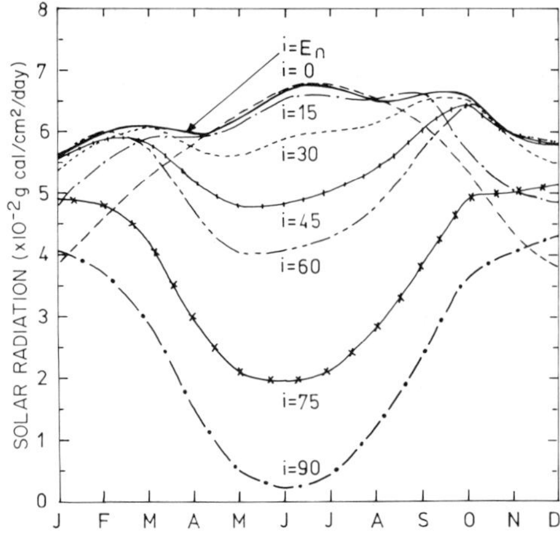

FIG. 8 Solar radiation falling on various sloped surfaces in Riyadh.

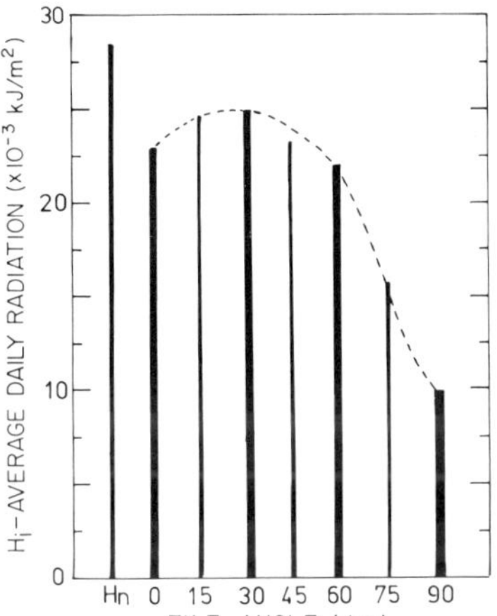

FIG. 9 Average daily solar radiation on sloped surfaces in Riyadh.

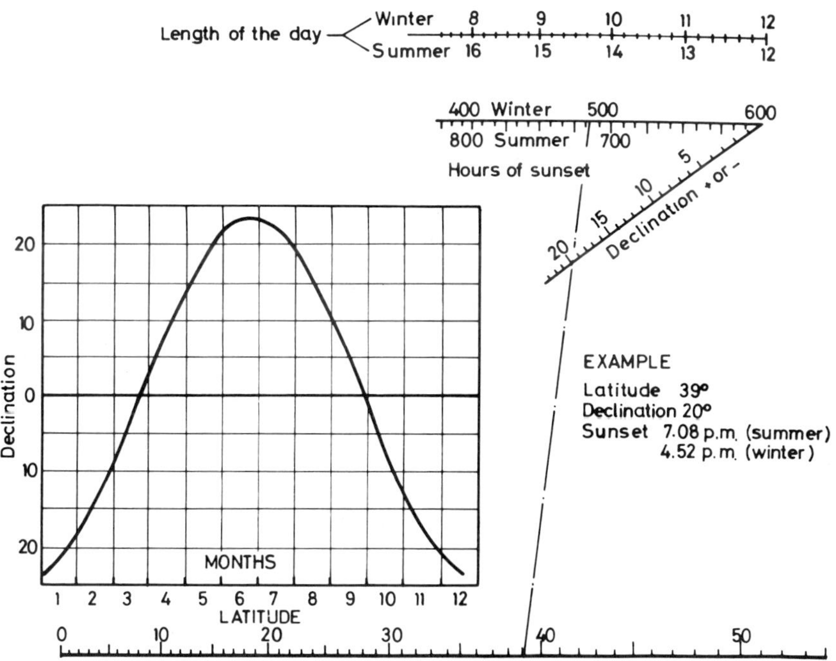

FIG. 10 A nomogram for calculating the length of the day (Z) in hours.

solar radiation falling on an inclined plane with orientation determined by angles b and p for any latitude L and at various moments of the day (the hourly angle h) or at any time during the year (the declination of the sun δ).

If the plane is horizontal, then $b = 0$ and H_n is the direct radiation on a horizontal plane. Equation (8) then becomes

$$H_0 = H_n \sin \alpha \tag{12}$$

Also

$$H_i/H_0 = \cos b + (\sin b/\tan \alpha)\cos \bar{Z} \tag{13}$$

where $\bar{Z}$ is the relative azimuth between the sun and normal to the receiving surface.

If Eq. (13) is rewritten as

$$\left(\frac{H_i}{H_0} - \cos b\right)\frac{\tan \alpha}{\sin b} = \psi(H) = \cos \bar{Z} \tag{14}$$

then the characteristic line is obtained by plotting $\psi(H)$ for given values of α and b versus $\cos \bar{Z}$. Figure 11 shows this plot. It is a straight line of slope unity and passes through the points (1, 1), (0, 0), and (−1, −1). This is true only if b is always less than α.

The scale for $\cos \bar{Z}$ is in the reverse direction to that normally used. This is justified if one starts with the plate facing the sun and progressive relative motion is assumed until the sun is behind the plate.

If b is greater than α, then the rays of the sun cease to strike the front surface of the plate. For example, if $b = 90°$, it means that the plate is vertical; the plate will be shaded when $\bar{Z} = 90°$ or $\cos \bar{Z} = 0$, irrespective of α.

For values of b less than 90° but greater than α, the ratio H_i/H_0 be-

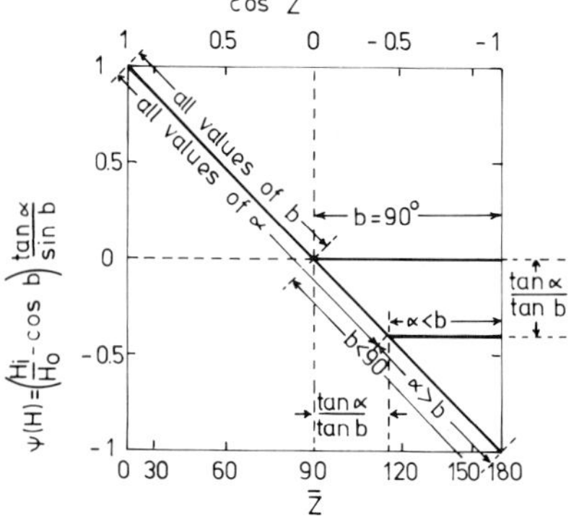

FIG. 11 Ideal radiation line for direct radiation only.

comes zero. This occurs when the rays of the sun pass tangentially to the place

$$\cos \overline{Z} = -\tan \alpha / \tan b.$$

1.5 DIFFUSE RADIATION

For a clear sky, the diffuse radiation component of solar radiation depends mostly on air mass m and atmospheric turbidity, water vapor, dust content, and aerosols. This component is equivalent to roughly 16% of the total radiation falling on a horizontal surface at noon. It rises to 25% five hours later. These values are for midsummer in a temperate climate (Heywood, 1965). If the distribution of diffuse radiation is uniform over the whole of the visible sky hemisphere, then the diffuse radiation falling on an inclined plane is

$$D_i/D_0 = (1 + \cos b)/2 = \cos^2(b/2) \tag{15}$$

For further work regarding diffuse radiation on inclined surfaces, see Kondratyev and Fedorova (1960). Figure 12 shows the ratio of diffuse radiation on an inclined surface to that on a horizontal surface for a solar altitude of 47°, at Woolwich.

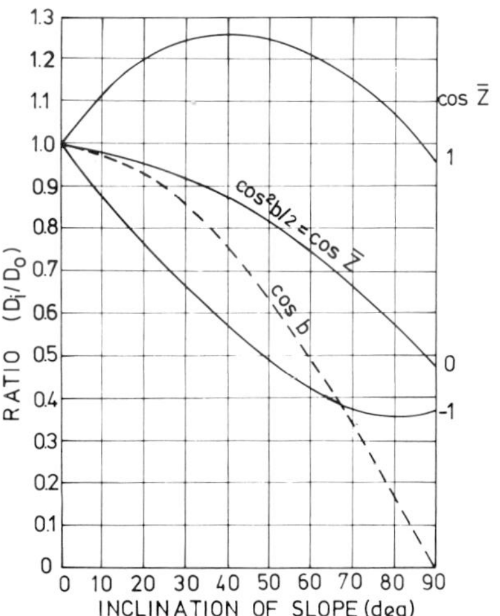

FIG. 12 Ratio of diffuse radiation on inclined and horizontal surfaces for Woolwich, June 26, 1964 (from Heywood, 1965) for a solar altitude of 47°.

1.6 REFLECTED RADIATION

Reflected radiation or that reradiated from the ground is in most cases neglected, especially from a surface with inclination b less than 60°.

Heywood thought this factor could be as large as 8% for a vertical surface. Care must be taken in using radiation equipment so that no reflected radiation from an adjacent wall or the ground is present. However, it is difficult to separate this component from the diffuse radiation component. An extensive analysis of diffuse radiation as well as reflected radiation is given by Kondratyev and Fedorova (1976). They expressed the diffuse radiation on inclined surface as

$$D_i = \int_0^{\pi} d\bar{Z} \int_{\alpha_1(\bar{Z})}^{\pi} J(\alpha_1, \bar{Z}) \cos i \cos \alpha_1 \, d\alpha_1 \tag{16}$$

where $J(\alpha_1, \bar{Z})$ is the intensity of radiation scattered in the direction determined by coordinates α_1 (angular height) and $\bar{Z}$ (relative azimuth direction); $\alpha_1(\bar{Z})$ is the minimum angular height of a point in the sky at azimuth $\bar{Z}$ relative to the plane of the horizon. In the case of isotropic scattering radiation, $J(\alpha_1, \bar{Z}) = \mathrm{J} =$ constant; therefore formula (15) will be

$$D_i = \tfrac{1}{2}(1 + \cos b)\pi J \tag{17}$$

Another relationship can be written for reflected radiation,

$$R_i = \int_0^{\pi} d\bar{Z}_1 \int_0^{\alpha_1} \bar{J}(\alpha_1, \bar{Z}) \sin \alpha_1 \cos \alpha_1 \, d\alpha_1 \tag{18}$$

where $\bar{J}(\alpha_1, \bar{Z})$ is the reflected radiation intensity,

$$\alpha_1(\bar{Z}_1) = \arccos \frac{\cos b}{\sqrt{1 - \sin^2 b \cos^2 \bar{Z}_1}} \qquad \text{with} \quad 0 \leqslant \bar{Z} \leqslant \pi$$

Assuming that the radiation reflected by the horizontal surface is isotropic $[\bar{J}(\alpha_1, \bar{Z}) = \mathrm{J}_{\mathrm{ref}} = \text{constant}]$, Eq. (18) becomes

$$R_i = 2J_{\mathrm{ref}} \int_0^{\pi/2} d\bar{Z}_1 \int_0^{\alpha_1(\bar{Z}_1)} \sin \alpha_1 \cos \alpha_1 \, d\alpha_1$$

$$= \pi J_{\mathrm{ref}} \sin^2 (\tfrac{1}{2} b) \tag{19}$$

1.7 TOTAL RADIATION OR GLOBAL RADIATION

It is of great importance in solar energy technology to know the total radiation falling on an inclined plane. In calculating the total flux, the non-isotropy of scattered radiation should be taken into account. It is apparent that when the sky is clear and the solar altitude small, a plane facing the sun will receive a substantial amount of scattered radiation as compared with total radiation. For a large solar altitude the relative contribution of scattered radiation is insignificant. Then the global radiation flux to the

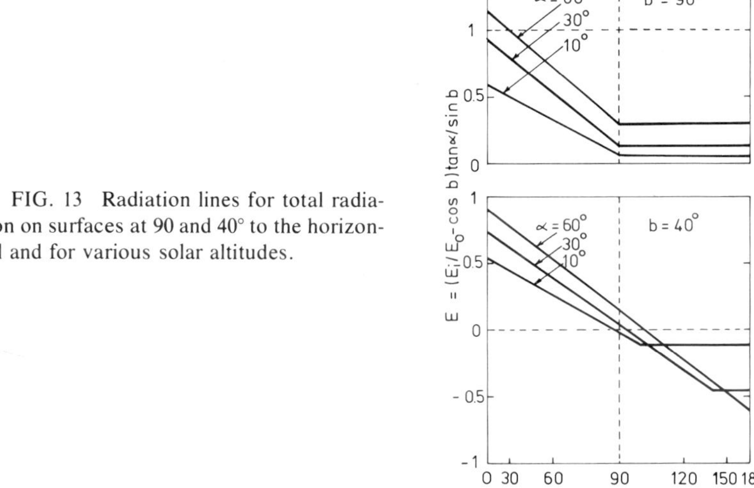

FIG. 13 Radiation lines for total radiation on surfaces at 90 and 40° to the horizontal and for various solar altitudes.

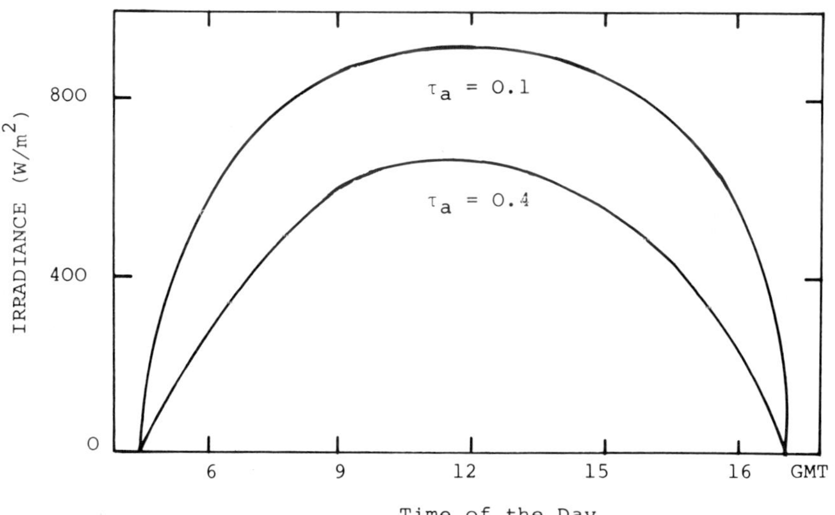

FIG. 14 Calculated direct irradiance at normal incidence on June 22 at latitude 52° N for turbidities τ_a = 0.1 and 0.4.

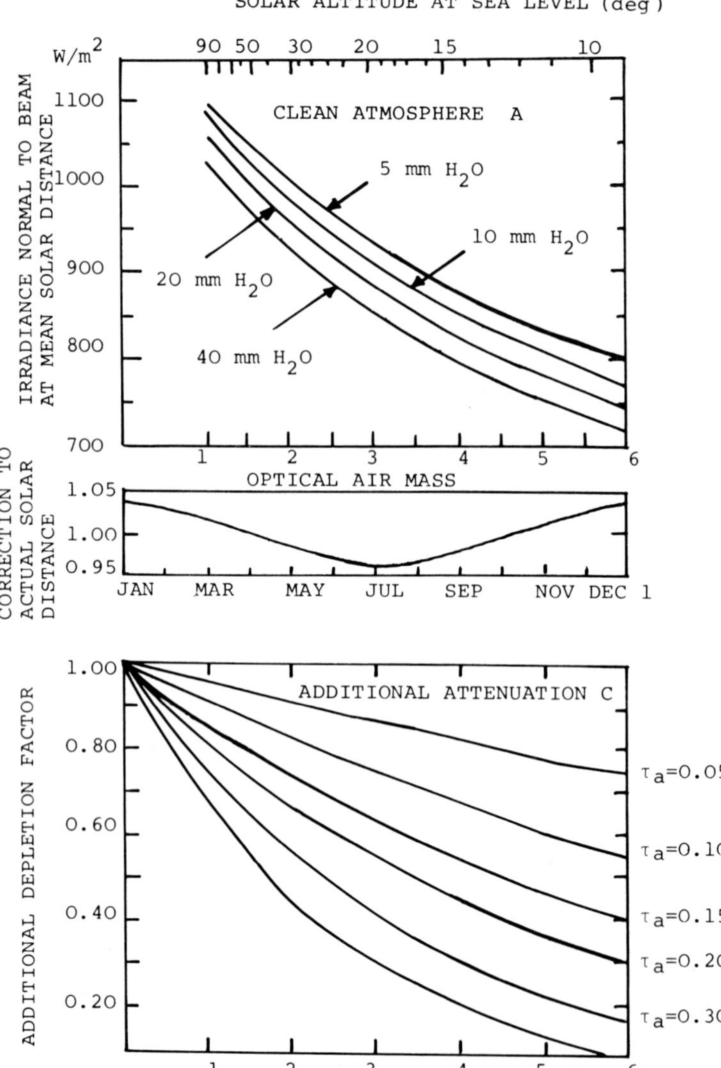

FIG. 15 The intensity of the direct beam for a given solar altitude.

slope may be calculated assuming the isotropy of scattered radiation. The experimental values of total radiation E_i and E_0 on cloudless days show a straight line, but this line differs from the ideal by having a slope less than unity and not passing through the point (0, 0). Numerous experimental readings have led to the relationship

$$\left(\frac{E_i}{E_0} - \cos b\right)\frac{\tan \alpha}{\sin b} = \psi(E) = C + M \cos \overline{Z} \qquad (20)$$

where $C = \psi(E)$ when $\cos \overline{Z} = 0$, and M is the slope of the characteristic line. Figure 13 shows typical characteristic lines for two angles of inclination of a plane and three altitudes. Estimation of total solar radiation on a horizontal surface has been carried out by Sayigh (1977c).

1.8 EFFECT OF TURBIDITY ON SOLAR INTENSITY

Turbidity (τ_a) effects are almost inversely proportional to the total solar radiation reaching the ground. This is clearly shown in Figs. 14 and 15. Unsworth and Monteith (1972) found from a long series of measurements of τ_a in Britain that τ_a depends on the airstream prevailing over the area. It was also stated that $\tau_a = 0.05$ in clear polar airstreams, while $\tau_a = 0.35$ in airstreams of continental origin and polluted urban areas. τ_a depends mostly on the amount of aerosol in the atmosphere.

1.9 THE SURFACE ALBEDO α_g

An important parameter in dealing with solar radiation intensity is the reflectance of various surfaces. This parameter is called surface albedo (α_g). It is not exactly uniform for a given surface, but for practical pur-

TABLE 3

The Albedo of Various Surfaces at Various Solar Elevations

Surface / Solar elevation: p	15°	30°	60°	90°
Fresh snow	0.80	0.75	0.71	0.70
White sand	0.68	0.65	0.61	0.60
Old snow	0.80–0.50	0.75–0.40	0.71–0.35	0.70–0.35
Desert sand	0.45	0.30	0.25	0.25
Dry grassland (semidesert)	0.30	0.23	0.16	0.16
Eucalyptus forest	0.27	0.19	0.15	0.13
Smooth sea	0.18	0.08	0.05	0.03

poses is assumed to be uniform. It is also a function of the air mass or solar zenith angle Z. Table 3 shows the albedo for various surfaces at various solar elevations (Paltridge and Platt, 1976).

1.10 CONCLUSION

It is essential to know how much solar energy is available at a given location before utilizing it. In buildings, the most effective absorber is the flat plate collector. This is because of its ability to utilize diffuse radiation as well as to direct radiation. The correct orientation of the solar absorber is also essential in obtaining maximum solar radiation intensities. Obviously, a sun-tracking absorber is best, but owing to the expense of such a device and to its impracticality in most buildings, a fixed orientation for the collectors in a building is preferred.

2

Materials for Solar Energy Collectors

ALBERT G. H. DIETZ

DEPARTMENT OF ARCHITECTURE
MASSACHUSETTS INSTITUTE OF TECHNOLOGY
CAMBRIDGE, MASSACHUSETTS

2.1 INTRODUCTION

This chapter describes briefly some of the principal requirements for and properties of materials employed in solar collectors used for the transformation of solar energy into thermal energy. Beginning with the solar spectrum as a reference point, the properties of diathermanous (radiant-energy-transmitting) materials, the requirements of absorbing surfaces, the properties of storage media, and the properties of thermal insulation materials are treated.

2.2 SOLAR RADIATION

Materials for solar energy applications must be considered in relation to the solar spectrum, the irradiance or intensity of solar radiation at various wavelengths.

In Fig. 1 (Yellott, 1976), three curves are given. Curve A is for the distribution of solar energy ($W/cm^2/\mu m^1$) versus wavelength outside the earth's atmosphere. Curve B is for the same distribution after passing through a single air mass, i.e., with the sun at its zenith above a point at sea level. Curve C is the distribution for a black body radiating at 5762°K, the temperature of the surface of the sun. By definition, a black body is a perfect radiator and a perfect absorber of radiant energy, the distribution of that energy depending on the temperature.

It is seen that the sun closely approximates a black body at its surface

ISBN 0-12-620860-3

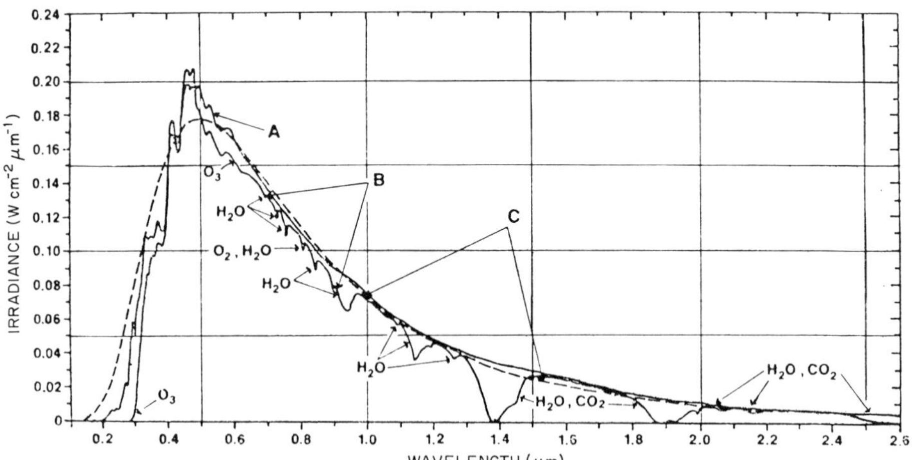

FIG. 1 Spectral distribution of solar radiation A: $m = 0$; B: $m = 1.0$; C: black body irradiance for $T_{bb} = 5762°K$ (Yellott, 1976).

temperature, but there are differences. Curve B shows that absorption of energy occurs because of constituents in the atmosphere, mainly O_3, O_2, H_2O, and CO_2, with sharp absorption occurring at various wavelengths, especially in the infrared (beyond 0.7 μm), with similar absorption caused by O_3 in the ultraviolet spectrum (below 0.4 μm).

Practically all of the sun's energy is found at wavelengths below 2.5 μm. Materials used to transmit solar energy should therefore be highly transparent over the range of wavelengths from 0.3 to 2.5 μm, but can be essentially opaque outside that range, without noticeably affecting solar transmission.

2.3 DIATHERMANOUS MATERIALS

The term diathermanous is applied to materials capable of transmitting radiant energy, including solar energy. From the standpoint of the utilization of solar energy, the important characteristics are reflection, absorption, and transmission. The first two should be as low as possible and the latter as high as possible for maximum efficiency.

A. Reflection and Transmission

Fresnel (Hardy and Perrin, 1932) derived a quantitative expression for the amount of light (or incident radiation) reflected at the boundary surface between two transparent media. In Fig. 2, let I_0 be the intensity of a

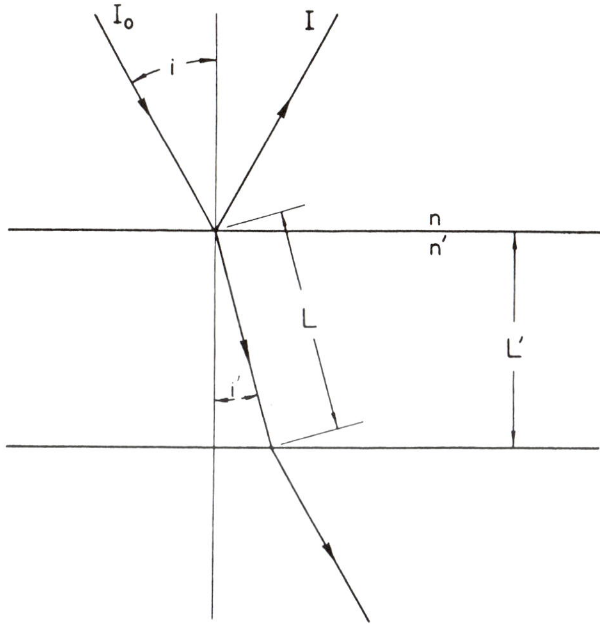

FIG. 2 Relationships between incident, reflected, and transmitted beams in media having two different indices of refraction. Snell's law: $\sin i' = (n/n') \sin i$ and $L = L'/\cos i'$.

beam of plane-polarized light incident at an angle i on the boundary between two media whose indices of refraction are n and n'. If the beam is plane polarized in the plane of incidence, the ratio of the intensity of reflected beam I to the incident beam I_0 is

$$\frac{I}{I_0} = \frac{\sin^2(i - i')}{\sin^2(i + i')} \tag{1}$$

in which i' can be computed from Snell's law, $n \sin i = n' \sin i'$. Similarly, if the incident beam is plane polarized in a plane perpendicular to the plane of incidence,

$$\frac{I}{I_0} = \frac{\tan^2(i - i')}{\tan^2(i + i')} \tag{2}$$

Natural or unpolarized light may be assumed to consist of equal amounts of the two components, and the proportion of natural light reflected at the boundary is

$$\frac{I}{I_0} = \frac{\sin^2(i - i')}{2 \sin^2(i + i')} + \frac{\tan^2(i - i')}{2 \tan^2(i + i')} \tag{3}$$

For light falling perpendicular to the surface, if one medium is air ($n' = 1$),

$$I/I_0 = [(n' - 1)/(n' + 1)]^2 \tag{4}$$

Figure 3 shows the relationship between percent reflected and angle of incidence for glass having an index of refraction $n = 1.526$. Curve C for natural light is of interest in solar heating. At normal incidence, reflection from each surface of the glass is about 4%; so the transmittance of the glass as a whole (except for transmission losses) is approximately $(0.96)^2$ or nearly 92%. Reflection does not become marked until the angle of incidence exceeds 60°.

To obtain the overall transmittance τ_r of one or more transparent sheets (Hottel and Woertz, 1942), allowing for reflection losses only, a unit beam of either of the plane-polarized components of light may be considered. A reflectivity equal to r exists for a given angle of incidence. Consequently, the fraction $1 - r$ penetrates to the second surface of which $r(1 - r)$ is reflected, and $(1 - r)^2$ is transmitted through the second surface. Part of the reflected beam $r(1 - r)$ is re-reflected at the first surface, and part is lost outward. The summation of reflections and transmissions results in the transmittance τ_1 for a single sheet.

$$\tau_1 = (1 - r)/(1 + r) \tag{5}$$

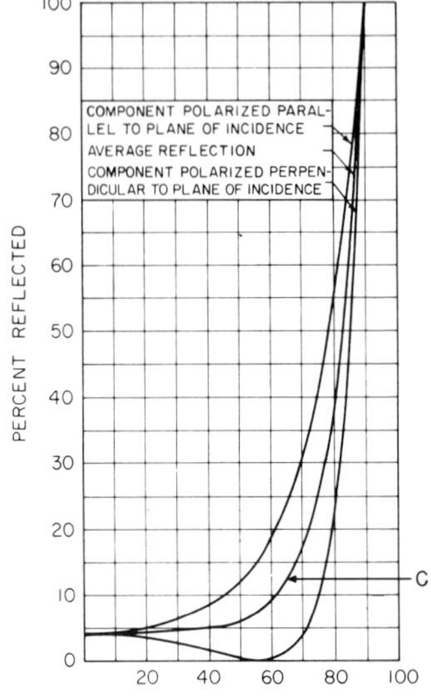

FIG. 3 Reflection versus angle of incidence for polarized and natural light. (Index of refraction equals 1.526.)

For n sheets the overall transmittance, allowing for reflection losses only, is

$$\tau_r = (1 - r)/[1 + (2n - 1)r] \tag{6}$$

This relation holds for either component of polarization. For natural light having equal values of the two components, the overall transmittance is the arithmetic mean of the two values of τ_r calculated for the two components.

The overall transmittance of a transparent sheet is dependent not only on the proportion of the incident light that is reflected, but also on the proportion absorbed in passing through the sheet. If the length of path through the sheet is L, and if this is visualized as divided into a number of layers dL, each of which reduces the intensity I by dI in proportion to the thickness dL and the absorption or extinction coefficient K of the sheet, then

$$-dI = IK\,dL \tag{7}$$

from which, integrating or summing over the distance L and the limits I and I_0, the transmittance τ_a due to absorption alone becomes

$$\tau_a = I/I_0 = e^{-KL} \tag{8}$$

In this expression L is the actual length of path traversed by the beam of light in passing through the sheet of thickness L'. It can be calculated from the geometrical relationship shown in Fig. 2.

Diathermanous materials used in solar collectors are characterized by small values of the extinction coefficient K and reflectivity r. For such materials the overall transmittance τ_1 for a single sheet can be simplified to

$$\tau_1 = [(1 - r)/(1 + r)]\,e^{-KL} = \tau_{1r}\tau_{1a} \tag{9}$$

the subscript 1 referring to a single sheet. For several plates, if τ represents the overall transmittance of the plate system, τ_r the transmittance of the same system due to reflection only, and τ_a the transmittance due to absorption only,

$$\tau = \tau_r\tau_a \tag{10}$$

to a high degree of accuracy. The simplified relations are sufficiently accurate for most problems of solar heat collector design.

Parmelee (1945) has calculated families of curves showing transmission of direct solar radiation by glass as a function of the KL of the glass and of the angle of incidence. A family of such curves for a single sheet of glass is shown in Fig. 4.

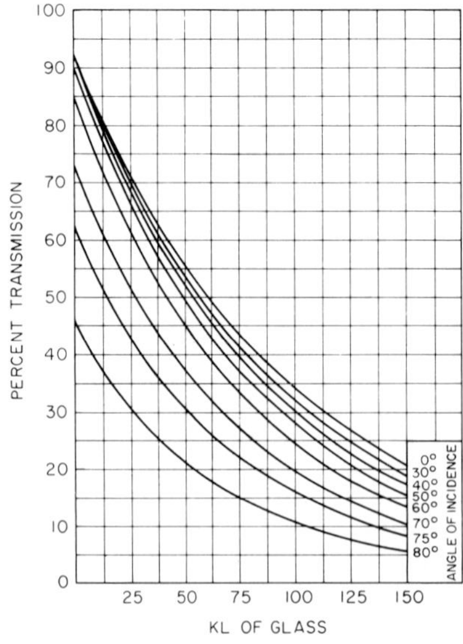

FIG. 4 Transmission versus *KL* values for various angles of incidence (Parmelee, 1945).

B. Classes of Diathermanous Materials

From the standpoint of windows and solar energy collectors, glass is the most important material, and principal emphasis is placed upon its properties. Transparent plastics and composites of glass fibers and plastics, as well as laminar plastics-based composites, are of increasing importance and are also reviewed.

Window glasses are primarily silica (SiO_2), soda (Na_2O), and lime (CaO) with minor amounts of magnesia (MgO), alumina (Al_2O_3), and iron oxide (Fe_2O_3).

Of these constituents, the iron content, expressed in percent Fe_2O_3, is of particular importance in its effect on spectral transmittance. Figures 5–8, from values supplied by Arner (1950), show this effect for four different percentages of Fe_2O_3 (radiation at normal incidence). For the lowest percentage (0.02) the transmittance rises to more than 90% in the near-ultraviolet region, remains high and uniform throughout the visible and well into the infrared regions, and then breaks off sharply at about 2.7 μm, beyond which it drops to zero (that is, becomes opaque) at about 4.7 μm. In the high-level region, reduction of transmittance is caused almost entirely by reflection. As the Fe_2O_3 content rises to 0.10 and 0.15%, the transmittance is decreased, although it remains at a high level in the visible region. The heat-absorbing glass (0.50 Fe_2O_3) exhibits a sharp decrease in the visible red and near-infrared regions. The typical bluish color of heat-absorbing glass is caused by this rather selective absorption.

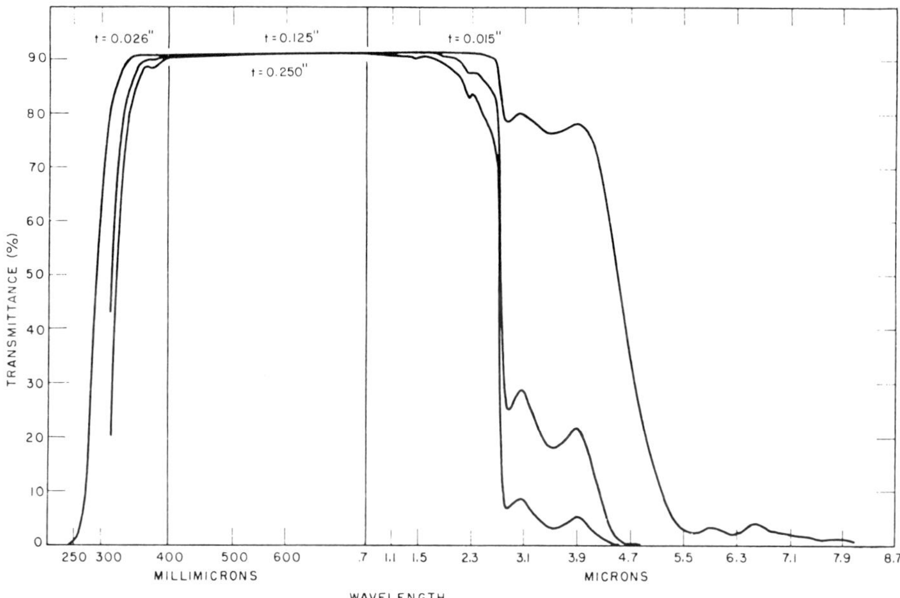

FIG. 5 Spectral transmittance of glass containing 0.02 Fe_2O_3.

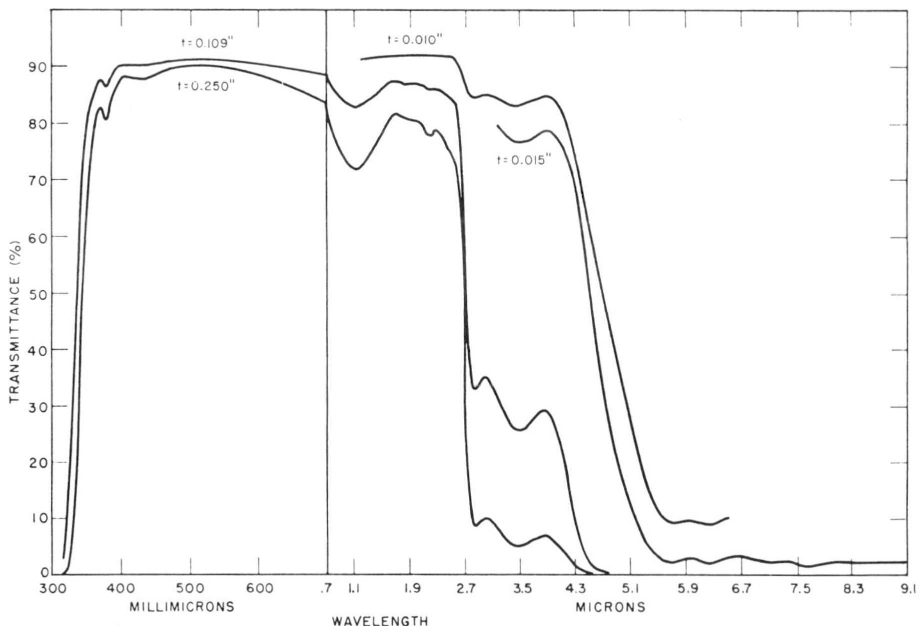

FIG. 6 Spectral transmittance of glass containing 0.10 Fe_2O_3.

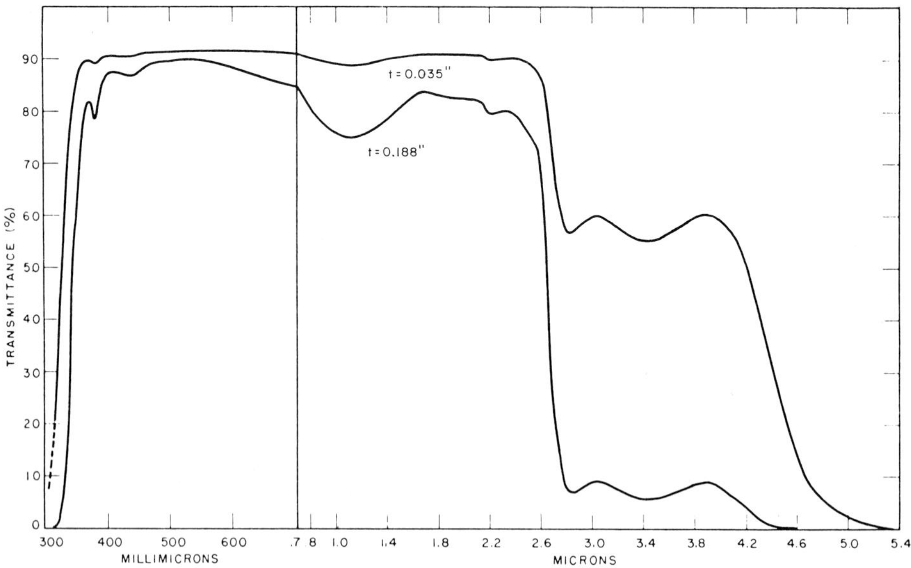

FIG. 7 Spectral transmittance of glass containing 0.15 Fe_2O_3.

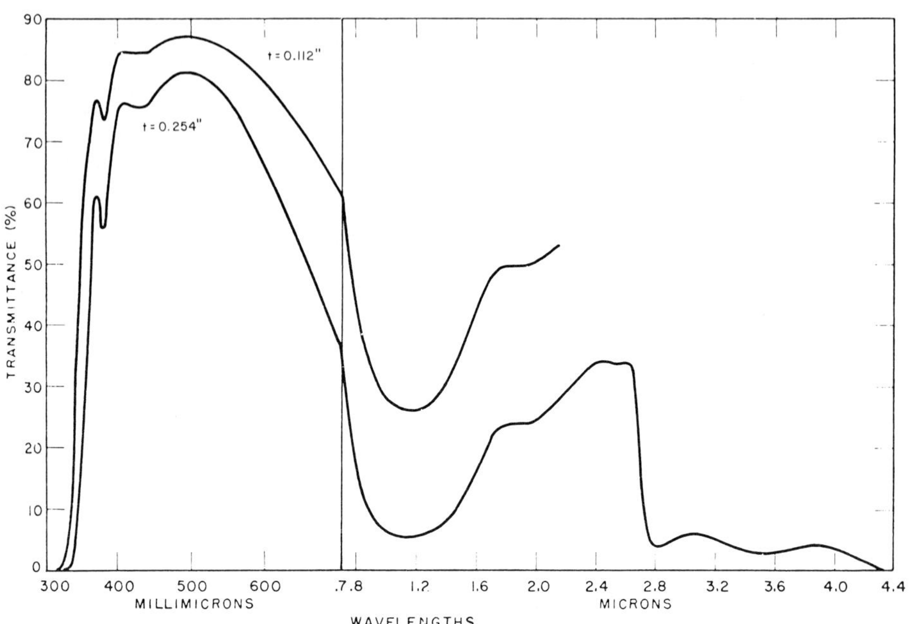

FIG. 8 Spectral transmittance of glass containing 0.50 Fe_2O_3.

For maximum solar energy collection it is, of course, desirable to have the highest possible transmittance throughout the solar spectrum, including the infrared to a wavelength of about 2.5 μm, beyond which the radiant energy of the sun drops to negligible proportions. In the very far infrared, corresponding to the temperatures to be found at the absorbing surfaces of solar heat collectors or the contents of dwellings, the transparent media of the collectors or of windows should be opaque, so as to act as barriers against radiation outward. Glasses used in building construction are completely opaque in these far infrared regions corresponding to low-temperature radiation. For example, radiation from a source at 38°C has its maximum intensity at 9.3 μm. The range 8–10 μm covers radiation at the temperatures of interest in solar space-heating applications (see Sections 2.3D and 2.4).

The effect of increased thickness of glass is shown in Figs. 5–8. In each instance increased thickness results in diminished transmittance, with the most marked effect in the infrared and ultraviolet regions. For low-iron glass increased thickness has an almost negligible effect in the visible region, showing that the extinction factor or *KL* value is low in that

TABLE 1

Characteristics of Some Glass Samples for Normal Incident Solar Radiation

Glass type	Thickness *L* (mm)	Percent transmission	*KL*	Absorption coefficient *K*	References
Single strength					
A quality	1.8	88.8	0.030	0.411	*a*
Window glass		86.0			*b*
Double strength					
A quality	3.2	89.2	0.025	0.194	*a*
Window glass		84.5			*b*
Clear plate	6.9	87.2	0.047	0.174	*a*
Regular	3.2	85.0			*b*
Regular	6.4	78.0			*b*
Grade A				0.432	*c*
Water shite		91.0		0.107	*c*
Tank plate	6.2	77.5	0.166	0.683	*a*
Water White	6.6	88.4	0.033	0.129	*a*
Heat absorbing, H_1	6.5	39.5	0.840	3.300	*a*
Heat absorbing, H_2	3.3	43.5	0.742	5.750	*a*
Heat absorbing, H_3	2.8	51.0	0.584	5.260	*a*
Heat absorbing, H_4	6.3	34.0	0.986	3.990	*a*
Heat absorbing, H_6	5.9	18.5	1.598	6.890	*a*

[a]Yellott (1976). [b]Parmellee (1945). [c]Hardy and Perrin (1932).

region. As the iron content increases, the effect of thickness also increases until it becomes marked in the heat-absorbing glass, particularly in the red and near-infrared regions.

Percent transmission values for solar radiation at normal incidence for a number of different glasses, as reported by various authorities and assembled by Hottel, are given in Table 1. Extinction coefficients K are also given.

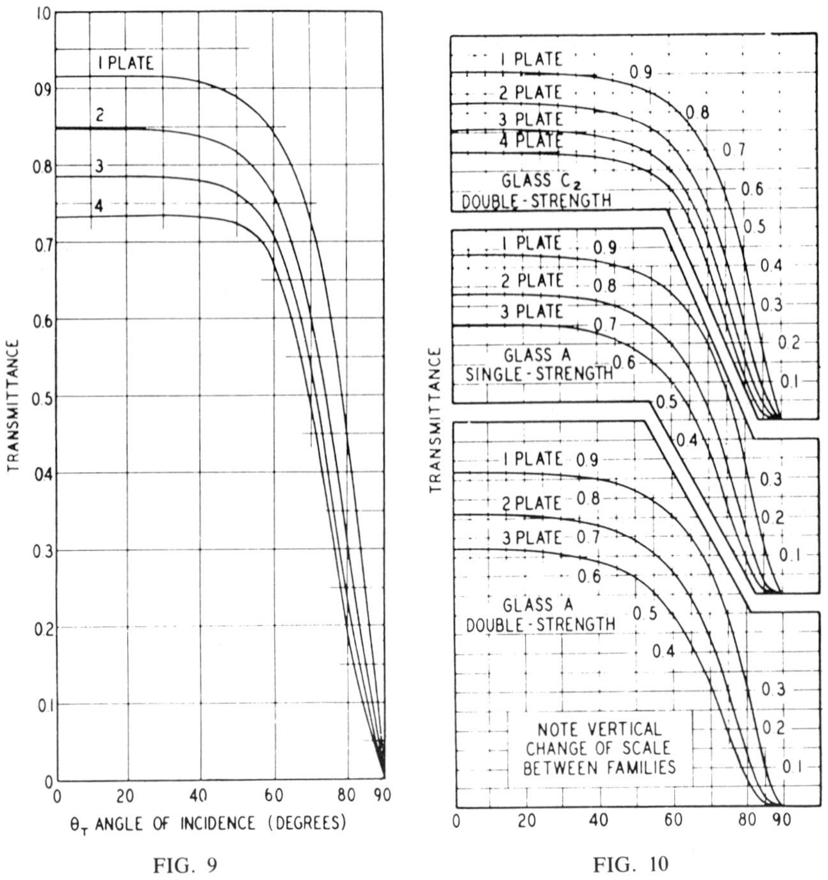

FIG. 9 Curves of calculated transmission versus angle of incidence, taking into account reflection losses only (refractive index = 1.526).

FIG. 10 Curves of calculated transmission versus angle of incidence, taking into account reflection and absorption losses. For top, middle, and bottom families, KL per sheet = 0.0125, 0.0370, and 0.0524, respectively.

C. Transmittance of Glass versus Angle of Incidence

With the equations for transmittance of one or more sheets of glass, and with measured values of reflectivity and extinction coefficients, it is possible to compute values of transmittance as a function of angle of incidence. Curves (Hottel and Woertz, 1942) of calculated transmittance τ_r of a system of glass plates having a refractive index of 1.526 (reflective losses only) are given in Fig. 9. Sets of curves (Hottel and Woertz, 1942) of calculated transmittances of systems of glass plates (reflection and absorption losses) are shown in Fig. 10. Top, middle, and bottom families are for glasses having *KL* values of 0.0125, 0.0370, and 0.0524, respectively.

Measurements have been made in connection with the MIT solar project to determine the transmittance of various glasses as functions of the angles of incidence. Some of the results of these measurements, correlated with calculated values for the same glasses (3), are shown in Fig. 11. Measurements were made utilizing natural sunlight. The curves shown are for three sheets of glass of type C_2, a water-white plate glass, each

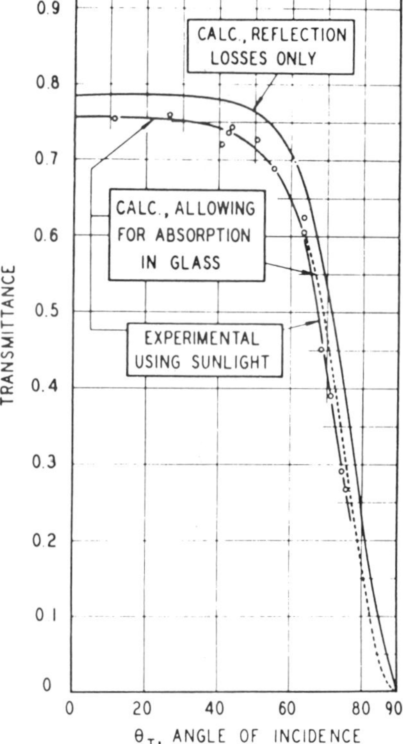

FIG. 11 Calculated curves of transmission versus angle of incidence, with experimental check points.

sheet 0.297 cm thick. The glass was maintained either horizontally or at an angle of 30° to the horizontal, and variations in angle of incidence depended on variations of the position of the sun in the sky. The small discrepancies between the calculated and measured values at angles larger than 65° are probably due to the changed solar spectrum when the sun is low in the sky and contains a larger fraction in the infrared, which is more readily absorbed by the glass.

D. Plastics and Composites

Some plastics and composites, such as glass-fiber-reinforced plastics and laminated plastics, are highly transparent to solar radiation. Some have low indices of refraction, leading to lower reflection losses than most glass. In general, weight is less than that of glass while resistance to breakage is greater. Durability over long periods of intense exposure to sunlight, characteristic of solar collectors, is not in every case established, although a few plastics do have long histories of exposure. Those that are thermoplastic should be installed in such a way as to avoid overheating and softening. All are softer and more prone to scratching than glass, but scratching and etching must be severe before there is appreciable loss of efficiency.

One of the most commonly used transparent plastics is polymethyl methacrylate (PMMA). Its transmission characteristics are shown in Fig. 12 (Dietz, 1969). Depending on composition, transmittance is high or low in the ultraviolet region. Transmittance is high (approximately 90%) in the visible region. As is true of most transparent materials, transmittance is variable in the infrared, and cuts off (becomes opaque) at approximately 2.8 μm, corresponding to the low-intensity end of the solar spectrum.

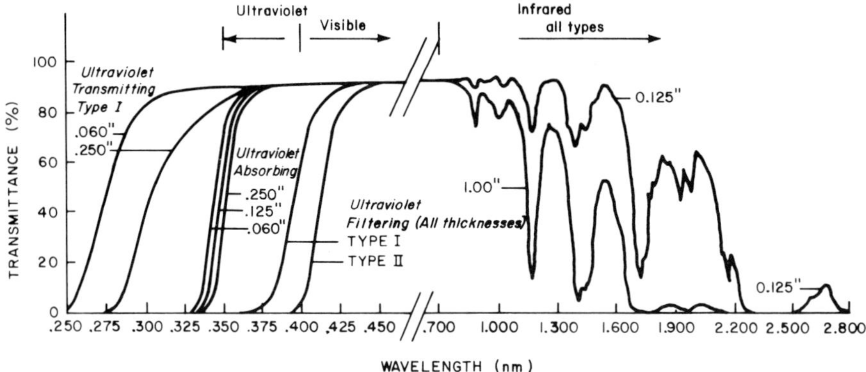

FIG. 12 Spectral transmittance of polymethylmethacrylate (Dietz, 1969).

Like the best glass, this material therefore exhibits high transparence to solar energy.

The index of refraction (1.50) of PMMA is similar to that of glass. Its reflectance is therefore also similar. Lowering the refractive index lowers the reflectance and, consequently, increases the transmittance. Some transparent plastics, notably the fluorocarbons, have lower refractive indices (approximately 1.34 to 1.4) and, therefore, lower reflectances (0.027 versus 0.04) at normal incidence [Eq. (4)] resulting in a net gain of solar transmission. This has an appreciable effect.

Figure 13 shows the transmission characteristics of polyvinyl fluoride (PVF) over the 0.2–15 μm range (Anonymous, 1978), in each case 1 mil (0.001 in. or 0.025 mm) thick. In Fig. 13a, for uv, transparent, and near-infrared radiation, transmittance is high. In the far infrared (Fig. 13b), transmittance is variable, reaching a low point at 9 μm, corresponding roughly to low solar collector temperatures. At the higher collector temperatures to be expected, where emission is in the 7–9 μm range, transmittance is higher.

Figure 14 shows transmittance curves for polytetrafluoroethylene (FEP), which also has a low refractive index of approximately 1.341–1.347 (Anonymous, 1978). Transmission is high in the uv, transparent, and near-infrared spectra, but variable in the far infrared, with a sharp cutoff (opacity) at the 7.5–9 μm range corresponding to temperatures likely to be found in solar collectors. Overall solar transmittance, as the curves show, is higher than for glass.

PVF and FEP are expensive and must be employed in thin films if costs are to be reduced. Furthermore, transmittance is considerably reduced, especially in the visible region, if thickness is increased much beyond 5–10 mils (0.125–0.250 mm). Such films are vulnerable to outdoor weather conditions and, if used as films, must be protected by a surface cover sheet such as glass or a thicker, tough, transparent plastic in a multicover collector. Alternatively, they may be laminated to a transparent sheet, thus providing a low surface refractive index. Furthermore, these materials are highly resistant to deterioration by sunlight and can screen uv effectively, thus protecting the substrate, such as a transparent plastic. Because of their higher transmittance than glass, more layers of film or laminate can be employed as cover sheets in a multilayer collector, thus markedly increasing the insulating value of the cover with no decrease in overall transmittance. Higher collector temperatures and efficiencies can therefore be achieved than with a smaller number of glass cover plates.

Great strength and toughness accompanied by high spectral transmittance can be achieved by incorporating glass fibers in a transparent plastic matrix as, for example, selected unsaturated polyesters. When there is a

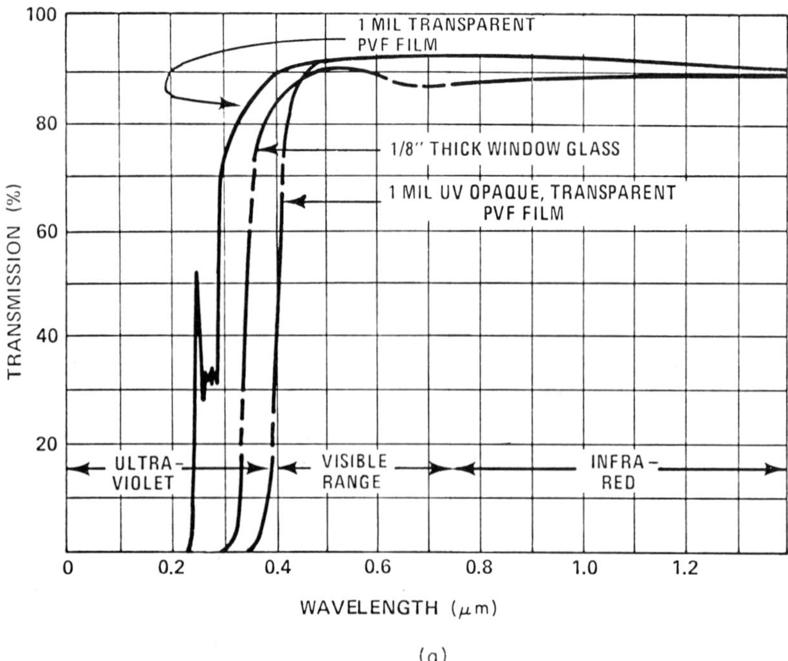

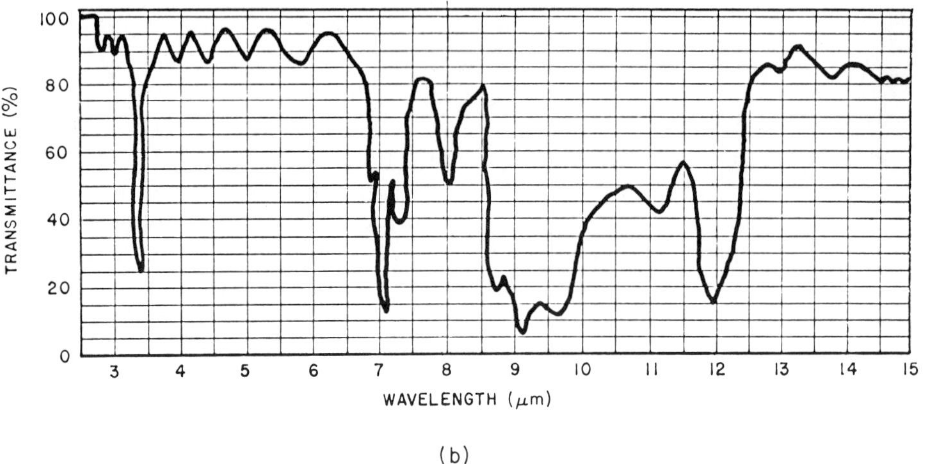

FIG. 13 Spectral transmittance of polyvinyl fluoride (PVF): (a) ultraviolet, visible, and near infrared; (b) far infrared.

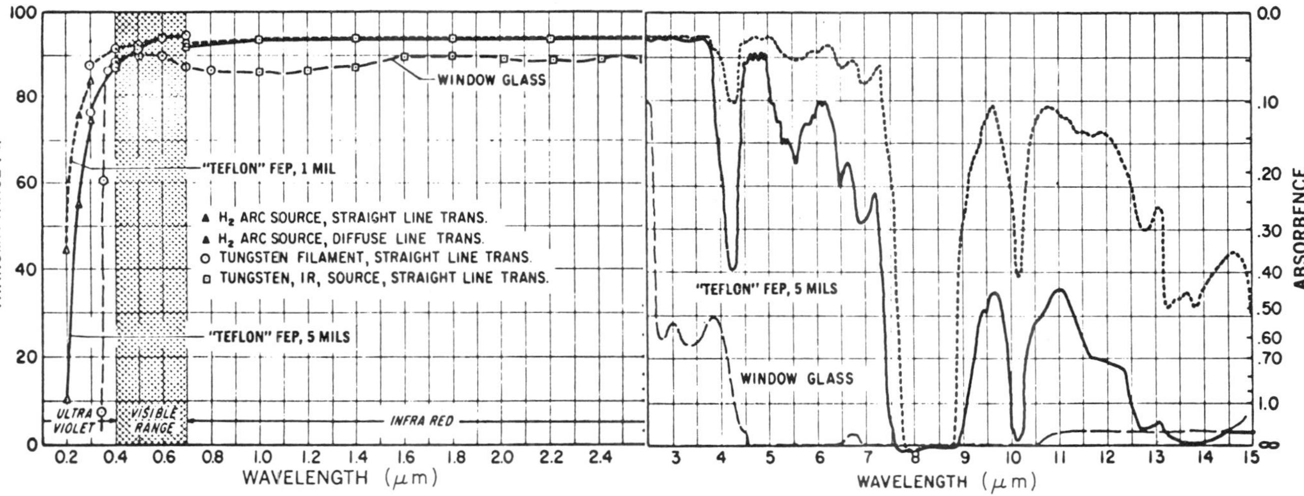

FIG. 14 Spectral transmittance of polytetrafluoroethylene (FEP). △, H_2 arc source, straight line transmission; ▲, H_2 arc source, diffuse line transmission; ⊙, tungsten filament, straight line transmission; ⊡, tungsten infrared source, straight line transmission.

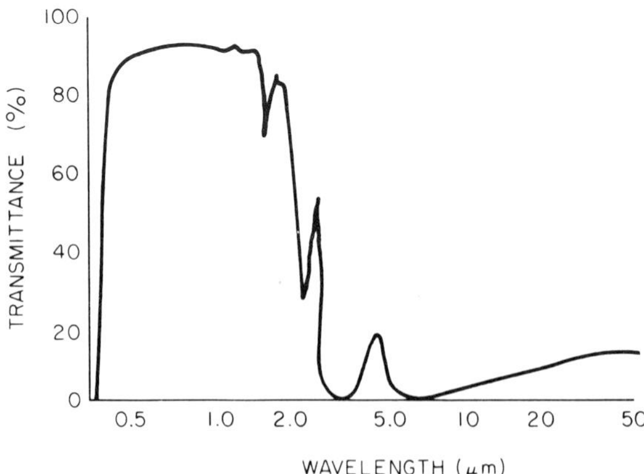

FIG. 15 Spectral transmittance of high-clarity glass fiber-reinforced plastic (Keller, 1977).

good match of refractive indices of glass fiber and matrix, and if the glass is thoroughly wetted by the matrix during fabrication, internal scattering of light is reduced, and overall transmittance of solar energy is high, as shown in Fig. 15 (Keller, 1977). Surface refractive index can be reduced by laminating a film of fluoroplastic, such as PVF or FEP. Because the material is tough and strong, it can be used in thinner layers than glass

TABLE 2

Infrared Transmission of Selected Plastics[a]

Material	Thickness (mm)	Percent transmission
Polyethylene	1 mil 0.025	85
	4 0.100	72
	7 0.175	62
Polyvinyl chloride	2 mil 0.050	40
	6 0.150	15
	16 0.400	5
	20 0.500	3
Polyvinylidene chloride	1 mil 0.025	65
Polyethylene terephthalate	½ mil 0.013	55
	2 0.050	28
	3 0.075	20
Polypropylene	4 mil 0.100	80
	50 1.250	21
Cellulose acetate	3 mil 0.075	12
	20 0.500	0

[a] Johnson *et al.* (1975).

(typically 0.625–1.00 mm), and its flexibility allows it to be bent into curved shapes if necessary. Being thermosetting, i.e., not softened appreciably by solar collector temperatures, it is less vulnerable to heat than many thermoplastics. Its higher inherent cost than glass requires it to be used in thin layers to keep costs down.

The foregoing are examples of plastics and plastics-based composites usable for solar installations. Many plastics are or can be transparent, and may be usable. Practically all have, or can have, high transmissivity in the visible spectrum, with variable infrared transmission depending on composition and thickness. Infrared transmission for a few films is listed in Table 2 (Johnson *et al.*, 1974/1975). Polyethylene, for example, is widely used in greenhouses where its low cost is attractive in spite of limited life.

2.4 ENERGY-ABSORBING AND -EMITTING MATERIALS

In a solar collector some surface must absorb the radiant solar energy falling on it and transform it into thermal energy, unless direct conversion to electrical energy or use in chemical or biological processes is intended. This discussion is limited to thermal energy. Absorption of solar energy impinging on a collecting surface should be as high as possible, but re-emission (loss) outward from the collector should be minimized.

A. Emission and Absorption

As already defined, a black body is a perfect absorber of radiant energy and is a perfect radiator; that is, it has an absorptance α and an emittance ϵ, each equal to unity. Actual surfaces do not behave like perfect absorbers or perfect radiators and have absorptances and emittances less than unity. Parenthetically, it may be pointed out that a black body is not necessarily nonluminous but may be as bright as the sun (which is not quite a black body). The term merely indicates a surface that is a perfect radiator and a perfect absorber.

According to Kirchhoff's law, at thermal equilibrium (when radiating and absorbing surfaces are the same temperature) the ratio of the emissive power of a surface to its absorptance is the same for all bodies. Another way of stating this law is that at thermal equilibrium the absorptance and emittance of a body are the same.

The emittance of a surface varies with its temperature and its roughness. If it is a metal, it depends also on its degree of oxidation. Highly polished metals have low emittances, provided oxidation and surface imperfections are kept to a minimum. The absorptance of a surface depends on the same factors as the emittance and, strictly speaking, on the distribution of wavelengths in the spectrum of incident radiation. If the character of the absorbing surface is such that its absorptance is independent of the distribution of incident wavelengths, it is called gray, and its absorptance and emittance are the same even though thermal equilibrium does not exist—that is, even though the temperatures of the radiator and the receiver are not the same. Evidently, if the difference in temperature between an emitting and an absorbing surface is small, gray body conditions can be assumed with little error. For example, room temperatures and the temperatures attained by solar collectors or by ordinary radiators are nearly enough alike to permit each to be considered "gray."

Table 3 (Anderson, 1976) gives values of emittances and absorptances for various materials. It also gives values of reflectance ρ, the ratio of reflected solar energy to incident solar energy. It is noteworthy that many

TABLE 3

Absorptance α, Reflectance ρ, Emittance ϵ, and Ratio α/ϵ for Various Materials[a,b]

Materials	α	ρ	ϵ	α/ϵ
White plaster	0.07	0.93	0.91	0.08
Fresh snow	0.13	0.87	0.82	0.16
White paint	0.20	0.80	0.91	0.22
White enamel	0.35	0.65	0.90	0.39
Green paint	0.50	0.50	0.90	0.56
Red brick	0.55	0.45	0.92	0.60
Concrete	0.60	0.40	0.88	0.68
Grey paint	0.75	0.25	0.95	0.79
Red paint	0.74	0.26	0.90	0.82
Dry sand	0.82	0.18	0.90	0.91
Green roll roofing	0.88	0.12	0.94	0.94
Water	0.94	0.06	0.96	0.98
Black tar paper	0.93	0.07	0.93	1.00
Flat black paint	0.96	0.04	0.88	1.09
Granite	0.55	0.45	0.44	1.25
Graphite	0.78	0.22	0.41	1.90
Aluminum foil	0.15	0.85	0.05	3.00
Galvanized steel	0.65	0.35	0.13	5.00

[a] Anderson (1976).

[b] $$\alpha = \frac{I_a}{I} = \frac{\text{absorbed solar energy}}{\text{incident solar energy}}$$

$$\rho = \frac{I_r}{I} = \frac{\text{reflected solar energy}}{\text{incident solar energy}}$$

$$\epsilon = \frac{R}{R_b} = \frac{\text{radiation from material}}{\text{radiation from blackbody}}$$

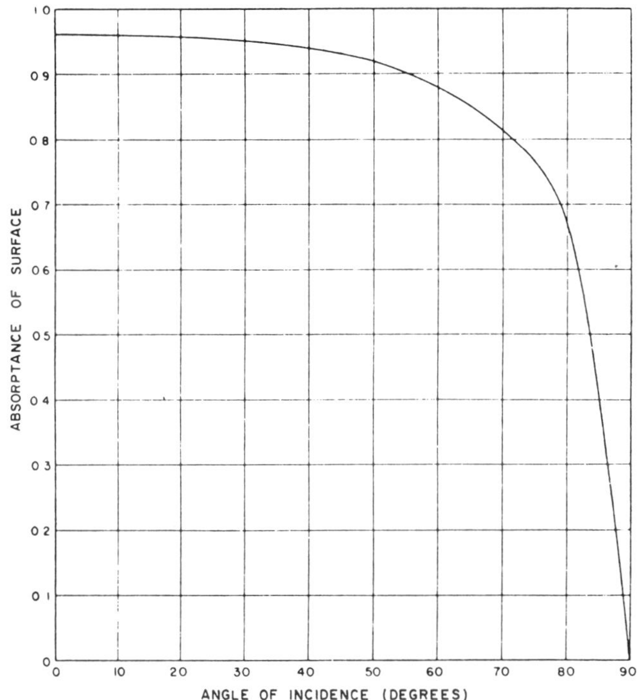

FIG. 16 Absorptance of a blackened surface for artificial sunlight transmitted through glass.

common building materials have excellent emitting surfaces for long-wave radiation.

Work carried out at the Building Research Station of the Department of Scientific and Industrial Research, Great Britain (Anonymous, 1933; Hottel and Woertz, 1942), relates the angle of incidence to the absorptance of a blackened surface for artificial sunlight transmitted through glass. This effect is shown in Fig. 16. At large angles of incidence the absorptance drops off markedly and approaches zero at the grazing angle.

B. Selective Surfaces

As shown previously, where absorptances are high, as in desirable surfaces for solar collectors, emittances also are generally high, so that such surfaces tend to reemit energy originally absorbed. Emission is at a much longer wavelength because the surface is emitting at a much lower temperature than the sun. Figure 17 (Anonymous, 1970), for example, shows the emission curve (B) for a black body at 35°C. For higher temper-

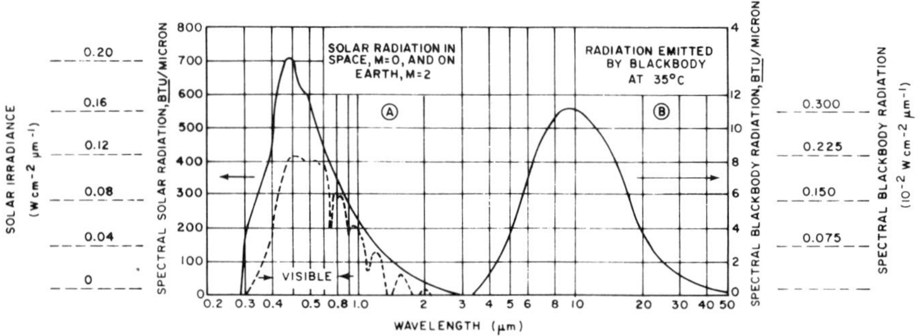

FIG. 17 Spectral transmittance of black body at 35°C compared with solar radiation in space and through air mass = 2.

atures the peak is moved to the left; for lower temperatures it moves to the right. This may be compared with the simplified solar curve (A).

Even though the emission is at a longer wavelength and the peak intensity is much less than in the solar curve, outward radiation losses, accompanied by heating and convection, are appreciable. It is, therefore, highly desirable to reduce the emittance as much as possible, while keeping absorptance high.

Much research has been carried out to develop such "selective" surfaces, i.e., surfaces with high absorptances for solar energy but low emittances at the temperatures reached in solar collectors. Several such surfaces are set forth in Table 4 (Duffie and Beckman, 1974). Absorptances are less than for the best flat black paints, but emittances are significantly lower.

TABLE 4

Properties of Selective Surfaces for Solar Energy Application[a]

Surface	α	ϵ
Chrome black	0.95	0.10
Nickel black on polished nickel	0.92	0.11
Nickel black on galvanized iron[b]	0.89	0.12
CuO on nickel	0.81	0.17
Co_3O_4 on silver	0.90	0.27
CuO on aluminum	0.93	0.11
Ebanol C on copper[b]	0.90	0.16
CuO on anodized aluminum	0.85	0.11
PbS crystals on aluminum	0.89	0.20

[a] Duffie and Beckman (1974).
[b] Commercial processes.

2.5 HEAT STORAGE

Once the solar energy has been converted to thermal energy, it is usually necessary to store at least part of it for use when the sun is not shining. Storage may be directly at the absorbing area ("passive" system), or the thermal energy may first be transported and then stored ("active" system).

Transportation of thermal energy is usually by airstream or by liquid, of which by far the most common is water. Where freezing is a problem, the flow of water can be shut off and the collector drained when freezing temperatures are approached or when the collector temperature is less than the storage temperature. Alternatively, use may be made of a non-freezing fluid such as an ethylene glycol–water mixture or other liquid that does not freeze at temperatures likely to be met in a collector.

Thermal storage may be by sensible heat or latent heat. Sensible heat storage involves a rise in temperature of the storage medium as heat is absorbed and a drop as heat is released. Latent heat involves a change of stage, usually solidification and liquefaction (freezing and thawing), with heat evolution and absorption at constant temperature.

A. Sensible Heat Storage

Sensible heat storage media may be liquid or solid. By far, the most commonly employed liquid storage medium is water. It has the highest specific heat per unit of weight of all materials, and is generally plentiful and cheap. Its properties are well known. It is easy to store and to circulate. The installation must be leak proof.

Various solids may be employed for sensible heat storage. Most common is a bed of fragmented rock or gravel through which warm air from the collector is blown for storage, and through which air from the building is blown to pick up heat. Although its density is higher than water, its specific heat is less, and when it is fragmented to allow air to pass through, its heat capacity per unit volume is considerably less than an equal volume of water. Higher-density materials, such as iron ore or scrap iron, may be employed to reduce this disadvantage. Table 5 lists specific heats, densities, and heat capacities of several sensible heat storage materials (Dietz, 1954, 1963; Telkes, 1974).

B. Latent Heat Storage

The principal drawbacks of sensible heat storage are varying temperature, causing difficulties in the design of heating and cooling systems, and the large storage volume needed. Latent heat storage has the theoretical advantages of constant temperature and smaller volume for a given heat

TABLE 5

Heat Capacities of Various Materials[a]

Material	Heat capacity (10^3 J/kg °K)	Density (kg/m^3)	Heat capacity (10^3 J/m^3) 1°K	Heat capacity (10^3 J/m^3) 20°K
Sensible heat storage				
Water	4.12	1000	4120	82,400
Concrete	0.88	2320	2042	40,800
Stone, various	0.84	2560	2150	43,000
Broken (40% voids)				25,800
Iron ore, Fe_2O_3	0.62	5290	3280	65,600
Broken (40% voids)				39,400
Scrap iron	0.47	7850	3690	73,800
Broken (40% voids)				44,300

Latent heat storage

	Melting point (°C)	Density (kg/m^3)	Heat of fusion (10^3 J/kg)	Theoretical heat capacity fusion only (10^3 J/m^3)
$CaCl_2 \cdot 6H_2O$	29–39	1630	174	285,250
$Na_2CO_3 \cdot 10H_2O$	32–36	1440	267	355,680
$Na_2HPO_4 \cdot 12H_2O$	36–48	1520	279	402,800
$Na_2SO_4 \cdot 10H_2O$	32	1470	244	358,680
$Ca(NO_3)_2 \cdot 4H_2O$	40–42	1830	209	254,800
$Na_2S_2O_3 \cdot 5H_2O$	48–51	1730	221	382,330

[a] Dietz (1954).

capacity. The advantage of volume is diminished in practice because, especially in "active" systems, some fluid such as air or water is needed to transport the heat from reasonably small containers of the latent heat medium, thus requiring additional space, similar to that needed in a rock pile. Water storage, on the other hand, can usually be compact, calling, at most, for an internal heat exchanger.

Several fusible salts, melting at temperatures useful for solar collectors, are listed in Table 5. Still lower temperatures can be achieved by various eutectic mixtures. In practice, difficulties have been encountered with some fusible salts after numerous cycles of freezing and thawing because of segregation and loss of efficiency. Thin layers, various crystallization promoters, and suspending media are commonly employed in an attempt to overcome this tendency. Nevertheless, latent heat storage materials have the attractive features, in theory at least, of constant temperature and small volume.

2.6 THERMAL INSULATING MATERIALS

Heat losses from the various parts of a solar collecting system must usually be avoided or, at least, controlled, although such losses into the building can be considered contributions to heating requirements. Thermal insulation, for example, behind the collector itself, around pipes or ducts, and around the heat storage container is usually needed.

Thermal insulating materials commonly employed in buildings generally are suitable for solar energy installations. They are well known and their properties need not be repeated here in detail. Nevertheless, a few common materials, together with their resistances R, are listed in Table 6 (ASHRAE, 1970). Conductances are the reciprocals $1/R$ of the resistances.

TABLE 6

Resistances of Common Building Materials[a]

Material	Thickness (mm)	R value (m^2, °K/W)
Hardwood siding	25	0.16
Softwood siding	25	0.22
Gypsum board	13	0.08
Wood shingles	lapped	0.15
Wood bevel siding	lapped	0.14
Brick, common	100	0.14
Concrete (sand and gravel)	200	0.15
Concrete blocks (filled cores)	200	0.34
Gypsum fiber concrete	200	0.85
Mineral Wool (batt)	89	1.92
Mineral Wool (batt)	150	3.31
Fiberglass board	25	0.77
Corkboard	25	0.68
Expanded polyurethane	25	1.04
Expanded polystyrene	25	0.73
Molded polystyrene beads	25	0.68
Loose fill insulation		
Cellulose fiber	25	0.65
Mineral wool	25	0.70
Sawdust	25	0.39
Flat glass	3	0.16
Insulating glass (6-mm space)		0.27
Vertical air space	19	0.15
Vertical air space	100	0.18

[a] ASHRAE (1970).

3

Systematic Classification of Climate for Solar House Design

*J. K. PAGE**

DEPARTMENT OF BUILDING SCIENCE
UNIVERSITY OF SHEFFIELD
SHEFFIELD, UNITED KINGDOM

3.1 INTRODUCTION

The meteorological influences that need to be considered in the systematic design of solar houses are extremely complex. It is difficult at the current stage of knowledge to produce simple global climatological classifications to assist architects in solar house design. Weather not only affects the day-to-day supply of solar energy, but also the magnitude and the temporal patterns of the different demands for the use of that energy, say for space heating or for cooling. Weather in many parts of the world is highly variable from day to day, and in all parts of the world, outside the equatorial regions. The climate also varies strongly from season to season. The need to match the supply of solar energy with diurnally varying energy demands usually creates a need for energy storage, but the establishment of appropriate energy storage design strategies is entirely dependent on the availability of adequate statistical descriptions of the probable relationships between energy demands and solar energy supplies. Figure 1 taken from the UK–ISES Report (Page and Hall, 1976) attempts to describe the complex interrelationships that exist between component design for solar houses and the multidimensional external meteorological environment. Figure 1 distinguishes between the passive collection of solar energy

* Sections 3.1–3.8 of this chapter are based on a paper presented to the North East London Polytechnic-UNESCO Conference on Solar Building held in London in July 1977.

ISBN 0-12-620860-3

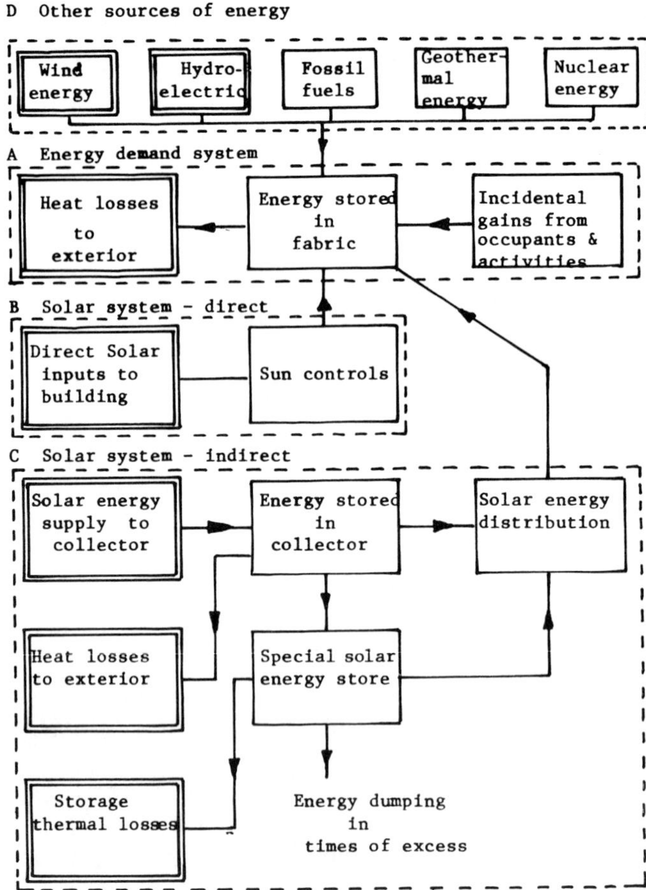

FIG. 1 The systematic design of solar building involves understanding the interactions between the energy demand system and the different energy supply systems, no less than three of which are used in a typical solar building. The solar systems interact with the wider energy supply system. Many of the factors are weather sensitive. These factors are shown in double boxes (Page and Hall, 1976).

through windows, etc., and the active external collection of solar energy using external solar collectors on the roof, etc. It indicates the problem of solar controls and energy dumping. It also distinguishes between energy storage in the building fabric itself and energy storage in separate, specially constructed energy stores. It assumes the potential for energy collaboration with other types of energy supply systems, ambiently based, fossil fuel based, and where appropriate, nuclear based. The numerous meteorologically sensitive aspects of system design are all ringed in double boxes. The large number of climatic interactions is the clearest feature of the diagram. It is clear that solar system design is difficult.

The design of solar houses needs to be strongly conditioned by the precise geographical location of the proposed project and its associated climate analyzed in detail. The principle weather variables that have to be considered are

(1) the incoming short-wave radiation on various slopes, which includes direct, diffuse, and ground-reflected components;

(2) the long-wave flux due to thermal radiation from the atmosphere, which is influenced by cloud as well as by atmospheric constituents and also varies with the slope angle;

(3) the external air temperature and wind speed, which affect the heat loses by convection both from solar collector devices and from buildings to which the collectors are attached.

The relationship between gains and losses must be understood. The demand patterns are frequently not in phase with the temporal patterns of solar energy supply, and the two patterns have to be matched when necessary through the installation of a solar storage system of appropriate capacity.

This chapter attempts to outline briefly some of the variations found across the world in the climatological variables relevant to solar house design. Schemes for particular areas must be seen to rest within a particular climatological context, and their success must be judged within that geographical context. The designer must seek an *appropriate* solar technology for *each geographical location.* The logical form of solar buildings is strongly conditioned by latitude and cloudiness characteristics. Architects must put aside any belief there can be any universal international solar style.

Accurate climatological analysis is the foundation of successful solar house design, and borrowing designs out of their climatological context can lead to disaster. For example, Fig. 2 compares the measured monthly mean daily irradiation on a vertical south-facing wall at Odeillo in the French Pyrenees with that measured at Bracknell, just west of London, England. The difference of latitude is just over 10°, but the energy availability in winter is vastly different. Odeillo has a clear sunny winter climate with an average of about $5\frac{1}{2}$ hours of bright sunshine in January; Bracknell has a typical UK winter climate with a monthly mean bright sunshine level of only about $1\frac{1}{2}$ hours a day. The high-level mountain climate of Odeillo is also much clearer than the low-level winter climate of the United Kingdom; so it was for a very good reason that Felix Trombe took the French CRNS Solar Energy Laboratory to Odeillo. The Trombe

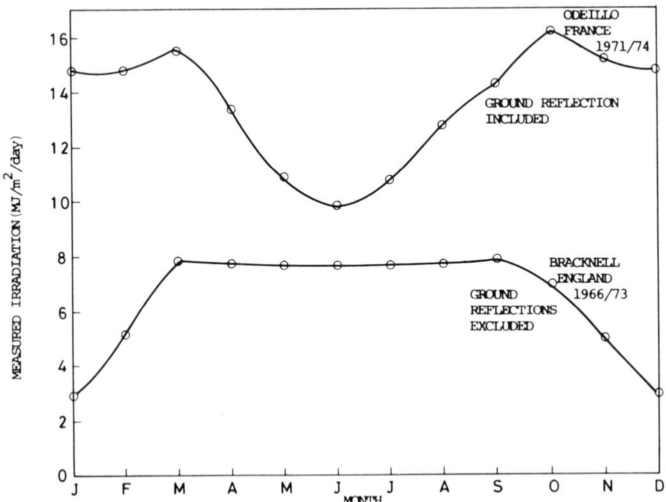

FIG. 2 Measured monthly mean daily irradiation on vertical south walls at Odeillo and Kew. Note large winter differences.

passive solar wall system may have proved itself successful in Odeillo (Trombe, 1974b), but Fig. 2 shows that one must not assume that it would necessarily be successful in midwinter in the United Kingdom, however attractive the concept might seem at a superficial glance.

3.2 GLOBAL PATTERNS OF SOLAR IRRADIATION

Figure 3, based on Budyko's world map, shows the annual mean global irradiance on a horizontal plane. The plot is a rough one and gives no detail; however, it does give a general impression of the range across the main inhabited parts of the world, from 100 W/m^2 in high-latitude climates to 300 W/m^2 in the Red Sea area. The great desert areas of the world typically located between 20 to 30° N and 20 to 25° S have mean irradiances of 250 W/m^2, over $2\frac{1}{2}$ times as great as countries such as the United Kingdom. The low mean irradiance values at the equator are also evident. The Congo Basin and Paris have very similar mean annual irradiances.

Figure 4 shows recorded monthly mean daily totals of the horizontal irradiation $\overline{H}_h$ for six stations of widely differing latitude, some north of the equator and some south. The monthly mean diffuse irradiation $\overline{H}_{dh}$ is also shown. The variation in monthly mean daily global irradiation with season increases with latitude. The even supply of solar energy in the equatorial region from month to month is very obvious at Kinshasa. The

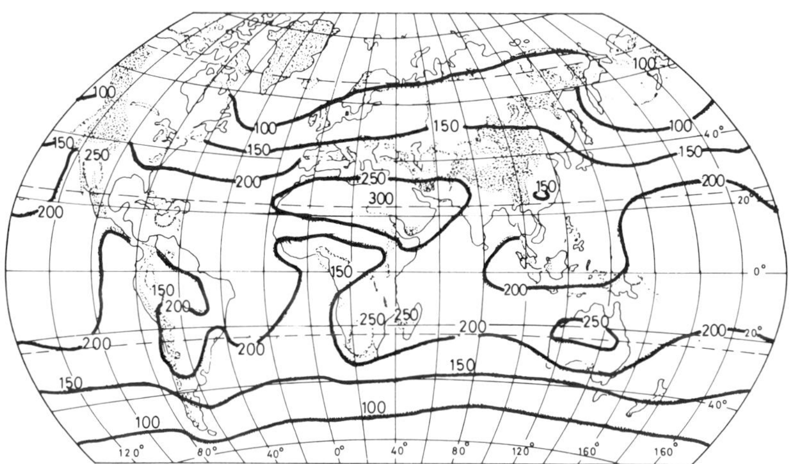

FIG. 3 Annual mean global irradiance on a horizontal plane at the surface of the earth (W/m^2 averaged over 24 hours 365 days of year) (Page and Hall, 1976).

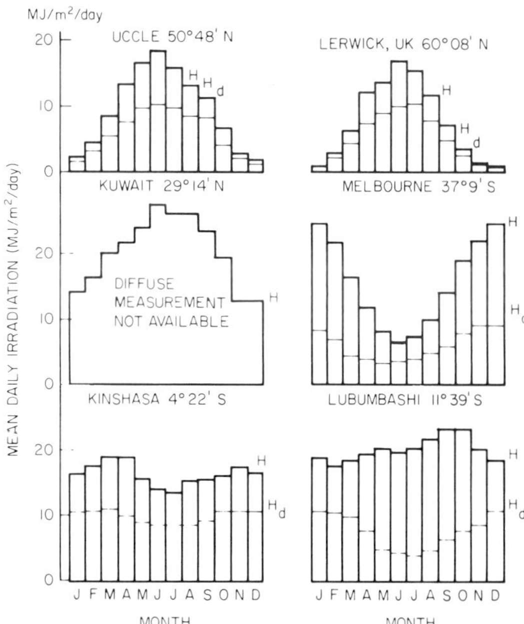

FIG. 4 Measured mean monthly global and diffuse irradiation on a horizontal surface for six stations of widely differing latitude. Note high proportion of diffuse irradiation at high latitudes and close to the equator as well.

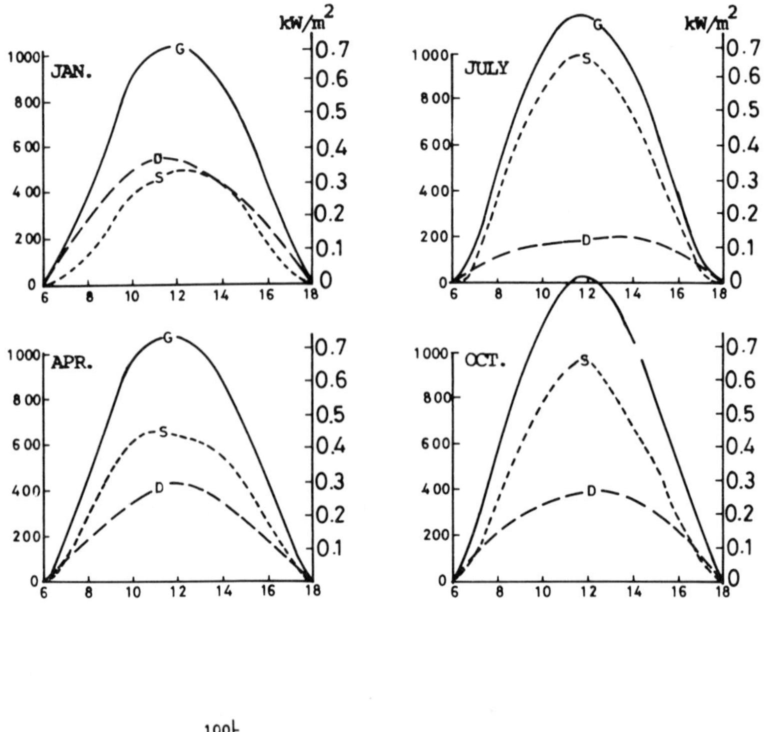

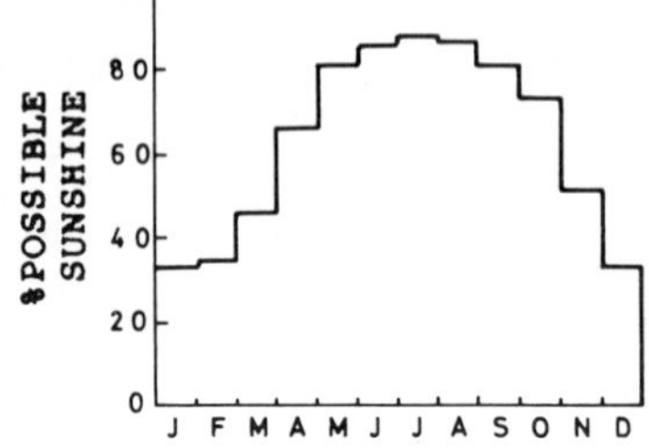

FIG. 5 Variations in mean global, direct, and diffuse irradiance on a horizontal surface in a low-latitude climate with a clear and a cloudy season: Lubumbashi, 11°39′ S, 27°28′ E, 1298 m, 1955–1960. Sunshine values are for 1950–1955 (Bultot, 1971).

diffuse irradiation forms a big proportion of the short-wave irradiation at Kinshasa in the Congo Basin. Lubumbashi a few degrees in latitude further to the south from the equator has a two-season climate, sometimes wet, humid, and cloudy, sometimes dry and sunny according to the position of the intertropical convergence zone. The proportion of diffuse radiation varies accordingly. Kuwait, a desert-type area, has a high insolation throughout the year, but, being situated 29° N of the equator, the horizontal monthly mean irradiation levels during the winter half of the year fall back as the sun moves to its southward position. Melbourne, Australia, at 37° S has the minimum monthly mean global irradiation level in June. The diffuse proportion at Melbourne is relatively high during the Southern Hemisphere winter, but this proportion decreases in the summer period. Uccle, outside Brussels, Belgium, shows a big range from summer to winter, with a large proportion of diffuse radiation throughout the year. The annual range is even greater at Lerwick in the Shetlands to the north of the Scottish mainland at latitude 60° N. The winter solar energy availability is very low indeed. The prospects for solar houses close to the Arctic Circle are poor.

It is useful to start by looking at the monthly changes of horizontal surface global irradiance G_h. Figure 5 presents variations in the measured mean monthly irradiance with time of day at Lubumbashi in Central Africa for four selected months, together with information about the associated percentage possible sunshine. In January with the percentage possible sunshine as low as 34%, the diffuse horizontal surface irradiation exceeds the direct. In July the percentage possible sunshine is about 87%, and every day is virtually cloudless. The direct horizontal surface irradiance dominates, and the proportion of diffuse is very small. In October the sun is overhead; so the mean global irradiance reaches its peak, but there is slightly more cloud and the diffuse has risen compared with July.

Solar building design involves understanding such radiation patterns in order to make best use of them. Knowledge of irradiation on sloping and vertical surfaces is critical for proper design, and it is essential to have methods for developing architectural design data from basic meteorological information. However, understanding the transmission properties of the atmosphere in relation to the astronomical movements of the earth about the sun is complex. This chapter considers the problem of modeling these important properties for architectural design purposes.

3.3 THE ABSORPTION AND SCATTERING PROCESSES IN THE ATMOSPHERE

The principal extraterrestrial variations in the solar energy supply arise from the variations in the distance between the earth and sun that are the consequence of the earth's elliptical orbit. The extraterrestrial solar irradiance normal to the beam at mean solar distance G_{sc}, known as the solar constant, is about 1353 W/m^2. This value is based on Thekaekara's survey of the observational data drawn mainly from US sources. There is still considerable discussion on the precise value to be adopted, and also whether significant variations in the value of G_{sc} occur because of variations in solar activity (see Chapter 1). Thekaekara's value is adopted consistently throughout this survey, and all ratios involving the extraterrestrial irradiation or irradiance in this chapter are based on this particular value at the mean solar distance (Robinson, 1966; Unsworth, 1975a).

The extraterrestrial intensity varies by ±3.3%, being greatest around January 1 and least around July 5. The direct beam intensity is substantially reduced as it passes through the atmosphere. The lower the sun, the greater the path length and the greater the reduction due to absorption and scattering. The path length through the atmosphere is normally described in terms of its optical air mass m. The standard atmospheric path at sea level is given a path length of unity with the sun at the zenith, dead overhead. As the sun gets lower, the path length will increase. Ignoring optical effects, which are important only at very low solar altitudes at sea level, $m \triangleq \sec \alpha$, where α is the solar altitude angle. The optical air mass decreases with increase of station height, and the effective solar path length at elevated sites is much less at a given solar altitude compared with sea level. The correction is proportional to the station atmospheric pressure p_{h} compared with p_0, the standard sea level atmospheric pressure.

Some radiation is absorbed by gases in the atmosphere, mainly ozone, oxygen, and carbon dioxide, some by water (vapor and cloud droplets); and some by aerosol, man-made and natural. In terms of absorption per unit air mass, ozone absorbs about 3% of the solar constant G_{sc}, oxygen and carbon dioxide absorb less than 2%, and water vapor some 10–15%, depending on its concentration. Variations in the surface irradiance due to absorption depend mainly on the water vapor distribution. Bannon and Steele (1960) have mapped global data on the amount of precipitable water vapor in the atmosphere. The precipitable water held gaseously in the atmosphere varies considerably from area to area and is greatest in the equatorial regions where it may reach 100 mm. In high latitudes about 5 mm is typical in winter. The water vapor absorption removes energy mainly from the infrared part of the spectrum. Air molecules scatter the blue light preferentially. Scattering by cloud droplets and aerosols depends strongly on particle size and on wavelength. Scattering by aerosol removes from 5 to 40% of G_{sc} in unit air mass, but a significant part of the scattered radiation still reaches the earth's surface as diffuse radiation. Figure 6 shows the main spectral features of the direct solar beam. The theoretical direct solar irradiance G_{bn}^{*} expected at the surface depends on

(1) the solar constant G_{sc},
(2) the sun–earth radius vector,
(3) the total water vapor content of the atmosphere expressed as precipitable water vapor,
(4) the air mass m, which is related to solar elevation α by the for-

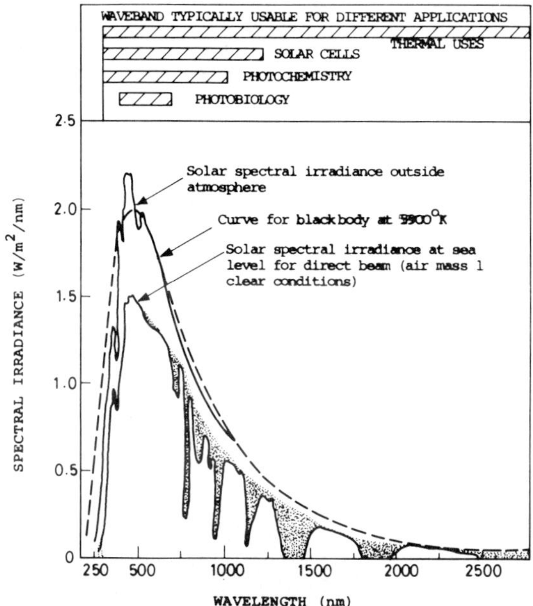

FIG. 6 Spectral irradiance curves for direct sunlight outside the earth's atmosphere and at sea level at air mass; shaded areas indicate absorption due to atmospheric constituents, mainly H_2O, CO_2, and O_3. Wavelengths potentially utilized in different solar energy applications are indicated at the top (Page and Hall, 1976).

mula $m = 1/\sin \alpha$ for $\alpha > 10°$ for stations at or close to sea level. For lower altitudes the Smithsonian Tables (List, 1951) provided suitable values. Clearly the major variations of G^*_{bn} are due to water vapor and air mass. The additional attenuation is ascribed to aerosol and can be described by a turbidity coefficient that relates G_{bn} to G^*_{bn}.

A variety of turbidity coefficients have been proposed, many of them depending on radiation measurements in a narrow waveband (for discussion refer to Robinson, 1966). For solar energy applications there are advantages in using the Monteith and Unsworth turbidity coefficient τ_{a} which is based on measurements of G_{bn} over the whole solar spectrum 0.3–3 μm. τ_{a} is defined by the relationship

$$\tau_{\mathrm{a}} = -\frac{1}{m}\log_e \frac{G_{\mathrm{bn}}}{G^*_{\mathrm{bn}}} \qquad (1)$$

A. Geographical Variations in Atmospheric Turbidity

There are very large geographical variations in mean annual atmospheric turbidity. Large seasonal differences are found as well. The clearest region is the Antarctic. The Arctic is also very clear, and air masses originating from these polar regions, having a low precipitable moisture content and also a low dust burden, have low turbidities. The solar beam intensity in such high-latitude regions may be surprisingly high in spite of the low solar altitude. In the great desert areas of the world, a considerable amount of dust enters the atmosphere, and the turbidity may be relatively high throughout the year. The high midday solar altitudes associated with lower latitudes mean that the midday air mass is low, so that the energy that penetrates is still very considerable. In the humid equatorial regions the turbidity is often lower than in the desert regions on either side, even though there is a lot of cloud, which substantially reduces the average irradiation by intercepting the direct beam for a high proportion of the day. The high absorption by gaseous water vapor gives a lower value of G^*_{bn}, and so does not affect the turbidity coefficient τ_a. However, in these areas there are also lots of small water droplets, which produce substantial forward scattering and a characteristically milky sky. In two-season, low-latitude climates where a rainy season intervenes into a clear season, the rainy season typically reveals lower atmospheric turbidities. The high rainfall tends to wash the dust particles out of the sky. Further, the fast-growing vegetation and the moist ground surfaces suppress surface dust generation due to turbulence. In temperate midlatitude climates the turbidity is characteristically lowest in the middle of winter and increases to a maximum around July. The atmospheric water vapor contents also increase from winter to summer; so the absorption in the infrared absorption bands increases, thus reducing summer direct beam irradiance.

Figure 7 shows typical month-to-month variations in the mean monthly "pseudo" turbidities for several Northern European sites. The pseudoturbidity was derived from the daily totals of direct and diffuse radiation using the Sheffield University computer solar radiation model, by dividing the derived mean monthly daily direct radiation on a horizontal surface by the percentage possible sunshine for the month in question, fixing the day lengths as the length of time during which the direct solar irradiance is above 200 W/m^2 (which is the intensity at which burning of the sunshine recorder first starts), assuming a Monteith and Unsworth turbidity $\tau_a = 0.2$. Figure 7 shows that the highest turbidities are associated with urban situations, i.e., with Kew and Hamburg. The remote seashore sites have substantially lower turbidities than the urban sites, the clearest conditions at sea level being found at Valentia in the far west of Ireland, which is a considerable distance from the main continental sources of pollution. Further, the air masses come mainly off the clean Atlantic Ocean. One high-level French mountain station, Odeillo, is in-

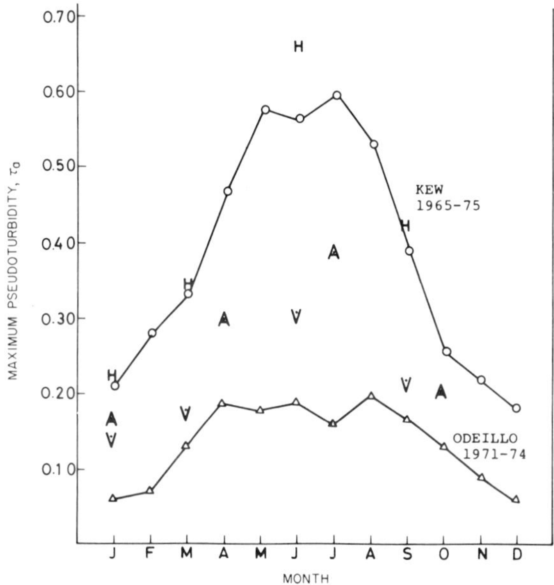

FIG. 7 Annual variations in daily maximum pseudoturbidity in Northern Europe; an average of all days. H, Hamburg; V̇, Valentia, W. Ireland; O, Kew; A, Aberporth; Δ, Odeillo.

cluded. The low turbidities associated with this elevated site are evident. At Odeillo some turbidity increase from winter to summer is present, but is more subdued than at Kew. Published data for the United States (Bilton *et al.*, 1969) bear out very much the same annual pattern.

B. Urban Influences on the Direct Radiation Climate on Cloudless Days

In addition to the general geographical factors already discussed, one also has to consider urban pollution influences, which may extend over very wide areas. Figure 8 gives data for the mean atmospheric transmission $\overline{H}_h/\overline{H}_{oh}$ around London, where $\overline{H}_h$ is the monthly mean daily global irradiation and $\overline{H}_{oh}$ is the corresponding monthly mean daily extraterrestrial irradiation. The falloff from the center of London to Cambridge some 100 km away is considerable. Fortunately, under the influence of clean air legislation, the smoke pollution burden of the UK atmosphere has been falling; so over the last decade the turbidities in many of the large towns have decreased. The opposite situation unfortunately holds in many other parts of the world. Typical values of τ_a for different types of air mass (Unsworth, 1975b) with corrections for urban and other polluting influ-

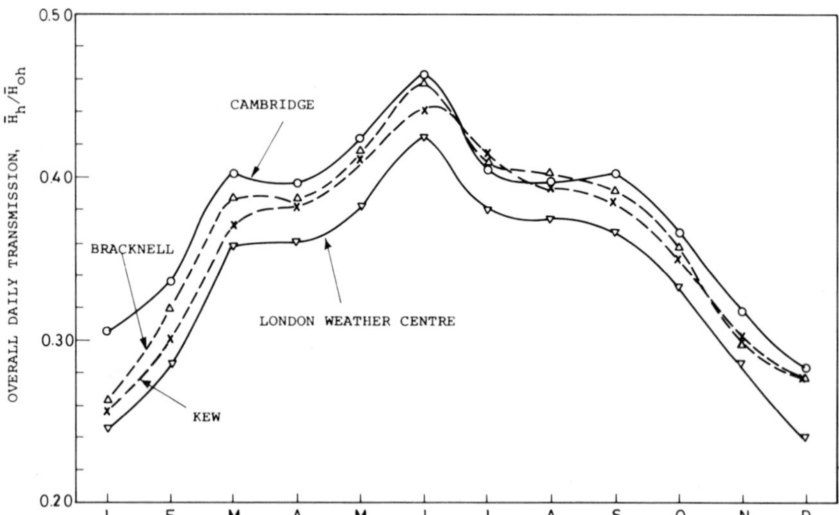

FIG. 8 Mean atmospheric transmission around London ($\bar{H}_h/\bar{H}_{oh}$) showing effects of increasing distance from city center. Monthly means averaged, 1965–1970.

ences in high latitudes are given in Table 1. Figure 9 shows the calculated effect of turbidity variations on vertical and horizontal surface irradiance for cloudless days on a south-facing vertical surface at latitude 50° N for March 21. The range of τ_a values selected were chosen to match the actual range found across the United Kingdom from the clearest conditions in the far west to the inland urban polluted situations in towns with poor control of smoke emissions, fortunately now a minority in the UK. The

TABLE 1

Characteristic Values of the Monteith and Unsworth Turbidity Coefficients τ_a for the UK Showing the Influence of Air Mass Type and Location in Relation to Pollution Sources

Location	Air mass type	τ_a
Northerly island site, minimum pollution from land sources	Polar	0.05
	Average	0.20
	Continental	0.35
Rural or coastal site exposed to natural aerosol pollution and small amounts of smoke	Polar	0.10
	Average	0.25
	Continental	0.40
Urban site within or close to a large town (say population exceeding 100,000)	Polar	0.25
	Average	0.40
	Continental	0.55

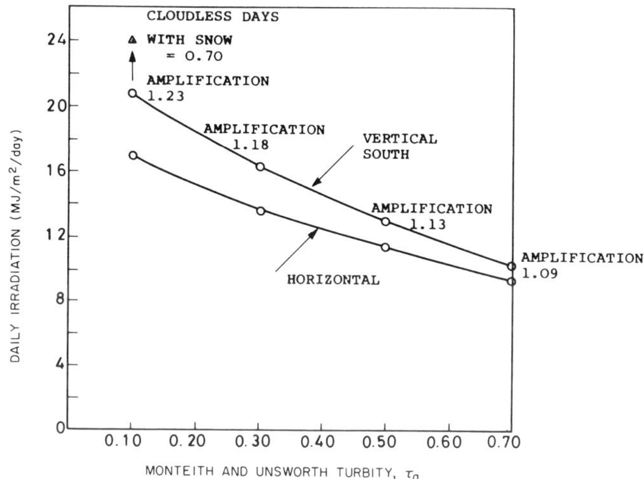

FIG. 9 The effect of changes in turbidity on horizontal and vertical surface irradiation on cloudless days. Latitude 50° N, March 21; ground albedo, 0.25; precipitable water vapor, 10 mm.

halving of available energy on cloudless days is a serious loss, and careful attention needs to be given to urban pollution effects in assessing the feasibility of solar houses in high altitudes. At lower latitudes a given pollution level has a smaller effect, as the path length through the atmosphere is shorter. The effects of ground albedo are also important, and a snow ground cover will substantially increase vertical surface irradiation. One point has been plotted in Fig. 9 to demonstrate the influence of increasing the ground albedo from 0.25 to 0.70. The ratio between the clear day, inclined surface, daily irradiation and the corresponding clear day, horizontal surface, daily irradiation may be defined as the clear day amplification factor. As the turbidity increases, the clear day amplification factor for vertical and tilted south-facing surfaces will decrease. The appropriate vertical surface amplification figures are included in Fig. 9.

3.4 THE DIFFUSE RADIATION CLIMATE

A. Cloudless Days

The amount of diffuse solar radiation on cloudless days also varies considerably from one part of the world to another, because dust particles, especially from deserts, and general atmospheric pollution from man's activities add substantially to the atmospheric scattering. The falloff in the direct beam irradiance due to scattering by dust and other pollution is partly compensated for by the associated increases in the scattered diffuse irradiation reaching the earth's surface. Approximately half the scattered energy moves toward the surface, the precise proportion depending on the nature of the scattering particles. Increase in station altitude tends to produce a falloff in both Rayleigh and atmospheric scattering due to dust; so the diffuse energy tends to be relatively low on cloudless days at high-altitude stations, as the typical deep blue skies on clear days clearly demonstrate. In contrast, in the low-level humid tropics, water droplets are present in large numbers, and a great deal of forward scattering takes place. The diffuse horizontal surface cloudless day irradiance levels are high in such areas, and on clear days the sky tends to have a bright milky appearance.

It is possible to find systematic links between the observed turbidity and the associated diffuse radiation on cloudless days. Studies at Sheffield University have made extensive use of the U.S. clear sky observations of Parmelee (1954), linking his original U.S. observations with observations in other countries, including the United Kingdom. Parmelee's clearness index approach has been interpreted into an overall approach based on the simultaneous use of the Monteith and Unsworth turbidity coefficient. Parmelee presented results of diffuse solar irradiance on both horizontal and vertical surfaces. His results were derived from measurements made at Cleveland, Ohio, on days that, although cloudless, differed in atmospheric clarity. A study of his results for a horizontal surface shows that for a fixed solar altitude α, a linear relationship exists between the diffuse horizontal irradiance and the direct horizontal irradiance. The relationship takes the form

$$G_{\mathrm{dh}} = a_0 - a_1 G_{\mathrm{bh}} \quad (\mathrm{W/m^2}) \tag{2}$$

where G_{dh} is the diffuse irradiance on a horizontal surface ($\mathrm{W/m^2}$), G_{bh} is the direct irradiance on a horizontal surface ($\mathrm{W/m^2}$), and a_0 and a_1 are constants for a particular solar altitude α and are given in Table 2.

Unsworth and Monteith (1972) report that for cloudless conditions, when the solar altitude is about 30°, the ratio of the diffuse irradiance on a horizontal surface G_{dh} to global irradiance on a horizontal surface G_{h} in Central England may be expressed as a function of turbidity τ_{a} as

$$G_{\mathrm{dh}}/G_{\mathrm{h}} = c + d(\tau_{\mathrm{a}})$$

where $c = 0.097 \pm 0.009$ and $d = 0.68 \pm 0.04$. Page has extended this work to lower altitudes. Page (1975) has shown that values obtained from Parmelee compare favorably with Unsworth's values for solar altitudes of more than 30° and with observations made by Blackwell (1954) at Kew.

TABLE 2

Constants in the Parmelee Formula $G_{dh} = a_0 - a_1 G_{bh}$ for Different Solar Altitude Cloudless Days[a]

Solar altitude (degrees)	a_0	a_1
0	(0)	(0.290)
10	(63.1)	(0.295)
20	134.9	0.314
30	222.1	0.360
40	284.3	0.362
50	383.0	0.424
60	484.6	0.492
70	(552.1)	(0.520)
80	(604.3)	(0.545)
90	(624.7)	(0.560)

[a] The figures in parentheses were obtained by extrapolation.

Similar data obtained by Dogniaux in Belgium also compare well with those derived by Page from Parmelee. Values of G_{dh}/G_h for cloudless days are tabulated for various values of τ_a in Table 3. Unsworth's values, which are independent of solar altitude, for air mass 1.15–2.0 in summer are also given in Table 3. They correspond reasonably well with values derived from Parmelee.

Equation (2) has been incorporated into the Sheffield University clear sky computer programs used to prepare the clear day diagrams in this chapter, the constants a_0 and a_1 being determined for a particular solar altitude by interpolation on the values given in Table 3. Thus, knowing the solar altitude, one can derive the direct irradiance on the horizontal surface G_{bh} from the Monteith and Unsworth turbidity calculation and esti-

TABLE 3

Values of G_{dh}/G_h for Different Turbidity Values and Solar Altitudes—Clear Days without Cloud

Turbidity (τ_a)	Air mass (m): Solar altitude (α):	1.5 42°	2 30°	3 20°	4 14.5	5 11.5	6 9.6	1.15–2.0 Unsworth
0.05		0.092	0.13	0.17	0.22	0.25	0.28	0.13
0.10		0.12	0.17	0.24	0.29	0.34	0.39	0.17
0.20		0.19	0.24	0.33	0.42	0.50	0.57	0.23
0.30		0.24	0.32	0.45	0.56	0.65	0.72	0.30
0.40		0.30	0.41	0.55	0.67	0.77	0.83	0.37

mate the corresponding associated diffuse clear sky irradiance G_{dh} on the horizontal surface.

B. Diffuse Radiation on Vertical and Inclined Surfaces on Clear Days

It is not possible to discuss in detail the relationships between vertical surface diffuse irradiation and horizontal surface diffuse irradiation under clear sky conditions. Figure 10, which is derived from Parmelee's observations, shows the importance of the precise orientation of the surface receiving the diffuse radiation relative to solar position. The ratio of the diffuse vertical surface irradiance to the diffuse horizontal surface irradiance is expressed in Fig. 10 as a function of the cosine of the angle of incidence of the direct sun on the surface for different altitudes of the sun. The large variations of diffuse irradiance with orientation are immediately apparent. This figure demonstrates forcibly that the isotropic approximation is very inaccurate for the estimation of incident, diffuse, cloudless sky radiation on vertical surfaces. Fuller discussion is given by Page (1976). It is desirable to model these nonisotropic features of the clear sky in clear sky computer programs.

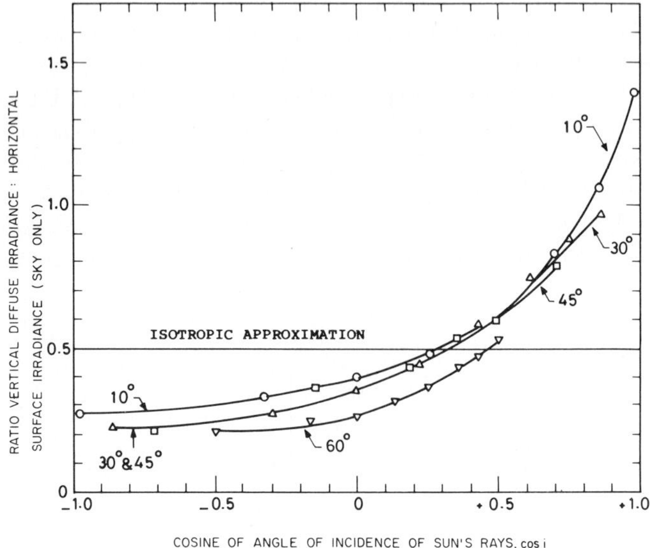

FIG. 10 This diagram, based on Parmelee's work, demonstrates the large errors that may result from using the isotropic approximation for diffuse radiation. Cloudless skies data derived from Parmelee; clarity, 0.8; estimated ground reflected energy subtracted; figures on curves refer to solar altitude.

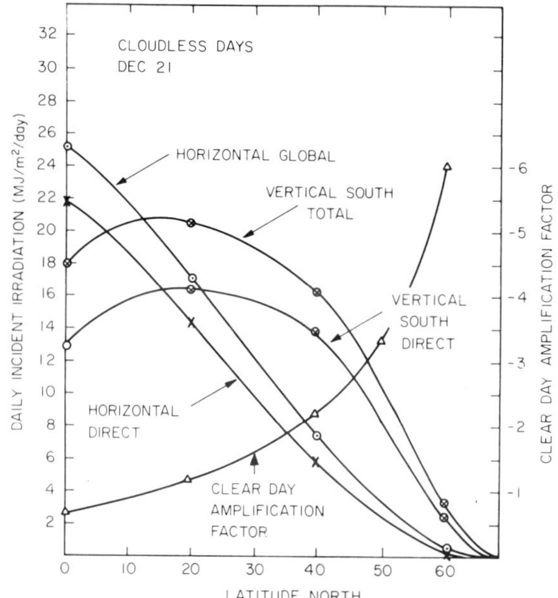

FIG. 11 Variations in the irradiation on a horizontal surface and on a vertical surface facing south with change of latitude; τ_a = 0.150; precipitable H_2O, 5.0 mm; ground albedo, 0.25; December 21.

Figure 11 shows the calculated effects of change of latitude on the cloudless irradiation of a horizontal surface and a vertical south surface, calculated for a fixed turbidity of $\tau_a = 0.150$ and a ground albedo of 0.25, for December 21 using Sheffield University's program SUN3. It shows that in December the most favored latitudes for passive solar houses lie between 30° and 45° N. Further north, much of the low sun's energy is lost in the atmosphere. The ratio of the irradiation on a vertical or inclined surface to the irradiation on a horizontal surface is defined here as the clear sky amplification factor, which may be greater or less than 1. It varies, of course, with orientation. The clear sky, south-facing, vertical surface amplification factor is quite large at high latitudes in midwinter owing to the favorable angle of incidence of the sun on the south wall, but the actual energy available close to the Arctic Circle on December 21 must necessarily be very low because of the long path length.

C. The Climatology of Diffuse Irradiation under Average Conditions

So far this chapter has considered clear sky conditions. However, in cloudy climates diffuse irradiation dominates. The study of the climatology of diffuse radiation is therefore very important in solar building applications, especially at high latitudes and also in the equatorial region where there is also a lot of cloud. Well over half the short-wave energy income may be diffuse in such areas, and it is important to have reasonably accurate methods for estimating the diffuse radiation available on building surfaces.

It is well known that the maximum amounts of diffuse radiation are received on days with partially clouded skies. Broken cumulus, in particular, can give very high values of diffuse radiation, and when the sun is shining through scattered broken clouds, instantaneous values of global radiation can be recorded that are higher than the solar constant.

The basic problem is how, in the absence of local measurements, to separate the monthly mean daily diffuse horizontal surface irradiation $\overline{H}_{dh}$ from the corresponding global irradiation $\overline{H}_h$. Page (1961) found it was possible, using data from stations where both diffuse and direct irradiation are observed, to set up reasonably reliable regression equations of the form

$$\overline{H}_{dh}/\overline{H}_h = c + d(\overline{H}_h/H_{oh}) \tag{3}$$

where $\overline{H}_{dh}$ is the mean monthly value of daily diffuse irradiation on a horizontal plane, $\overline{H}_h$ the mean monthly value of daily global irradiation on a horizontal plane, $\overline{H}_{oh}$ the mean monthly daily irradiation on a horizontal plane in the absence of any atmosphere, and c and d climatically determined regression constants that vary significantly from area to area throughout the world. Liu and Jordan (1960), in contrast, put forward a single formula based on Blue Hill Observatory, which they considered universally applicable, but no shade ring corrections were made.

Typical values of c and d for northern Europe and the United States are given in Table 4, while Table 5 gives some values calculated by the author for the African continent. Corresponding data are available for Canada (Tuller, 1976).

The mean regression equation for the 10 stations scattered across the world, studied by the author in 1961, was found to be

$$\overline{H}_{dh}/\overline{H}_h = 1.00 - 1.13H_h/\overline{H}_{oh} \tag{4}$$

An important feature of this formula for estimating diffuse radiation is that it mathematically predicts the occurrence of a high-level diffuse irradiation with relatively modest amounts of sunshine.

Rearranging Eq. (3), one gets

$$\overline{H}_{dh} = c\overline{H}_h + d(\overline{H}_h/\overline{H}_{oh})^2 \tag{5}$$

This is a parabolic curve, and one may find the maximum value by simple

TABLE 4

Regression Equations for $\overline{H}_{dh}/\overline{H}_h$ against $\overline{H}_h/\overline{H}_{0h}$ for Stations, Based on Period 1965–1970, except Belfast, 1969–1970, Using Monthly Mean Daily Values of $\overline{H}_h$ and $\overline{H}_{dh}$ over Period, and Diffuse Multipliers from Kew

Station	Latitude	Height (m)	Values of c and d	Correlation coefficient	$\frac{-c^2}{4d}$	Diffuse multiplier from Kew
London Weather Centre (1965–1970)	51°31′ N	77	c 0.990 d −1.103	−0.98	0.222	0.949
Kew (1965–1970)	51°28′ N	6	c 0.980 d −1.026	−0.97	0.234	1.000
Bracknell (1965–1970)	51°21′ N	73	c 0.995 d −0.990	−0.95	0.250	1.068
Cambridge (1965–1970)	52°12′ N	13	c 0.937 d −0.841	−0.91	0.261	1.115
Aberporth (1965–1970)	52°08′ N	115	c 1.064 d −1.140	−0.97	0.248	1.061
Lerwick (1965–1970)	60°08′ N	82	c 1.078 d −1.140	−0.96	0.254	1.089
Hamburg (1964–1973)	53°30′ N	—	c 1.043 d −1.038	−0.96	0.262	1.119
Valentia (1964–1974)	51°56′ N	20	c 0.958 d −0.850	−0.94	0.270	1.153
Uccle (1951–1965)	50°49′ N	120	c 0.971 d −0.936	−0.96	0.252	1.076
Blue Hill, U.S.A. (Nov 1945–Oct 1949)	42°13′ N	205	c 0.72 d −0.67	−0.15	0.193	0.827

differentiation; thus it is easy to show that

$$\overline{H}_{dh(max)} = -c^2\overline{H}_{oh}/4d \tag{6}$$

and that this value occurs when $\overline{H}_h/\overline{H}_{oh} = -c/2d$.

Observed relationships between daily irradiation and percentage possible sunshine for two stations close to the equator in Africa are given in Fig. 12. Both global and diffuse daily irradiation on a horizontal surface are plotted against percentage possible sunshine. In the equatorial climate of Kinshasa, the maximum diffuse occurs with percent possible sunshine around 30%. Lubumbashi has a very sunny climate for part of the year. During this period the diffuse radiation is very low compared with Kinshasa at the same percentage possible sunshine, where humid conditions prevail throughout the year. In the wet season at Lubumbashi, the characteristics are similar to Kinshasa. Figure 12 shows that the diffuse radiation

TABLE 5

Values of Constants c and d in Regression Equation $\overline{H}_{dh}/\overline{H}_h = c + d(\overline{H}_h/\overline{H}_{0h})$ *for Stations in the African Continent*

Station	Latitude	Height (m)	Values of c and d	Correlation coefficient	Range of monthly mean values of $\overline{H}_h/\overline{H}_{0h}$ (%)	$\frac{-c^2}{4d}$	Diffuse multiplier	Comments
Kisangani (Sept 1952–Aug 1953)	0°31′ N	437	c 1.07 d −1.16	−0.93	39–53	0.247	1.055	Hot humid climate
Kinshasa (Feb 1951–Dec 1952)	4°22′ S	450	c 1.08 d −1.21	−0.96	32–52	0.241	1.032	Hot humid climate
Windhoek (Aug 1951–Feb 1954)	22°34′ S	1728	c 0.88 d −0.95	−0.95	57–78	0.204	0.871	Hot dry climate, high
Pretoria (Jan 1951–Feb 1954)	25°45′ S	1369	c 0.98 d −1.16	−0.93	55–69	0.207	0.885	Hot dry climate, high
Tananarive (Feb 1953–Feb 1954)	13°53′ S	1310	c 1.20 d −1.39	−0.89	48–65	0.259	1.107	Hot humid climate, high
Durban (Jul 1951–Feb 1954)	29°50′ S	5	c 1.10 d −1.43	−0.97	46–61	0.212	0.904	Hot humid climate
Capetown (Sept 1951–Feb 1954)	33°54′ S	17	c 1.07 d −1.26	−0.93	56–68	0.227	0.971	Mediterranean-type climate

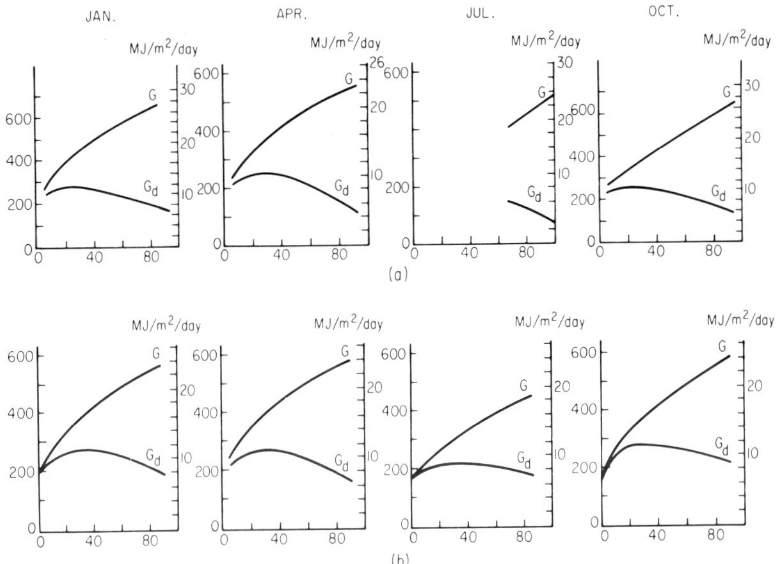

FIG. 12 Relationship between percentage possible sunshine and global and diffuse irradiance on a horizontal surface for two low-latitude stations; (a) Lubumbashi, (b) Kinshasa for the months of Jan., Apr., July, and Oct. (Bultot, 1971).

cannot be obtained by linear interpolation between clear and overcast day observations. Kasten (1977) has prepared comparable data for Hamburg, given in terms of cloud cover and solar altitude for different seasons. The maximum diffuse typically occurs with cloud cover between 4 and 6 oktas, when the diffuse irradiance is typically about 40% greater than the diffuse irradiance for cloudless sky conditions.

Tables 4 and 5 imply that the relative amount of diffuse energy for a given value of the monthly mean atmospheric transmission coefficient $\overline{H}_h/\overline{H}_{oh}$ varies from climate to climate. Some measure of the proportional variation from place to place may be obtained from values of $-c^2/4d$, which defines the maximum values of $\overline{H}_{dh}$ as a ratio to $\overline{H}_{oh}$. Values of $-c^2/4d$ are included in Tables 4 and 5 and are also expressed as a ratio to the corresponding value of $-c^2/4d$ for Kew, England. This figure is used in the average day programs in Sheffield, as a diffuse multiplier to correct the mean diffuse irradiance figures on the computer modeled on Kew observations to match actual observations at other stations more accurately. The multiplier may be greater than 1, for example, in clean, high-latitude, relatively overcast climates, such as Valentia, Ireland; or less than 1, for example, in clear sunny climates such as those found in the inland parts of southern Africa. These corrections appear to hold over quite large regions, provided proper account is taken of local urban influences. The urban influence appears to depress the diffuse multiplier within a specific region by some 5–10%. Experience in the United Kingdom indicates that the reduction in pollution that is being achieved in major cities has had a bigger impact on diffuse energy availability than on direct energy availability.

3.5 HOURLY RELATIONSHIPS FOR AVERAGE DIFFUSE IRRADIANCE ON HORIZONTAL SURFACES

The author has studied climatologically the observed relationships between monthly mean hourly diffuse horizontal surface irradiance and the mean monthly solar altitudes in the middle of each observational hour recorded for several northwestern European stations. Provided long-term means are taken, there is a very high correlation between monthly mean hourly diffuse horizontal surface irradiance and solar altitude for a number of stations in western Europe. The relationship function is remarkably simple:

$$\overline{G}_{\mathrm{dh}} = a' + b'\alpha \tag{7}$$

where $\overline{G}_{\mathrm{dh}}$ is the monthly mean hourly diffuse irradiance for a particular hour when the altitude of the sun at the midpoint of that hour is α. Table 6 presents some values of a' and b' found for northwestern Europe. When the daily irradiation data were studied on a monthly annual basis for a particular station for a run of individual years, it was found that as the average monthly global irradiation increased, the average diffuse irradiation also simultaneously increased in low-sunshine climates. If, however, one was dealing with a very sunny climate, the opposite relationship was found to hold. This observational result is, of course, implied by the parabolic nature of the relationship function already discussed:

$$\overline{H}_{\mathrm{dh}} = c\overline{H}_{\mathrm{h}} + d(\overline{H}_{\mathrm{h}}^{2}/\overline{H}_{\mathrm{oh}}) \tag{8}$$

This relationship was, of course, derived from the study of whole day data rather than from hourly values. The rising or falling characteristic simply depends on whether $\overline{H}_{\mathrm{h}}/\overline{H}_{\mathrm{oh}}$ is above or below the peak value of $-c/2d$. By incorporating this additional relationship function into any suitable computer program, one is able to predict more accurately annual variations of annual monthly mean diffuse radiation from annual monthly mean sunshine data. The monthly mean hourly diffuse relationship functions for the Southern Hemisphere have not been studied in the same detail. The relationship function discussed certainly holds approximately for the Australian data reported by Paltridge and Proctor (1976) and is more accurate

TABLE 6

Constants in Monthly Mean Hourly Diffuse Horizontal Surface Irradiance Formula, $G_{dh} = a' + b'\alpha$ (W/m²), *Where* α *is the Solar Altitude in the Middle of the Hourly Period Considered*

Station	Comment	a'	b'	Correlation coefficient
Data derived from hourly observations				
Kew (1959–1968)	Suburban London	2	4.532	0.996
Eskdalemuir (1959–1968)	Inland site in hills—possibly some pollution influence from Glasgow	2	4.798	0.997
Lerwick (1959–1968)	Northern exposed coastal position	2	5.068	0.997
Aberporth (1959–1968)	Coastal site West Wales	2	5.176	0.994
Hamburg (1964–1973)	Airport site	2	5.36	Fitted by eye; preliminary figure based on four months—Jan, Mar, June, Sept
Valentia (1964–1974)	Extreme westerly coast of Ireland; pollution levels very low	2	5.60	Fitted by eye; preliminary figure based on four months—Jan, Mar, June, Sept
Data derived from daily observations of $\overline{H}_h$ and $\overline{H}_{dh}$				
Kew (1965–1975)	Suburban London with reduced pollution	2	4.804*	
Bracknell (1965–1975)	New town in countryside outside London	2	5.068*	

* Used computer radiation model to determine correction to 1959–1968 period.

than Paltridge's method of estimation in spite of its far greater simplicity. Nevertheless, the relationship appears to hold somewhat less accurately than in the Northern Hemisphere, and it is possible that the remarkably simple result found for temperate areas of the Northern Hemisphere depends on the fact that the influence of the summer increase in turbidity on diffuse irradiation coincides with and precisely balances the summer decrease due to changes in the mean solar distance, while just the opposite situation will exist in the Southern Hemisphere. Pending further studies, Eq. (4) should be used with caution in the Southern Hemisphere. However, the real climatological issue in solar design is how to predict the irradiation of slopes.

3.6 CLIMATOLOGICAL VARIATIONS IN THE MEAN IRRADIATION OF SLOPING SURFACES

At high latitudes the levels of irradiation on horizontal surfaces even on cloudless days are very low in the winter half of the year. In summer the longer day compensates to a considerable extent for the lower midday sun. The low winter horizontal surface values are due partly to the unfavorable angles of incidence. Fortunately the solar energy density falling on a collector may be substantially increased by tilting the collector at an appropriate angle toward the equator. Table 7 presents data on observed values of the mean irradiation incident on vertical surfaces at different stations as a ratio to the mean irradiation simultaneously incident on a horizontal surface. If the mornings are clearer than the afternoons, an appropriate offset toward the easterly direction will give higher incident energy, and if the afternoons are clearer, vice versa. Figure 13 for Kinshasa shows that in some climates very strong asymmetry may occur. The high humidities in the equatorial zone produce heavy cloud obstruction during the morning, which lifts to some extent as the day progresses. Pro-

TABLE 7

Observed Ratios of Mean Monthly Vertical Surface Irradiation to Mean Monthly Horizontal Surface Irradiation

Month	South Kinshasa, 4°22′ S, 1954–1960	South Blue Hill, 42°13′ N, 1952–1956	South Odeillo, 42°29′ N, 1971–1974	South Bracknell,[a] 51°21′ N, 1967–1973	South Hamburg, 53°30′ N, 1952
Jan	0.46	1.80	1.76	1.38	1.40
Feb	0.37	1.38	1.33	1.26	1.30
Mar	0.27	0.93	0.90	1.00	1.15
Apr	0.26	0.61	0.67	0.73	0.80
May	0.30	0.44	0.49	0.60	0.62
June	0.29	0.39	0.43	0.52	0.59
July	0.31	0.42	0.44	0.55	0.60
Aug	0.31	0.54	0.59	0.66	0.75
Sept	0.44	0.79	0.85	0.87	1.00
Oct	0.39	1.23	1.24	1.20	0.96
Nov	0.44	1.60	1.69	1.59	1.30
Dec	0.49	1.94	2.09	1.65	1.50

[a] Original observations excluded ground reflections and are adjusted to a ground albedo of 0.20.

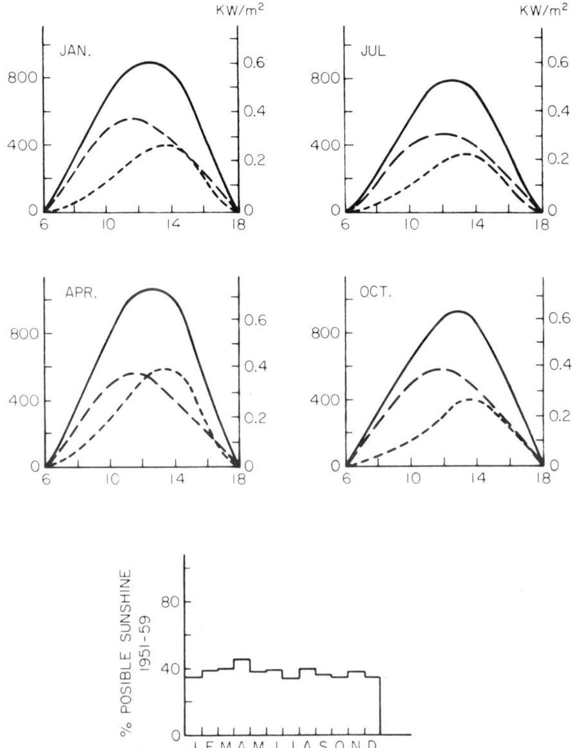

FIG. 13 Variations in mean global, direct, and diffuse irradiance on a horizontal surface in a cloudy equatorial climate. ——, global; – – –, diffuse; - - -, direct. Note the asymmetry of direct irradiance about solar noon: Kinshasa, 4°22′ S, 15°15′ E, 438 m, 1954–1960 (Bultot, 1971).

viding one can estimate the direct beam irradiance, which is possible if the hourly variations of turbidity and sunshine are known, it is fairly easy to calculate the tilt that will receive the greatest amount of direct solar energy on clear and partially clouded days, using trignometrical methods.

However, the estimation of the mean diffuse irradiation on slopes is more difficult because the radiance of the sky is not isotropic. One must also consider the effects of ground albedo, which are greatest for the steepest slopes. Furthermore, the ground albedo may vary strongly with the actual angle of incidence of the direct beam on that surface (Page, 1977). The author presented a fairly simple approach for estimating the available energy on slopes to the Rome conference (Page, 1961), and an updated version has been recently issued (Page, 1976).

At Sheffield University this problem is now handled in more detail on the digital computer in our average day solar radiation program. Hourly data for slopes of any inclination and any orientation are generated from mean monthly values of the sunshine recorded. At present the accuracy is fully tested for the United Kingdom, but at this stage the program appears reliable between latitudes 40 and 60° N. If appropriate values of the observed turbidity are available, it is certainly possible to estimate the mean slope irradiation within ± 10%. The current climatological model adopted considers the diffuse sky irradiation to be made up of two parts, one part coming from the clear blue sky and the other part from the overcast sky. The division of the diffuse into two parts is done on the basis of possible sunshine amount. The clear sky part is distributed in such a way that the area of high radiance around the sun is correctly allowed for, as described by Rodgers *et al.* (1977). The radiance of the diffuse overcast sky part is assumed to follow the Moon and Spencer distribution adopted by the CIE (Hopkinson *et al.*, 1966). The radiance of the overcast sky L_θ at an angle θ from the horizontal plane is related to the zenith radiance L_z by the relationship function

$$L_\theta = L_z(\tfrac{1}{3} + \tfrac{2}{3}\sin\theta) \tag{9}$$

This formula indicates the radiance at the zenith of the overcast sky is three times the radiance at the horizon. The diffuse sky irradiance on a vertical surface under overcast sky conditions is 0.3955 of the diffuse irradiance on a horizontal surface (Walsh, 1961). As a horizontal surface receives the greatest overcast diffuse sky irradiation, so, as the proportion of overcast diffuse radiation rises, the optimal tilt for maximum surface irradiation decreases. However, solar collectors are not linear systems; therefore it may be better to collect energy efficiently on the few clear days using a steeper tilt. The answer to such complex questions, however, can be accurately determined only by detailed simulation.

The Sheffield University average day program has been used to carry out a preliminary systematic study of the effects of sunshine hours and the diffuse multiplier on the monthly mean irradiance amplification factor for vertical south-facing surfaces at midwinter. Preliminary results for December 15 are presented in Fig. 14 for latitudes between 30 and 60° N. As the sunshine hours and atmospheric transmission decrease, so the amplification factor falls; thus the benefits of southerly steep tilts are very small in situations in which the sunshine hours are low. December, of course, is the most extreme month in radiation terms. Figure 14 brings out the enormous differences between sunny and cloudy climates. One must stress the dangers about being too optimistic concerning the average amounts of solar energy available. Clear day analysis can lead to excessively optimistic estimates of solar energy prospects.

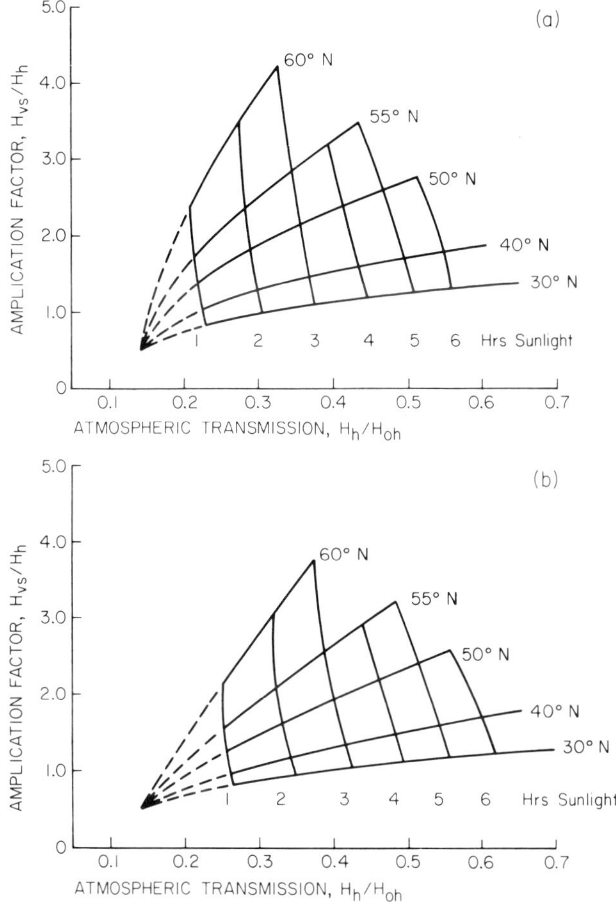

FIG. 14 Multiplication factors for estimating the irradiation of south-facing surfaces from horizontal surface irradiation H_h in different months from sunshine hours: latitudes 30–60° N December 15. This is close to the winter solstice, when amplification factors are greatest. (a) Diffuse multiplier = 0.9, (b) diffuse multiplier = 1.1. Refer to Tables 4 and 5 for typical values of the diffuse multiplier.

Even in low latitudes poor energy availability may be experienced in climates where there is a lot of cloud. Figure 15 compares the mean monthly variations in the vertical surface irradiation for the two low-latitude African sites already discussed. The lower the latitude, the lower the optimal tilt for efficient collection, as a comparison between Figs. 15 and 4 shows.

The broad implications of tilt are clear:

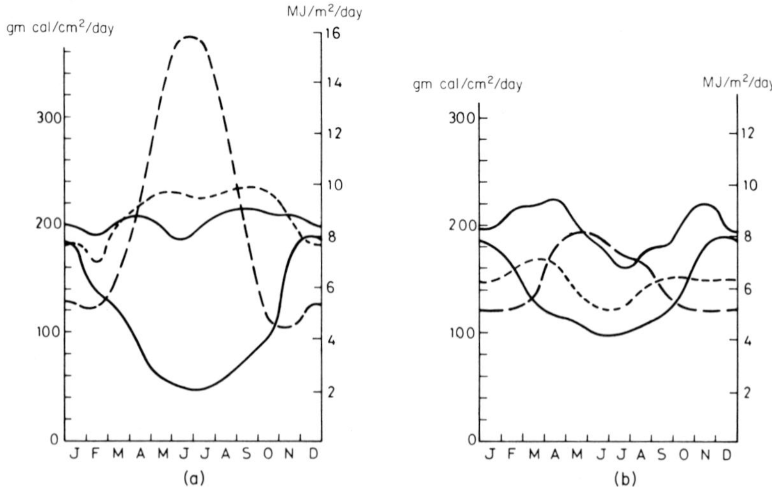

FIG. 15 Mean monthly irradiation falling on vertical surfaces facing two cardinal points at (a) Lubumbashi and (b) Kinshasa. Note roughly equal irradiation of all surfaces in the humid equatorial climate of Kinshasa and the marked differences that appear in the sunny season at Lubumbashi with the sun standing north of the elevation (Bultot, 1971).

(1) Tilting toward the equator reduces the winter–summer energy density range, especially at high latitudes. Steep tilts toward the equator at high latitudes increase winter energy availability and decrease summer energy availability.

(2) The optimum tilt depends on the purpose for which the solar energy is to be used.

(3) The corrections to horizontal surface data for increases in available energy density due to favorable tilt are greatest at high latitudes and least in equatorial regions.

(4) The more diffuse energy there is, the shallower the tilt required for maximum incident energy.

(5) There may be big differences between mean insolation in the morning and the afternoon, which may make it desirable to twist the solar collector in the more favored direction to take account of differences in cloudiness between morning and afternoon.

3.7 CLIMATOLOGICAL FACTORS AFFECTING HEAT LOSSES FROM FLAT PLATE SOLAR COLLECTION SYSTEMS

A. Long-Wave Radiation Exchanges

The overall thermal efficiency of collection of solar energy depends on the balance between energy gains and thermal energy losses. If the collection system is properly insulated, most of the heat loss will occur from the front face. The long-wave radiative losses will depend on the equivalent radiative temperature of the sky, which is strongly influenced by the amount of cloud cover.

Unsworth and Monteith (1975) have shown that measurements of ϕ_{th}, the long-wave flux density on a horizontal surface under cloudless conditions, in the Sudan and also at Sutton Bonington, United Kingdom, are well represented by the straight line.

$$\phi_{th} = c' + d'\sigma T_a^4 \tag{10}$$

where T_a is the screen air temperature in degrees Kelvin and σ is the Stefan-Boltzmann constant. The constants c' and d' were found to have values of -119 ± 16 and 1.06 ± 0.04 W/m², respectively. When the screen air temperature is expressed in degrees Celsius, the cloudless sky relationship reduced to

$$\phi_{th} = 213 + 5.5t_a \quad (\mathrm{W/m^2}) \tag{11}$$

The downward radiation is increased by the presence of cloud, and the flux of downward radiation below different types of cloud can be estimated from Kondratyev (1969). The difference between the outgoing long-wave radiation from a horizontal surface and the incoming short-wave radiation is known as the net long-wave radiation. The balance is of course affected by the emittance of the surface. The balance normally has a negative value. When the radiation temperature of the ground is equal to the screen air temperature, the ratio of the net long-wave radiation under cloudy skies to the corresponding net flux under clear skies is equal to $1 - 0.84c_c$, where c_c is the fraction of sky covered by cloud. Thus, the net long-wave radiation exchange under overcast conditions typically falls to 16% of the clear sky value for a surface close to outside air temperature. The long-wave radiance of the sky is not uniform, and Unsworth (1975b) has dealt in detail with the geometry of interception of long-wave radiation by slopes. Long-wave energy exchanges with the ground become increasingly important as the slope angle increases. Unsworth's paper has been used to prepare an interactive program for predicting long-wave radiation exchanges. Some predicted results are given in Fig. 16, which shows that steeply sloping collectors have a somewhat more favorable long-wave energy balance than collectors with small tilts, especially in clear weather. Clearly, more account needs to be taken of long-wave radiation exchanges in assessing collector performance.

B. Convection Exchanges

The convection losses for collectors and buildings are determined by external wind speeds and air temperatures. The wind speed, of course, in-

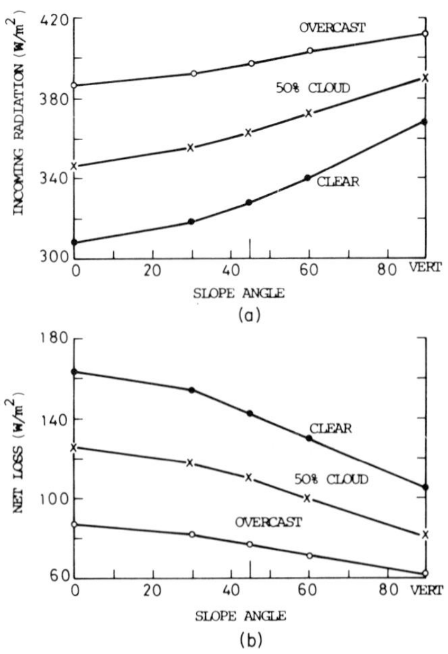

FIG. 16 Effect of slope angle and sky clarity on (a) total incoming long-wave radiation (sky + ground reflectance) and (b) net long-wave radiation loss from surface to sky and ground with air temperature 17°C, collector surface temperature 30°C, collector surface emittance 0.95, ground emittance 0.90, and ground surface temperature 20°C.

creases substantially with height. The aerodynamic factors that influence the detailed flow over the collection surface are important but poorly considered at present, and there is considerable scope for further study of how to reduce the energy losses from collectors due to wind flow. The heat loss per unit temperature difference from the flat glass surface exposed to the outside wind is often calculated from the expression

$$h_{\mathrm{cw}} = 5.7 + 3.8v$$

where h_{cw} is the surface conductance for heat transfer by wind (W/m^2/°C), and v is the wind speed over the surface (m/s).

Wind tunnel and full-scale work on energy transfer being carried out in the author's laboratory in Sheffield, which has so far been confined to the study of the actual energy losses from the windows of a 19-story building, has indicated that the actual heat transfer coefficients are different from those used in conventional building engineering calculations. For example, in one series of runs on the center of the eighteenth floor on the north side of a building, the mean relationship for all wind directions was found to be

$$h_{cw} = 4.4 + 1.4v$$

Wind effects are, of course, particularly important in the case of unglazed collectors. Low mounting positions on sites well sheltered by low trees but not overshadowed are likely to be best for solar collection when using unglazed collectors for swimming pools. In contrast, exposed sites on the tops of poles are favorable for the efficient operation of solar cells whose performance is adversely affected by significant rises of temperature.

Increasing attention is being paid to microclimatic factors in building design (for a general review refer to Page, 1976), but so far little systematic attention seems to have beeen given to the importance of microclimatic considerations in the solar house design field. For example, Fig. 17

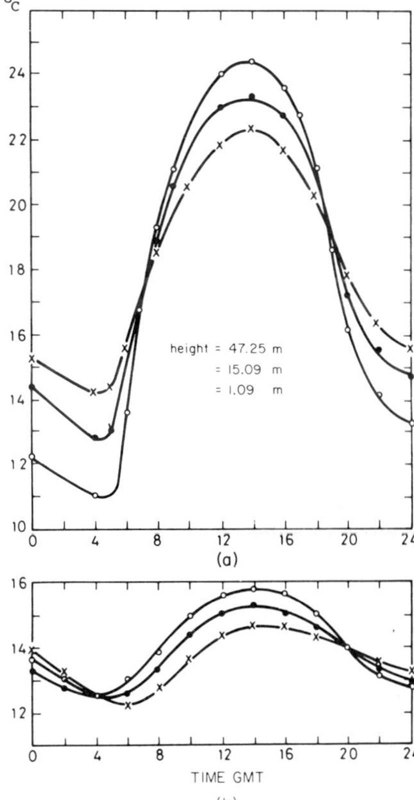

FIG. 17 Effect of height and insolation on air temperature regimes on (a) sunny days and (b) overcast days in southeastern England, summer.

shows typical relationships found between air temperature and height in the United Kingdom for clear and overcast weather.

3.8 METEOROLOGICAL DATA NEEDED TO ASSESS STORAGE NEEDS

The pattern of day-to-day weather is also important, as it affects storage needs. It is not possible to discuss here the best ways of describing the temporal patterns for design purposes. Simple statistical radiation

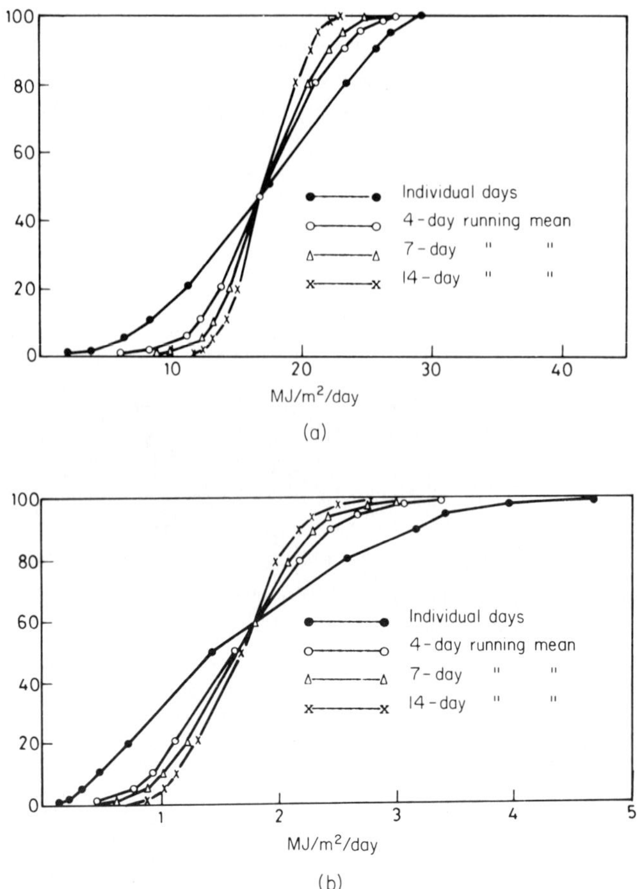

FIG. 18 Cumulative frequency distribution of average values of total daily global solar irradiation based on running means over different periods for (a) June and (b) December at Kew, 1951–1970. ●——●, individual days; ○——○, △——△, ×——×: 4-, 7-, and 14-day running means (Page and Hall, 1976).

data for the United States may be found in Jordan and Liu's recent review (1977). The UK Meteorological Office has compiled statistics on the variations in solar energy on a horizontal surface using running means for different periods of 1 to 14 days. Data presented from the UK–ISES Report (1976) shown in Fig. 18 demonstrate the very large variability in irradiation for single days, especially in December. The variability is substantially reduced if a 4-day period is considered, but there is little further change for the 14-day running means. If short-term storage is to be used in such a climate, the benefits of 14-day storage capacity will be modest compared with those of 4-day storage over 1-day storage. If really long-term storage is available, as in the Danish Zero Energy House (Esbensen and Korsgaard, 1976), then very long-term running means over months can be considered, and the principles of storage design are fundamentally changed in nature. Chapter 4 deals exclusively with thermal storage in buildings.

3.9 SUMMARY OF CLIMATOLOGICAL FACTORS AFFECTING SPACE-HEATING AND SPACE-COOLING DEMANDS

In assessing the climatic feasibility of using solar energy for space heating and cooling, it is obviously important to study simultaneously the impact of climate on the actual energy demands in the building in which the system is to be used. In space cooling, especially in situations in which the radiant heat load through the windows and/or the rest of the structure dominates over the latent heat load, the solar radiation supply will be closely linked to the energy demand patterns, and the problems of solar energy storage will be minimal. A few hours' storage may suffice. In contrast, in space heating the opposite situation will usually prevail, especially at very high latitudes, and the solar supply will be six months out of phase with the heating energy demand.

The solar heat gain variables and the heat loss variables are, of course, statistically interlinked, and the temperature regimes on sunny days are quite different from those on overcast days. The nights may be much colder after sunny days than when the weather is overcast, as the clouds associated with overcast days substantially reduce the outgoing radiation and so stabilize the night–day temperature differences. Figure 19 illustrates the screen air temperature regimes found for days of different radiation classes at Kew, England. Statistically, overcast days in winter are warmer throughout the 24 hours than clear days. In the United Kingdom it is normally windier on overcast days than on clear days. In other parts of the world this relationship could well be reversed, as high insolation in dry desert areas tends to stir up daytime surface winds because of strong

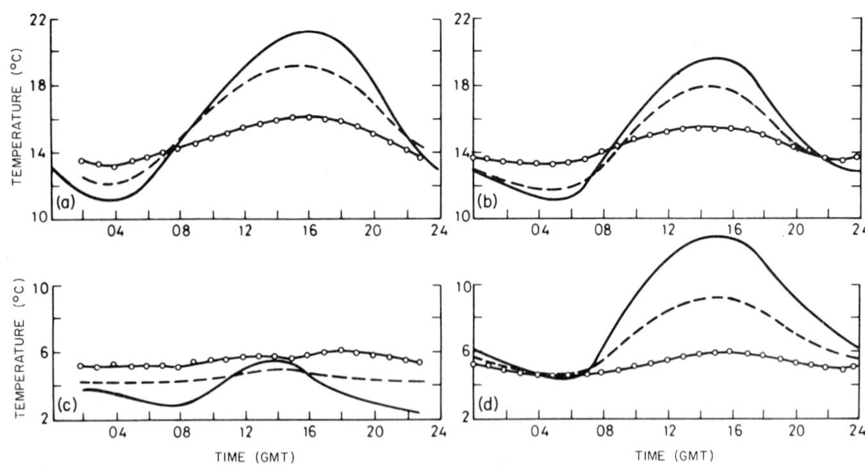

FIG. 19 Diurnal variations in shade air temperature at Kew for three classes of daily global irradiation (—, days of high radiation; —, average days; ○, days of low irradiation) for the months of (a) June, (b) September, (c) December, and (d) March (Page and Hall, 1976).

vertical convection. In the humid tropics wind speeds are often very low on overcast days, and the weather may become very oppressive indeed.

A. Design Implications of Geographical Variations in Radiation Climates

Solar climates, which may be relatively sunny or relatively overcast, can be broadly classified into four latitudinal bands for discussion of the broad impacts on design:

(1) high latitude, 50°+,
(2) middle latitude, 30–50°,
(3) lower latitude, 15–30°,
(4) equatorial, 0–15°.
These bands are necessarily approximate.

One can also distinguish upland and lowland climates. In upland climates the solar beam intensity is greater, but the associated air temperatures are lower than at sea level sites, especially at night.

B. High-Latitude Climates, 50°+

The main features that need to be considered in solar building design are

(1) The irradiation on horizontal surfaces is very low in winter. Steeply tilted surfaces are essential for effective energy collection on

sunny winter days, which may be rare. Figure 20 shows typical radiation impacts on differently orientated vertical facades at Bracknell, U. K., latitude 51°21′ N.

(2) The solar altitude is very low; therefore obstruction by trees and neighboring houses occurs easily. Solar collectors should be mounted as high as possible, especially on low structures, so that obstructions are cleared by the sun.

(3) In sunny climates steep tilts improve performance most in winter. In overcast climates, results of tilting may be disappointingly small.

(4) Typical tilts for optimum winter collection in sunnier high-latitude climates are latitude + 10°, giving tilts of 60 to 70°. However, there is usually very little energy to collect in the middle of winter, and it may be better to optimize performance for the end of the heating season, i.e., mid-February through May, using a tilt equal to the latitude or the latitude − 5°.

(5) As diffuse proportion is usually very high in winter, focusing collection devices are not suitable.

(6) In the Northern Hemisphere solar energy increases rapidly around mid-February onward. Increased energy can be used to shorten the heating season substantially. Winter tends to be long, 6–8 months or more.

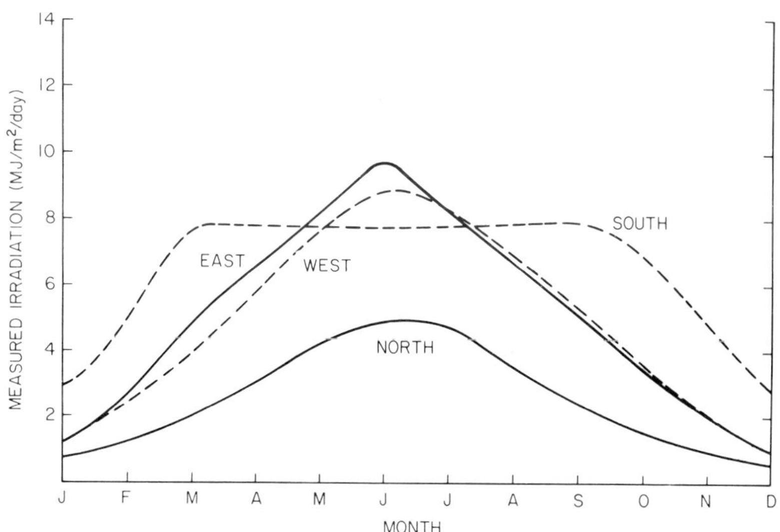

FIG. 20 Measured mean monthly daily irradiation on vertical surfaces facing south, north, east, and west at Bracknell, England, latitude 51°21′ N, 1966–1973. Ground reflections are excluded.

(7) Summer radiation availability is higher than many people realize. The long length of day helps.

(8) Relatively flat slopes give good summer collection. Summer water heating is feasible. Low tilts should be used for summer water heating.

(9) In the absence of adequate storage facilities for long-term storage of solar energy, it is difficult to meet winter needs for space heating.

(10) Short-term storage lasting 2–3 days is essential to meet high intermittency of solar energy supply.

(11) Summer cooling is rarely necessary. Adequate summer comfort can be achieved by appropriate architectural design.

(12) Passive solar houses must take account of potential adverse impacts of summer weather. Provision for sun exclusion is essential.

C. Middle Latitudes, 30–50°

The main features that need to be considered in solar building design are

(1) The irradiation on horizontal surfaces in winter is appreciably higher than at higher latitudes.

(2) The length of the heating season is usually appreciably shorter, especially as one moves toward latitude 30°.

(3) There are generally more sunshine hours in winter and the prospects for winter space heating are often quite good.

(4) The best tilts for winter space heating are typically around 55° at latitude 50° falling to around 35° at latitude 30° with active collectors.

(5) The irradiation on vertical south-facing surfaces falls back from winter into the summer; such facades are to some extent self-protected by their orientation from excessive summer heat gains.

(6) The summer overheating problem is much more acute, and solar houses must be protected against excessive heat gains.

(7) Because the midday sun gets progressively higher as one moves toward the equator, shading of south-facing surfaces by horizontal projections becomes progressively easier. West- and east-facing surfaces are, however, very exposed to excessive insolation, and heavily glazed areas should be avoided on these facades as they are difficult to shade.

(8) The better pattern of winter insolation compared with higher latitudes means that the need for long-term energy storage is much less acute. Short-term solar energy storage may suffice to produce an acceptable solution.

(9) Solar water heating is feasible for a bigger proportion of the

year. In midsummer the irradiation is high, and it may be better to tilt to collect energy most effectively around the equinoxes, i.e., tilt = latitude.

(10) Diffuse proportion is usually less than at higher latitudes, and use of focusing systems becomes more attractive.

(11) Summer cooling becomes increasingly desirable as one moves toward latitude 30°. While climatically appropriate architectural design can substantially improve summer building comfort, a desire for cooling is likely to be expressed.

(12) The tilt of solar systems to collect solar energy for summer cooling should be relatively flat—say latitude minus 15°, i.e., 35° at latitude 50° and 15° at latitude 30°.

D. Lower Latitudes, 15–30°

The main climatic features that need to be considered in solar building design in these regions are

(1) There is usually a high level of irradiation throughout the year. Dust, however, may impair collection efficiency.

(2) The proportion of direct radiation is usually relatively high, though it may fall if there is a marked wet season, say a monsoon period.

(3) The temperatures, except at elevated sites, usually are high enough to make solar space heating unnecessary.

(4) Conditions in midsummer are usually extremely oppressive, and a strong desire for cooling exists.

(5) It is critical to design buildings so that they can be kept cool economically. External shading devices are extremely desirable.

(6) In the heat of the summer, temperatures are very high, and the demand for hot water at that time is usually minimal.

(7) Solar water heaters should be tilted to provide the most energy in the cooler part of the year, say latitude minus 10 or minus 15.

(8) The prospects for focusing devices are extremely good.

(9) Flat plate collectors for cooling systems should be mounted at low tilt angles. Their sensitivity to orientation is not very great in view of the high solar position.

(10) As there is typically a shortage of rain, the ground albedo is frequently very high (0.4–0.5). The reflected radiation from the ground tends to make the building hotter and is also visually unpleasant. A view of the sky is usually preferable to a view of the ground in window design.

E. Equatorial Regions, 0–15°, Not Humid Regions

The main climatic factors influencing solar building design decisions in these regions are

(1) The weather has a high degree of repeatability from day to day, and the annual variations of irradiation are minimal.

(2) There is usually considerable broken cloud, and the diffuse radiation proportion is high.

(3) Concentrating collectors are ineffective because of the low proportion of direct radiation.

(4) There is often a strong asymmetry between morning and afternoon irradiation. Misty mornings may occur, which may clear later. Alternatively, cloud may build up progressively through the day and heavy rain storms occur in late afternoon. Collector orientation should take account of such facts if collectors are placed on vertical or steeply inclined surfaces.

(5) The high sun, which moves on different sides of the building at different times of the year, indicates a preferred use of almost horizontal collectors, but some small tilt helps drainage and reduces maintenance problems due to water.

(6) The regular pattern of weather makes storage problems minimal.

(7) The high humidities cause discomfort owing to difficulty in sweat evaporation. There is a strong demand for air movement in buildings. Dehumidification of the very moist air is needed to promote comfort rather than a drastic drop in temperature. Solar cooling design must recognize this need.

(8) The sweat-producing climate makes bathing a pleasant relief. There is a demand for solar water heating, which is relatively easy to meet, as the temperature rises needed are small.

3.10 CONCLUSIONS

The complex climatological interactions in solar building design can be properly assessed only by appropriate local meteorological studies and observations. There is a need to persuade meteorological services to prepare meteorological summaries for solar house design that respect a design logic that needs the simultaneous assessment of solar heat gains and collector heat losses. Guidance is also needed on the effects of height and shelter.

Further collaboration between meteorologists, building scientists, and architects will be needed to produce a sounder climatological foundation for solar house design. The computer can play an important role in assisting developments in this area, because many of the calculations are relatively long and tedious to perform. Considerable further detailed meteorological study is still needed to make design procedures for solar houses more reliable for various parts of the world. For example, more study is needed of the links between meteorological factors influencing the demands for energy and the factors that influence the supply of solar energy. The microclimatological considerations need greater attention. As solar house design is a very difficult systems engineering problem, collaboration in systems analysis will be vital to success, and the will to achieve effective interdisciplinary communication essential.

ACKNOWLEDGMENTS

The computer-based research project on solar radiation estimation reported here was partly financed by the Science Research Council. The help of my collaborators in this project, G. G. Rodgers and C. G. Souster, has been invaluable in the task of preparing the tables and diagrams. Figures 1, 3, 6, 18, and 19 have been drawn from the UK Section of the International Solar Energy Society's comprehensive report, "Solar Energy—a UK Assessment," and I am grateful to the Society for letting me reproduce them. Figures 5, 12, 13, and 15 have been taken from the excellent "Atlas Climatique du Bassin Congolais" which provides a fund of good information on how to present radiation and other meteorological data for a whole territory (Bultot, 1971). I am grateful to Dr. Bultot for permission to reproduce them.

4

Solar Thermal Energy Storage Systems

MEHDI N. BAHADORI

DEPARTMENT OF MECHANICAL ENGINEERING
SCHOOL OF ENGINEERING
PAHLAVI UNIVERSITY
SHIRAZ, IRAN

A. A. M. SAYIGH

MECHANICAL ENGINEERING DEPARTMENT
COLLEGE OF ENGINEERING
UNIVERSITY OF RIYADH
RIYADH, SAUDI ARABIA

4.1 INTRODUCTION

One of the drawbacks of solar energy is that it is not a continuous source of energy; it varies annually and diurnally. It also varies randomly owing to variable cloud cover. Therefore, it is an intermittent source of energy (Sayigh, 1977b). Nature has solved the storage problem of solar energy through the process of photosynthesis. It converts solar energy into a chemical form that can be used directly (combustion) or indirectly by fermentation process that lead to other forms of fuel. Therefore the basic energy problem is to devise ways of obtaining energy whenever and in whatever form available, then converting it (if necessary) to forms best suited for storage, and reconverting it, with minimal loss, to a useful form at the time it is needed.

The idea of thermal storage goes back to the caveman, when he used to choose his cave facing south so that solar radiation would supply him with heat during the day; and because of minimum conduction–convection exchange with the surroundings, the inside of the cave would

ISBN 0-12-620860-3

still be warm during the night (Sayigh, 1976f). For storing ice from wintertime to summertime, the Romans used to dig long, deep cellars in the side of the hills, which they filled with snow and ice during the winter. The snow and ice remained frozen all through the hot summer, because the rock and earth acted as insulators from the outside as well as being good storage materials. Another early example of rock storage is that of Montezuma's Castle, located about 120 km north of Phoenix, Arizona. The castle was completed in 1300 AD. In wintertime, because of the rock thermal storage properties of this masonry building and because of the time lag imposed by its thick walls, when the sun shone upon the south-facing masonry during the day, it provided sufficient heat for the night (Kreith, 1973). Although man has devised a number of systems for solar energy storage, none of them is as efficient as nature. Figure 1 shows solar conversion and storage in general.

The selection of a storage material must be based on the following criteria:

(1) The material should not be toxic.

(2) It should be neither flammable nor combustible.

(3) The material should have a long lifetime, approximately 10 to 15 years, and be noncorrosive.

(4) It should be inexpensive and available in abundance.

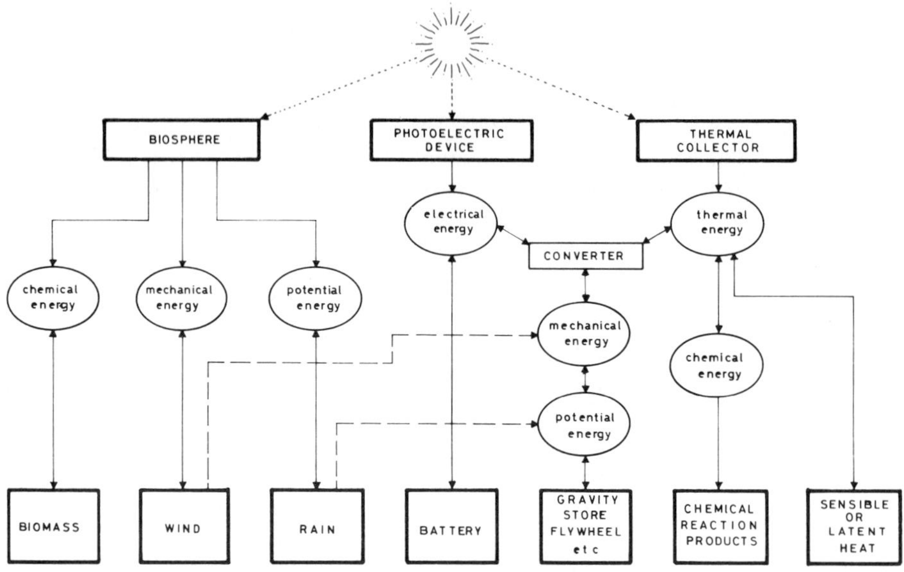

FIG. 1 Conversion and storage of solar energy.

(5) The material should have a high-energy storage capacity (Bahadori, 1976a).

4.2 SOLAR STORAGE REQUIREMENTS

Solar energy is not available every day nor does it have a constant flux when it is available. Also, various buildings require different heating and cooling loads. Therefore, if the heating or cooling requirements for a building are known, the thermal storage capacity can be assessed so that the rate of solar energy charging such storage (Q_u) should be equal to or greater than the load rate requirement (Q_L) plus the rate of heat loss to the surroundings (Q_S), i.e.,

$$mC_p(dT/dt) = Q_u - Q_L - Q_S \tag{1}$$

where m is the mass and C_p the specific heat of the material.

For short-term storage, three hypothetical cases for solar radiation absorbed by given collectors are now considered:

(a) Variations of absorbed solar energy and building heating load for a 24-hr period, as shown in Fig. 2. Here it is assumed that the amount of utilized solar energy is less than that required to heat the building, and therefore an additional source of heat must be brought to meet such requirements.

(b) Variations of utilized solar energy and building heating load with time for a 72-hr period, where two consecutive cloudy days prevail. Here, the collector is assumed to be large enough to meet the 3-day heating requirements (see Fig. 3).

(c) The solar energy absorbed by the collector is sufficient for the

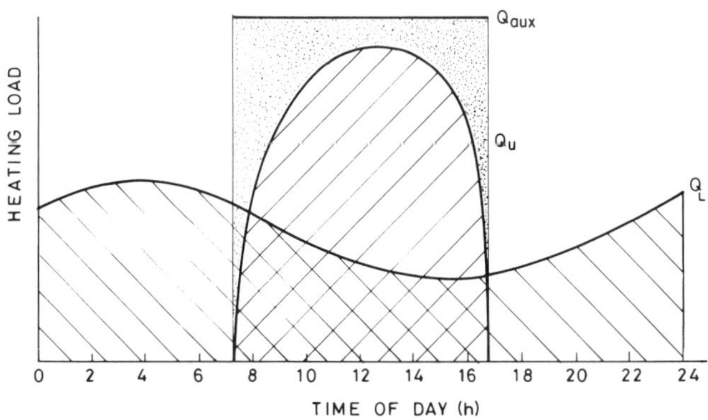

FIG. 2 Variations of utilized solar energy and building heating load with time.

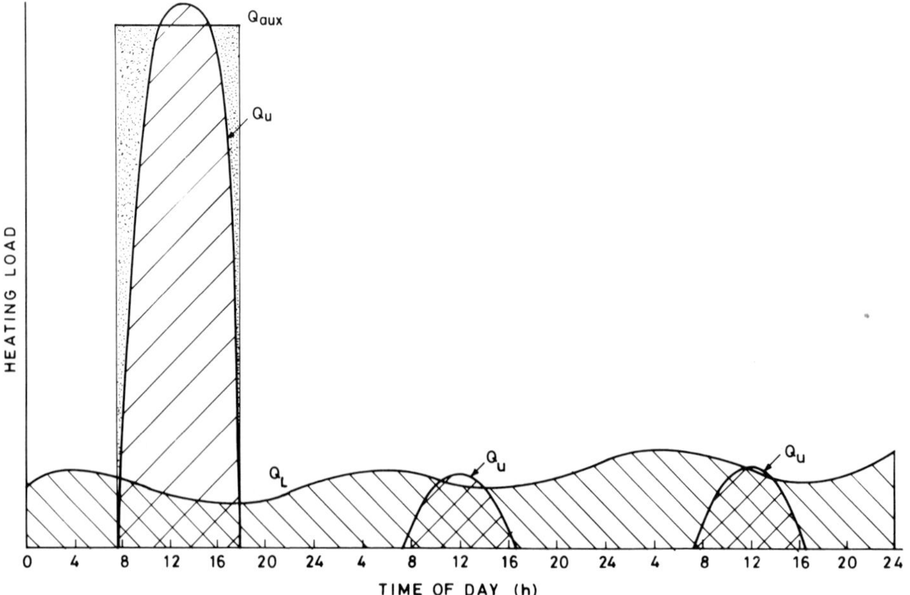

FIG. 3 Variations of utilized solar energy and building heating load with time for a 72-h period.

production of cooling load for a given building (see Fig. 4) (Bahadori, 1976a).

It is worth noting that the maximum cooling load requirement almost coincides with the maximum solar energy utilization during the day. Here again proper selection of the collector and proper storage will ensure the balance between solar energy utilization and cooling load.

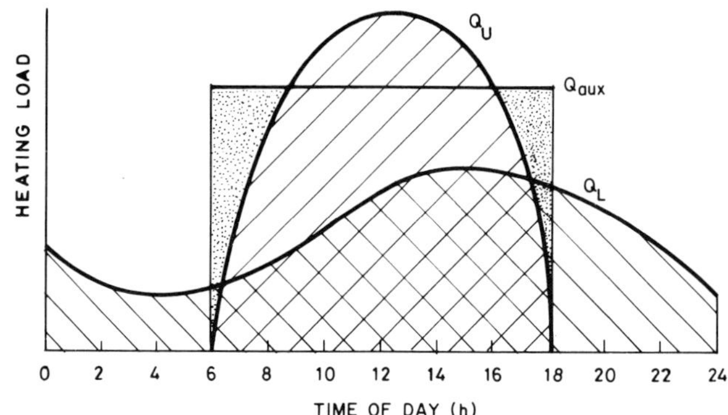

FIG. 4 Variations of utilized solar energy and building cooling load with time.

4.3 SENSIBLE HEAT STORAGE

Sensible heat storage is the most obvious method of heat storage and hence the most common. It can be summarized by the equation

$$q = mC_p \, \Delta T \tag{2}$$

But

$$m = v\rho \tag{3}$$

Therefore

$$q = v \, \rho C_p \, \Delta T \tag{4}$$

For a given volume and given temperature difference, the storage material is characterized by ρC_p, and the thermal diffusivity $k/\rho C_p$; the larger is ρC_p for a specific material, the greater the storage capacity. Table 1 shows a comparison of various common storage materials. From this it can be seen that the storage can be a water tank, a swimming pool, a solar pond, a container with gravel or pebbles, or part of the building itself, such as walls, roofs, or floors.

TABLE 1

Properties of Some Common Heat Storage Materials

Storage material	Specific heat C_p (Kcal/kg °C)	Density ρ (kg/liter)	Temperature T (°C)	ρC_p (kcal/liter °C)	Thermal diffusivity $k/\rho C_p$ (cm²/s)
Air (at P_0)	0.2415	0.0012	20	0.0003	—
Asbestos	0.2500	0.1500	—	0.0375	0.0083
Alcohol	0.55–0.65	0.8000	0–40	0.4800	0.0009
Brine	0.7090	1.2000	0	0.8508	0.0026
Brick	0.2000	1.7000	21	0.3400	0.0033
Charcoal	0.2420	0.4000	0–70	0.0968	0.0170
Concrete	0.2000	2.2000	21	0.4400	0.0082
Glass	0.16–0.20	2.515	—	0.4527	0.0045
Ice	0.5000	0.9180	−1 to −21	0.1836	0.0292
Porcelain	0.2550	2.3000	15–1000	0.5865	0.0041
Rock	0.2100	2.5600	20	0.5400	0.1416
Sand (dry)	0.1910	1.6000	—	0.3056	0.0026
Steel	0.1200	7.8500	0–200	0.9420	0.1243
Water	1.000	1.0000	4	1.0000	0.0014
Wood (oak)	0.5700	0.7700	0–70	0.4389	0.00113

A. Characteristics of Water as a Storage Material

Water, so far, has proved to be one of the best materials for solar thermal storage, mainly because of the following:

(1) It is nontoxic and nonflammable.
(2) It is cheap and abundantly available.
(3) It is a substance for which physical, chemical, thermodynamic, and technical data are well known.
(4) It can be used as a storage material as well as an energy collection medium.
(5) Water has the best storing properties for a given volume, as shown in Table 1.
(6) It has very suitable liquid–vapor equilibrium temperature ranges, of interest in heating and cooling.
(7) Its heat transfer and fluid dynamic properties are good.
(8) Its corrosion inhibitor technology is well advanced.

The disadvantages are

(1) Damage can occur as a result of expansion when the water is frozen.
(2) Using water will encourage electrolytic corrosion.
(3) Water is a good solvent for oxygen gas and hence promotes corrosion.
(4) A pollution problem arises with respect to the disposal of water containing toxic anticorrosion materials (Pickering, 1975).
(5) The heat recovery from water is not isothermal, as it could be in the case of phase change materials.

B. Hot Water Storage System

Now let us consider Eq. (1) and the calculations of its various terms for a given storage. The storage system considered here is water tank storage without stratification effects to simplify the calculations. Therefore,

$$Q_S = WA(T - T_a) \tag{5}$$

$$Q_u = \dot{m}_c C_p (T_{co} - T_{ci}) \tag{6}$$

$$Q_L = \dot{m}_L C_p (T_{Li} - T_{Lo}) \tag{7}$$

where T is the average water temperature in the tank, W the average overall heat transfer coefficient between the water in the tank and the ambient temperature, A the heat transfer area of the storage tank, T_a the ambient temperature, $\dot{m}_c$ the mass flow rate of water through the collector, $\dot{m}_L$ the mass flow rate of water through the load, T_{co}, T_{ci} the outlet and inlet water temperatures of the collector, and T_{Li}, T_{Lo} the inlet and outlet water temperatures of the load.

In order to calculate the various temperatures, the following procedure can be used. From Eq. (1),

$$T_{t+\Delta t} - T_t = (\Delta t/mC_p)[Q_u - Q_L - WA(T_t - T_a)] \tag{8}$$

For convenience, take $\Delta t = 1$; therefore at $t = 9$;

$$T_{10} - T_9 = (1/mC_p)[Q_u - Q_L - WA(T - T_a)]_{t=9} \tag{9}$$

T_a, T, and Q_L are known; W can be evaluated by knowing the film coefficients for the water inside the tank, T_a, and the type of insulation used on the tank (Krieth, 1973). Q_u can be determined from the collector's performance. The mass of water in the tank m and also C_p can be considered constant. Figure 4 shows hypothetical values of Q_u and Q_L for constants T_a, W, m, and C_p (Duffie and Beckman, 1974).

If stratification in the storage tank exists, then the tank can be divided into several zones, each with a constant temperature. By applying the first law of thermodynamics to each zone and then solving these equations numerically (Klein *et al.*, 1975; Gutierrez *et al.*, 1974), the temperature distribution or profile in the storage tank can be evaluated.

If the fluid in the collector and that of the load are not the same as that in the storage tank, then the energy balance equation is still given by Eq. (1). But Q_u and Q_L can be written

$$Q_u = \int_{T_{ci}}^{T_{co}} W_1(T_1 - T)\, dA_1 \tag{10}$$

$$Q_L = \int_{T_{Lo}}^{T_{Li}} W_2(T - T_2)\, dA_2 \tag{11}$$

where W_1 and A_1 are the overall heat transfer coefficient and area of the collector heat exchanger, respectively, and W_2 and A_2 the corresponding values for the load heat exchanger in the storage tank. T_1 and T_2 are the variable water temperatures through the collector and tank load heat exchanger, respectively. Combining Eqs. (6) and (10) and, similarly, Eqs. (7) and (11) (Duffie and Beckman, 1974), we get

$$(T_{co} - T_{ci})/(T_{co} - T) = 1 - e^{-(W_1 A_1/\dot{m}_c C_p)} \tag{12}$$

$$(T_{Li} - T_{Lo})/(T - T_{Lo}) = 1 - e^{-(W_2 A_2/\dot{m}_L C_p)} \tag{13}$$

Figures 5 and 6 show various systems for water storage.

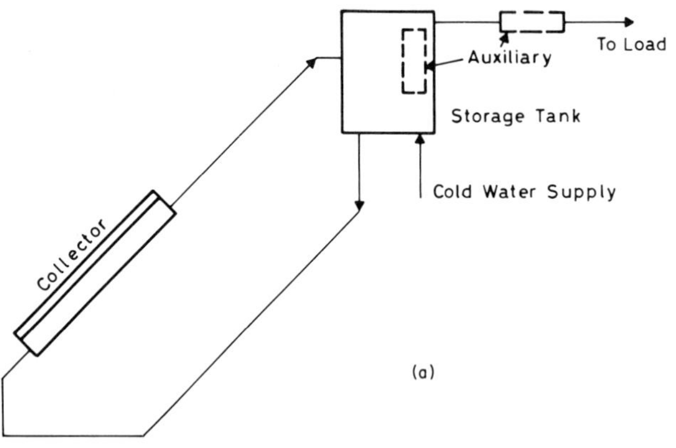

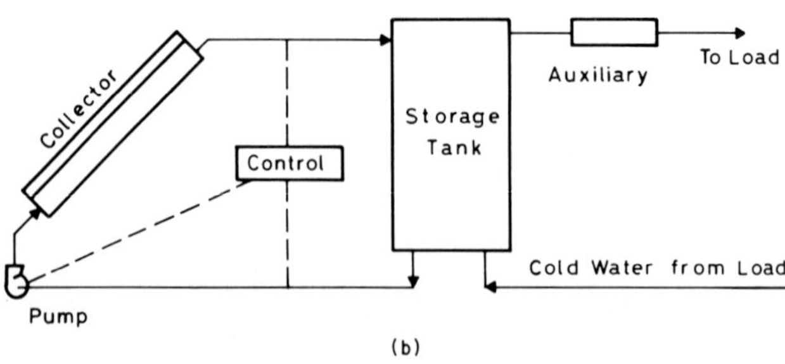

FIG. 5 Schematics of solar collector and thermal storage in water: (a) natural circulation, (b) forced circulation. Auxiliary heating may be provided in the storage tank or as a separate unit.

C. Water Storage Tank Materials and Insulations

The storage tank should be made of material capable of withstanding two atmospheric pressures, or preferably more, and a temperature of 95°C. The tank should be corrosion-free and have a lifetime of no less than 10 years. Additionally, it should have poor thermal conductivity and low cost. The common materials used for water storage tanks and the various insulations are shown in Table 2.

D. Thermal Energy Storage in Buildings

Buildings store thermal energy in their walls, roofs, floors, etc. In hot and dry areas of the world with a high daily temperature range, one often finds buildings built with massive walls of brick, stone, or adobe, and

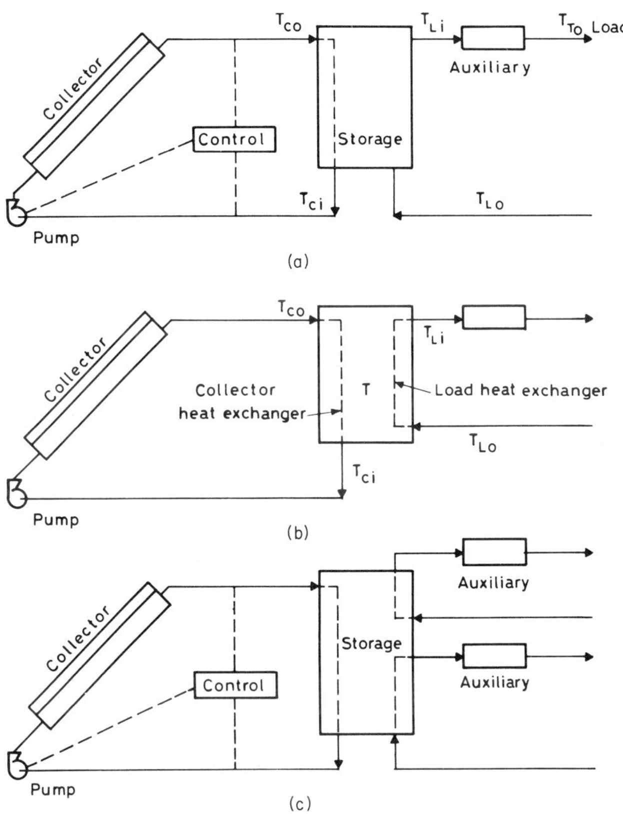

FIG. 6 Schematics of thermal storage in water: (a) collector water separated from storage and load; (b) storage thermal analysis; (c) one storage tank used for all of the loads.

roofs with more than 30 cm of sod on top. The summer performances of these buildings are excellent. At night with cool air allowed into the structure and with the thermal radiation loss to the clear sky, the building's mass is cooled appreciably. The building's thermal capacity absorbs the heat gained during the day, and a pleasant low temperature is maintained during the summer. In winter the thermal inertia of the building again plays an important role in maintaining a moderate daily temperature.

There have been a number of modern designs that make use of solar thermal storage (Barkmann and Wessling, 1975). One such design (Walton, 1973), in which the solar collector and thermal storage are incorporated to provide solar heating in winter and ventilation in summer, is shown in Figs. 7 and 8.

TABLE 2

Common Water Tank Storage Materials and Insulations

Storage material	Density ρ (g/cm³)	Thermal conductivity k (J · cm/cm² s °C)	Insulation	Density ρ (g/cm³)	Thermal conductivity k (J · cm/cm² s °C)
Glass	2.50		Asbestos cloth	0.15	0.0013
Aluminum	7.70	2.340	Asbestos wool	0.15	0.0006
Steel	7.85	0.490	Cork	0.25	0.0004
Concrete	2.200	0.015	Cotton wool	small	0.0003
Brick	1.700	0.012	Felt	small	0.0004
Copper	8.93	3.830	Glass wool	small	0.0004
Porcelain	2.300	0.010	Polystyrene	1.050	0.0004
Rock	2.56	0.320	Polyurethane	0.920	0.0003
Wood	0.77	0.0015			
Fiberglas	1.80	0.0035			
Teflon	2.200	0.0025			

E. Thermal Energy Storage in Rock Beds

The second most common material used for thermal energy storage in buildings is rock. This extensive use of rock is due mainly to the following characteristics:

(1) It is nontoxic and nonflammable.
(2) Rock is plentiful and cheap to purchase.
(3) Rocks act as both heat transfer surfaces and storage media.
(4) The thermal energy is transmitted to rock beds primarily by air circulation through the beds. Therefore the heat transfer between air and rock beds is good mainly because of the large heat transfer area. Also, the

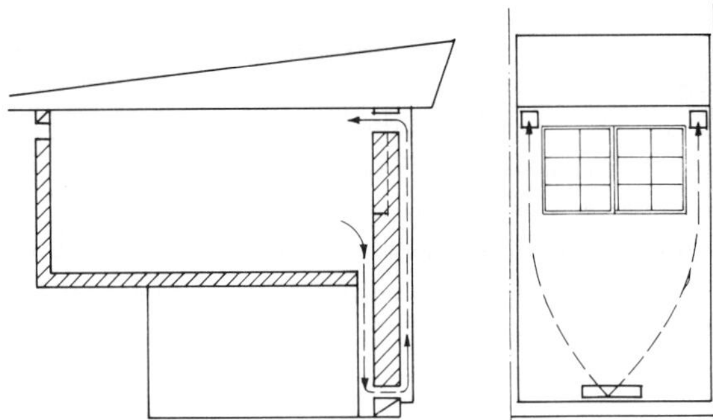

FIG. 7 Winter operation: schematics of solar collector and thermal storage in a building.

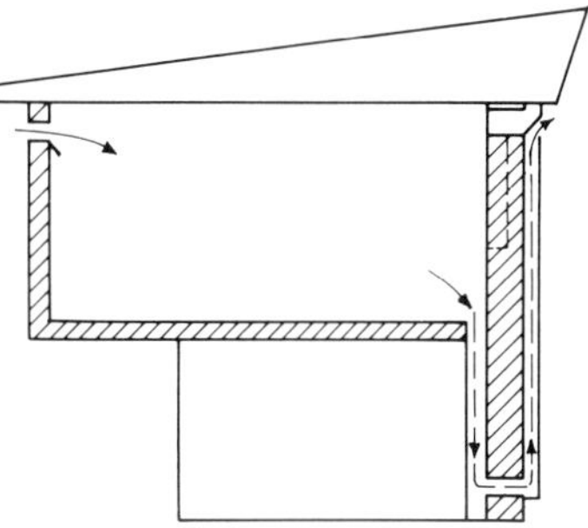

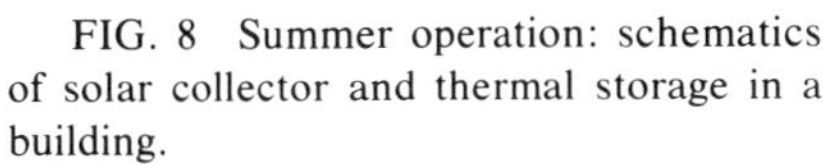

FIG. 8 Summer operation: schematics of solar collector and thermal storage in a building.

heat conductance of the rock pile is low because the area of contact between the rocks is small. This leads to low heat loss from the pile. These advantages make rock pile storage fairly efficient because air can leave a rock bed at a temperature nearly equal to that of the rocks at that point, which in turn is nearly equal to the temperature of the hot air entering the pile. This makes it possible to deliver heat from the storage at almost the maximum temperature of operation, which is independent of the amount of the stored energy.

The disadvantages of rock pile storage are

(1) It requires a larger volume than water storage (about double) (see Table 1).

(2) Greater power is required to circulate the air through a rock bed than to circulate water through water storage systems.

Rocks can be divided into two types: igneous rocks and sedimentary rocks. Table 3 shows the various densities for the two types.

A rock pile bed used for thermal storage should consist of a container that is normally cylindrical in shape, a porous structure to support the pile, and dividing fins for achieving uniform flow of air in both directions and hence minimizing air channeling.

In order to take advantage of the buoyancy effects and to ensure a satisfactory operation of the rock pile, the hot air should be admitted from the top and the cold air from the bottom of the pile, which is similar to a hot water storage system (Bahadori, 1976a; Sayigh, 1976f).

F. Theoretical Analysis of Rock Pile Thermal Storage

Several authors have studied the flow of air through packed beds (Carman, 1956; Leva *et al.*, 1951; Scheidegger, 1957; Ranz, 1952) by assuming the particles are made of spheres with void fractions of 30 to 60%.

TABLE 3

Types of Rock and Their Densities

Igneous rocks		Sedimentary rocks	
Name	Density ρ (g/cm^3)	Name	Density ρ (g/cm^3)
Granite	2.61–2.75	Limestone	2.6–2.7
Syenite	2.61–2.75	Dolomite	2.6–2.7
Ryolite	2.61–2.75	Sandstone	2.5–2.7
Trachyte	2.55–2.75	Conglomerate	2.5–2.7
Diorite	2.80–2.90		
Quartz diorite	2.70–2.85		
Dacite	2.70–2.85		
Andesite	2.70–2.85		
Basalt	2.90–3.10		
Obsidian	2.30–2.70		

For a bed with N particles, the hydraulic radius is

$$r_H = \frac{\text{void volume of bed}}{\text{surface area of packing}} \tag{14}$$

If v_p is the volume of a particle, and S_p the surface area of a particle, then the specific surface of a particle is

$$S_v = S_p/v_p \tag{15}$$

For a spherical particle,

$$S_v = 6/D \tag{16}$$

where D is the sphere diameter. If the particle is not of a spherical shape, then

$$D = 6/S_v \tag{17}$$

can be used. If e is the void fraction, then substituting in Eq. (14); one obtains

$$r_H = \frac{ev_pN/(1-e)}{S_pN} = \frac{ev_p}{(1-e)S_p} \tag{18}$$

or

$$r_H = e/(1-e)S_v = \tfrac{1}{6}[e/(1-e)]D \tag{19}$$

But

$$\mathrm{Re} = 4 r_{\mathrm{H}} u \rho / \mu$$

Therefore

$$\mathrm{Re} = [4e/6(1 - e)]((Du\rho/\mu) \tag{20}$$

where u is the average interstitial velocity at any cross section in the bed. It is better to use a superficial velocity u_{s} based on the cross section of the empty container; i.e.,

$$u_{\mathrm{s}} = eu \tag{21}$$

Therefore

$$\mathrm{Re} = [4/6(1 - e)](Du_{\mathrm{s}}\rho/\mu) \tag{22}$$

Hence, the friction factor f will be

$$f = \frac{4gr_{\mathrm{H}}(lw)}{2Lu^2} = \frac{gD(lw)}{3Lu_{\mathrm{s}}^2}\frac{e^2}{(1 - e)} \tag{23}$$

where lw is the friction loss. Another definition for Re and f are

$$\overline{\mathrm{Re}} = Du_{\mathrm{s}}\rho/\mu(1 - e) \tag{24}$$

$$\bar{f} = gD(1w)e^3/Lu_{\mathrm{s}}^2(1 - e) \tag{25}$$

Therefore

$$\bar{f} = (150/\overline{\mathrm{Re}}) + 1.75 \tag{26}$$

For laminar flow, $\overline{Re} < 1.0$ and $\bar{f} = 150/\overline{Re}$, while for turbulent flow, $\overline{\mathrm{Re}} > 10^4$ and $\bar{f} = 1.75$. For $1.0 < \overline{\mathrm{Re}} < 10^4$, Eq. (26) can be used. If the particles are not of the same sizes, then

$$S_v = \Sigma x_i S_{v_i} \tag{27}$$

where x_i is the volume fraction. Then

$$D = \frac{6}{S_v} = \frac{6}{\Sigma x_i(6/D_i)} = \frac{1}{\Sigma(x_i/D_i)} \tag{28}$$

Knowing e, one can calculate S_v and then D. Then, knowing ρ and u_{s}, one can calculate $\overline{\mathrm{Re}}$ and then $\bar{f}$; therefore the pressure drop can be calculated.

In order to calculate the average heat transfer coefficients for spheres (h), the Froessling correlation can be used (Ranz, 1952):

$$hD/k = 2.0 + 0.6(C_{\mathrm{p}}\mu/k)^{1/3}(Du\rho/\mu)^{1/2} \tag{29}$$

A concise approach for determining the coefficients of heat transfer for forced convection through packed beds is given by Bird *et al.* (1960); this is in fact

$$dQ = h(aS\,dz)(T_0 - T_b) \tag{30}$$

where $S\,dz$ is the bed volume (solid plus fluid) between two given cross sections distance dz apart in the flow direction, and a is the solid particle surface area per unit volume. A large number of experiments were carried out for various beds and the following empirical correlation was deduced:

$$j_H = 0.91\ \mathrm{Re}^{-0.51}\ \phi, \qquad \textit{for} \quad \mathrm{Re} < 50 \tag{31}$$

$$j_H = 0.61\ \mathrm{Re}^{-0.41}\ \phi, \qquad \textit{for} \quad \mathrm{Re} > 50 \tag{32}$$

where the Colburn factor j_H and Reynolds number Re are defined by

$$j_H = h/C_pG(C_p\mu/k)_f^{2/3} \tag{33}$$

$$\mathrm{Re} = G/a\mu_f\phi \tag{34}$$

The subscript f denotes properties evaluated at the film temperature $T_f = \frac{1}{2}(T_0 + T_b)$, where T_0 is the container temperature T_b the fluid temperature, and $G = \rho u_s$ the superficial mass velocity. ϕ is an empirical coefficient that depends on the shape of the particle (see Table 4). Now, if air is flowing through the storage tank, which is divided into several elements, say n, each of thickness dz, and if the radial temperature gradient is negligible, then for a section i with area A and temperature T_i, for which ρ and C are the apparent density and specific heat of the rock, and the air temperatures at the ith and $(i + 1)$th sections are $T_{a,i}$ and $T_{a,i+1}$, then

$$(\rho CA\ \Delta z)\frac{dT_i}{dt} = h_vA\ \Delta z(T_{a,i+1} - T_i) - q_i \tag{35}$$

$$(\dot{m}C_p)_a(T_{a,i+1} - T_{a,i}) = h_vA\ \Delta z(T_{a,i+1} - T_i) \tag{36}$$

where q_i is the heat loss from section i, and $(\dot{m}C_p)_a$ the product of the mass flow rate and the constant pressure specific heat of air going through the rock bed. h_v is the volumetric heat transfer coefficient and is given by

$$h_v = 650(G/D_i)^{0.7} \tag{37}$$

TABLE 4

Particle Shape Factors for Packed Bed Correlations[a]

Shape:	Sphere	Cylinder	Flake	Rasching ring	Partition ring	Berl saddle
ϕ:	1.00	0.91	0.86	0.79	0.67	0.80

[a] From Bird *et al.* (1960).

where G is the mass velocity (kg/m^2 s), D_i the equivalent spherical diameter (m) (Löf and Hawley, 1948), and

$$D_i = \frac{6}{\pi} \left(\frac{\text{net volume of particles}}{\text{number of particles}} \right)^{1/3} \tag{38}$$

Equations (35) and (36) can be written for n sections of the bed. This provides $2n$ equations. When energy is withdrawn from the storage, a similar set of equations can be written. Owing to different air mass flow rates and hence different h_v because of different temperatures in the charging and withdrawal processes, different sets of equations will result. When h_v is large and the Biot number is less than 0.1 ($k/h\mathrm{r}_0 < 0.1$, where r_0 is the radius of each rock), the air temperature leaving section i will be T_i or $T_i = T_{\mathrm{a},i}$ and $T_{i+1} = T_{\mathrm{a},i+1}$; therefore Eqs. (35) and (36) can be combined as

$$(\rho C A \ \Delta z) \frac{dT_i}{dt} = (\dot{m} C_{\mathrm{p}})_{\mathrm{a}} (T_{i+1} - T_i) - q_i \tag{39}$$

This leads to n equations instead of $2n$ equations (Löf and Hawley, 1948).

Rock sizes of 1 to 7 cm have been used for thermal storage, and the re-

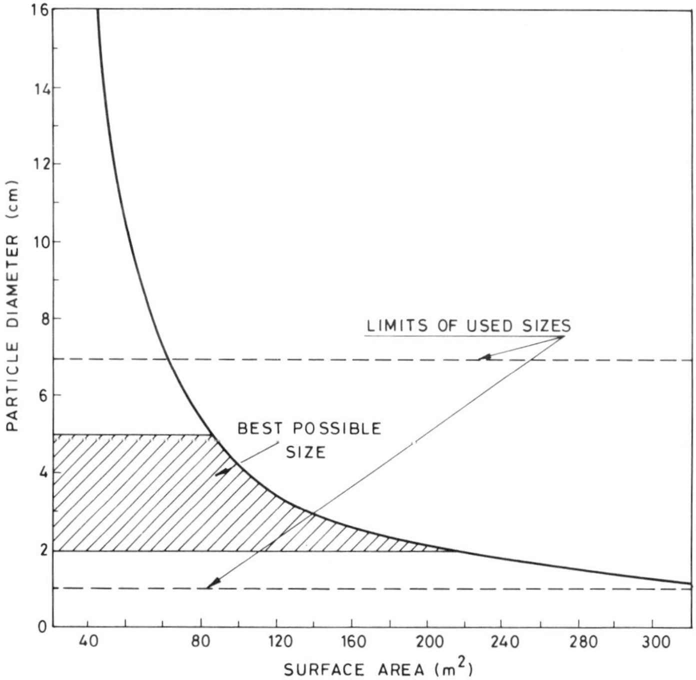

FIG. 9 Particle diameter and surface area in rock storage.

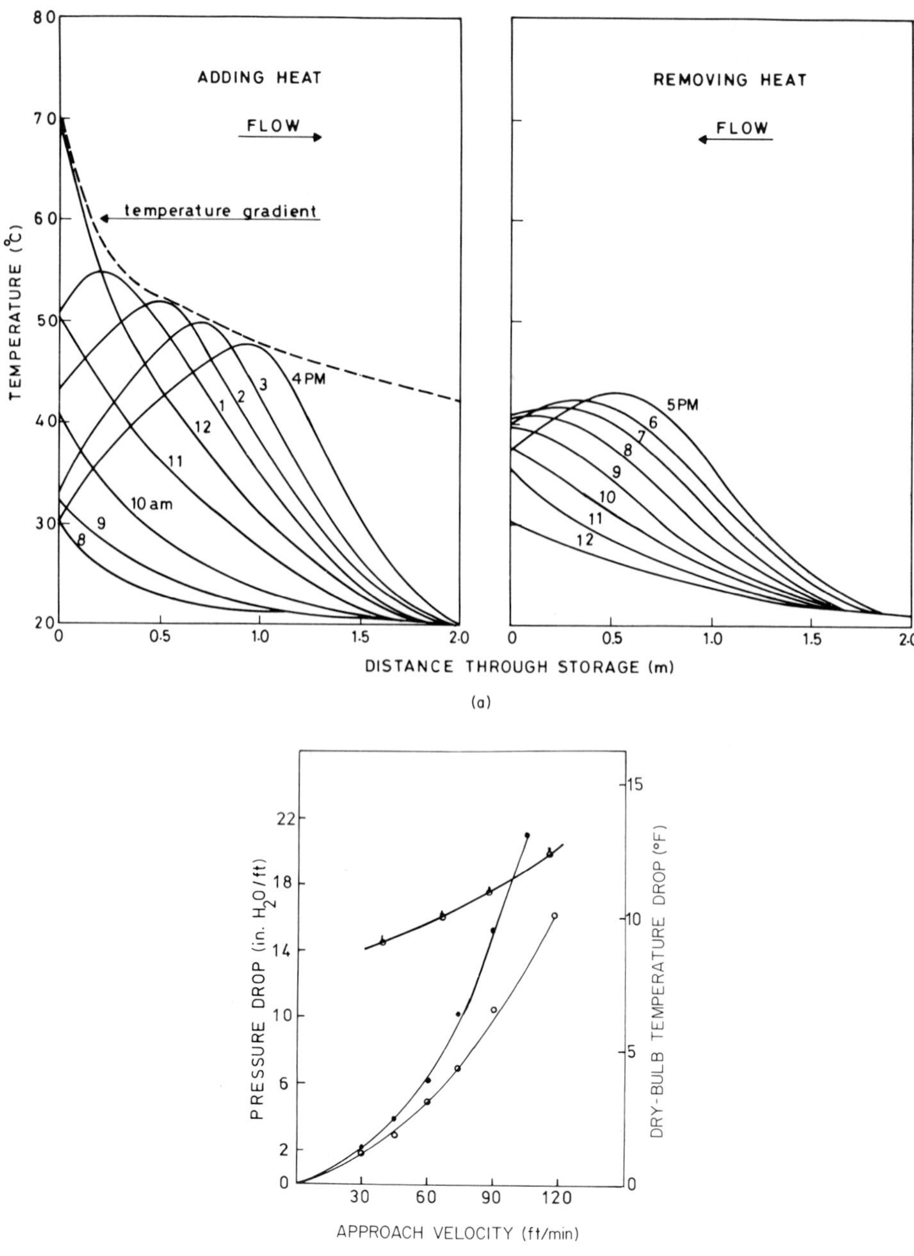

FIG. 10 (a) Air solar heater and limestone storage. (b) Pressure drop and dry-bulb temperature drop in rock beds. ɸ, temperature; ●, $\frac{1}{2}$–$1\frac{1}{2}$-in. rocks; ○, 2–3-in. rocks.

quirements are that they be of uniform size to obtain large void fractions, and have large surface area and low pressure drops. Figure 9 shows the surface area of various spherical particles in a given storage container. The container has a 30% void fraction. Obviously, with a fully packed container the void fraction does not change with different particle sizes, provided each packing has particles of the same diameter. But surface area and hence a larger heat transfer area will result from using small diameter spheres (Sayigh, 1977). McCormick (1975) has shown (Fig. 10a) temperature variations across a packed bed during the day. Figure 10b shows the pressure drop and air temperature drop in two sizes of rock beds (Bahadori, 1973).

4.4 THERMAL ENERGY STORAGE BY PHASE CHANGES

Most solar energy storage has been accomplished by sensible heating of either water or rocks with no phase or chemical changes taking place within the system. Nevertheless, storage for long periods requires a fairly large volume of rock bed or water tank, which at times is neither practical nor available. Therefore, using phase change processes will decrease storage size and hence the cost of storing energy by reducing the quantity of insulation materials and/or the space required. An example of such saving is given by Telkes (1974) for storing 250,000 kcal: 28.32 m^3 of water, 60.8 m^3 of rock, or 3.54 m^3 of phase-changing salt is needed; 25% of the volume was used for passages.

However, the choice of phase-changing material should be based on the fact that the substance should not be corrosive, toxic, flammable, and expensive. Additionally, it should be suitable for the temperature ranges required and suffer no degradation. Lane *et al.* (1975) gives a list of inorganic and organic compounds as well as eutectic mixtures for thermal storage in the 10–90°C range.

A. Solid–Solid Phase Change

The solid–solid phase materials generally have lower latent heat than those used for other forms such as solid–liquid phase change. Therefore, they can be used in the form of pebble beds (Bahadori, 1976a) and hence make use of the low heat loss pertinent to rock bed thermal storages.

B. Solid–Liquid Phase Change

Several inorganic, organic, and eutectic compounds undergo a fusion process that provides a fairly high-energy storage capacity for relatively low volume changes. For example, ice was used for many centuries as a storage material, but only around the 0°C limit. Hydrates provide good storage properties because they are cheap and plentiful if the associated difficulties can be eliminated, that is, incomplete melting and precipitation of some of the solids and supercooling. Several such materials are mentioned in Table 5.

C. Liquid–Vapor Phase Change

This type of phase changing is associated with much higher storage capacity than the solid–liquid phase, but it has the disadvantage of requiring a larger volume. A good way of solving such a problem is to employ two tanks, so that as the vapor is generated in one tank, it can be absorbed by the other, and the heat of vaporization given off to the environment from the second tank. If the vapor pressure of the volatile liquid in the first tank is lowered owing to a temperature drop, then by connecting the two tanks for the reverse process, the liquid in the second tank is evaporated, taking its heat of vaporization from the environment, and is condensed in the first tank. This is similar to absorption refrigeration systems.

In the case of lithium bromide–water solution the volatile material may be water; in the case of water–ammonia solution, the volatile material may be ammonia. Other systems may include sulfuric acid and water, silica gel and water, silica gel and nitrogen dioxide, silica gel and alcohol,

TABLE 5

Properties of Some of the Salt Hydrates Compared with Water

Chemical compound	Formula	Melting point (°C)	Density (kg/liter)	Heat of fusion (kcal/kg)	Heat of fusion (kcal/liter)
Calcium chloride hexahydrate	$CaCl_2 \cdot 6H_2O$	29–39	1.63	42	68.5
Sodium carbonate decahydrate	$Na_2CO_3 \cdot 10H_2O$	32–36	1.44	59	85
Disodium phosphate dodecahydrate	$Na_2HPO_4 \cdot 12H_2O$	36	1.52	63	96
Calcium nitrate tetrahydrate	$Ca(NO_3)_2 \cdot 4H_2O$	39–42	1.82	33	60
Sodium sulfate decahydrate	$Na_2SO_4 \cdot 10H_2O$	31–32	1.55	60	93
Sodium thiosulfate pentahydrate	$Na_2S_2O_3 \cdot 5H_2O$	48–49	1.66	50	83
Chloral hydrate	$C_2H_3O_2Cl_3$	51.7	1.90	33.2	62.7
Water (ice)	H_2O	0	0.914	79.8	72.9

and nickel chloride and ammonia (Daniels, 1964). This reaction is

$$NiCl_2{\cdot}6NH_3 \rightleftarrows NiCl_2{\cdot}2NH_2 + 4NH_3$$

As $NiCl_2{\cdot}2NH_3$ crystals are cooled, NH_3 is absorbed and $NiCL_2{\cdot}6NH_3$ is re-formed with the liberation of 250 kcal/kg of $NiCl_2{\cdot}6NH_3$. Other forms of storage using silica gel especially are mentioned by Close and Dunkle (1970) and Close and Pryor (1975). Figures 11 and 12 show 52 different

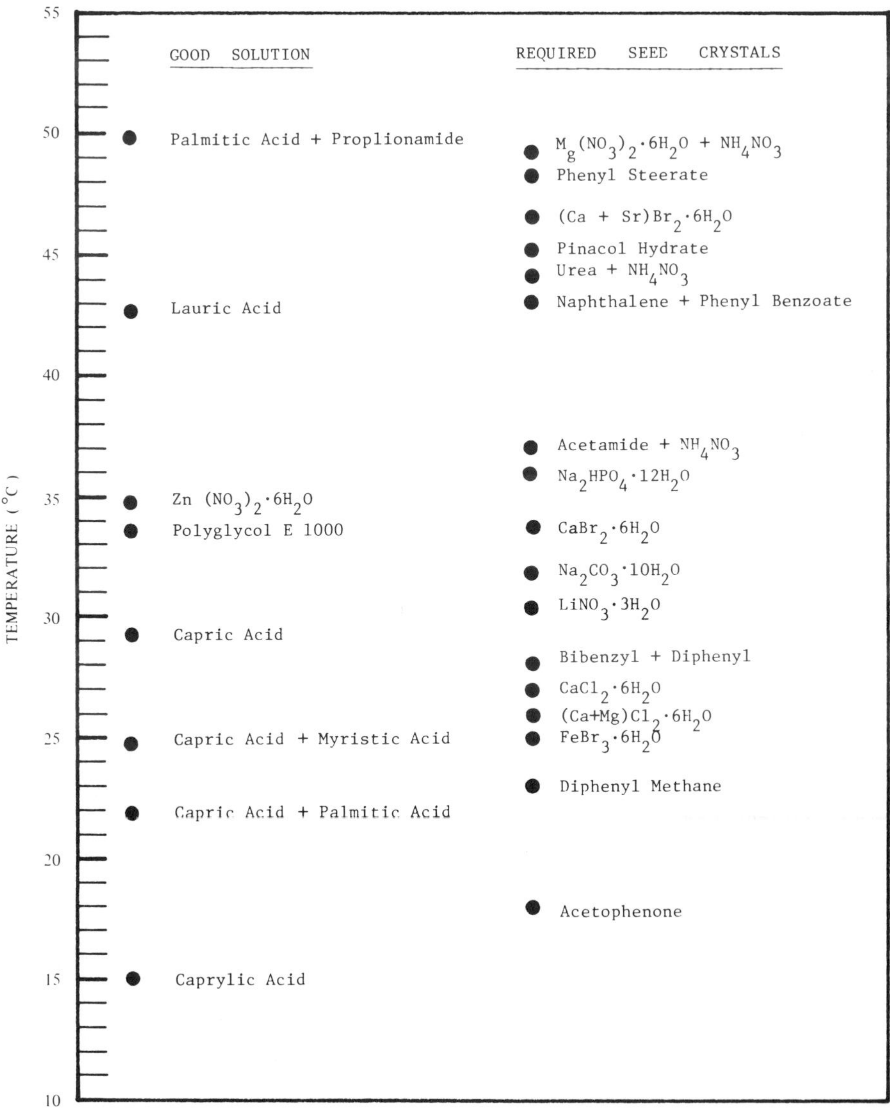

FIG. 11 Selected materials for heat storage, 10–50°C.

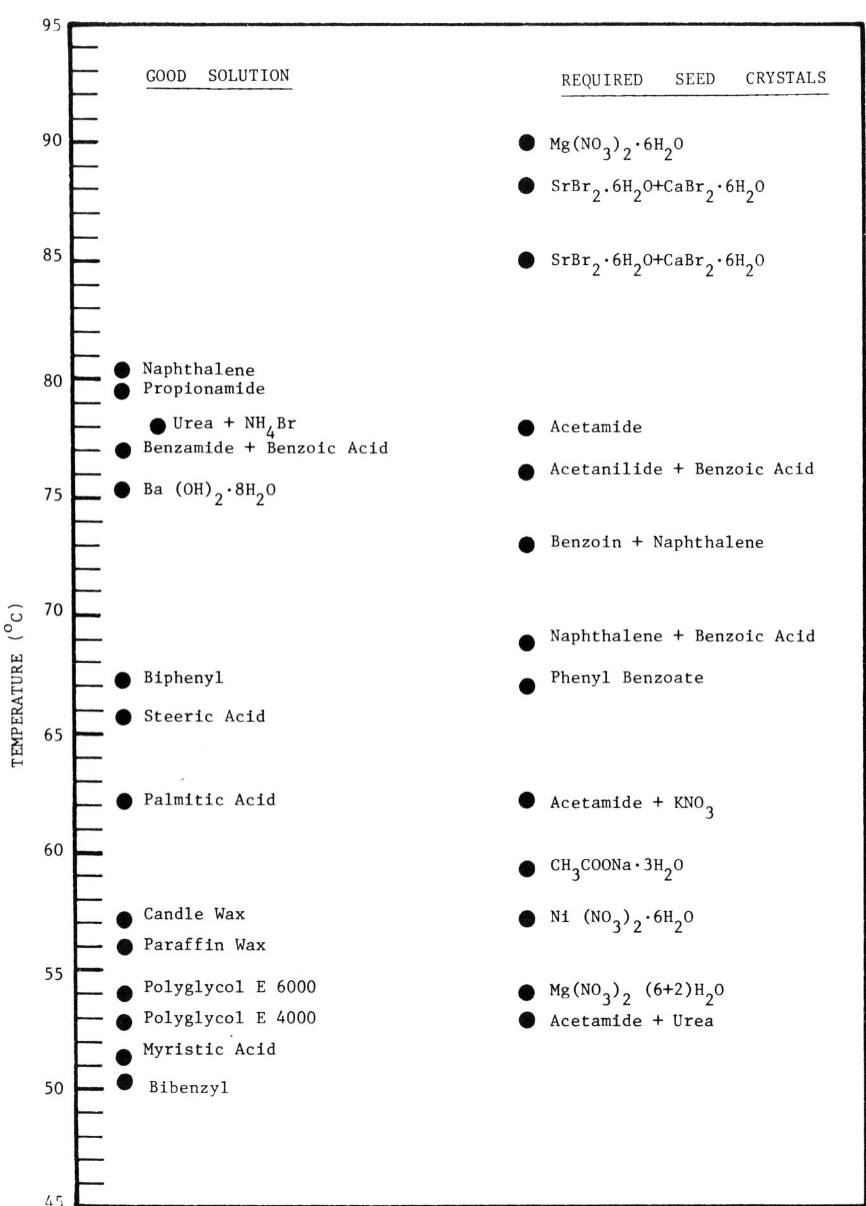

FIG. 12 Selected materials for heat storage, 50–90°C.

materials that can be used for heat storage by the principle of heat of fusion. Some of them do not require any seeding agent (Lane *et al.*, 1975).

4.5 OTHER FORMS OF ENERGY STORAGE

Energy can be stored by chemical solution properties. For example, energy can be used to dissolve such solids as ammonia nitrate, potassium nitrate, and silver nitrate. Heating the solution when it reaches saturation will increase its solubility.

Another form of energy storage is by chemical reaction. This is similar to the preceding method; i.e., if an equilibrium is reached between reactants and products, then raising the temperature will lead to a new equilibrium state and hence energy can be stored. An example of such a reaction is

$$Ca(OH)_2 \rightarrow Ca(OH) \rightleftarrows CaO + H_2O$$

This releases about 200 kcal/kg at 580°C.

Energy can be stored in a mechanical form, as either kinetic or potential energy. For kinetic energy, flywheels can be used as the kinetic energy storage mechanism. This depends heavily on sophisticated materials and maintenance capabilities.

Potential energy can be realized by pumping water to a reservoir at a high level and then allowing it to flow into a lower reservoir whenever energy is required. This is conventionally known as a pump-back system. Usually the kinetic energy of the flowing water is converted into electrical energy and can be transported from place to place. To use the energy directly as mechanical energy, a simple waterwheel can be placed at the location required. A difficulty of the pump-back storage system is that relatively large reservoirs are required.

It is better to try to find a reasonably large basin for an upper reservoir located near an existing lake or river. The latter then becomes the lower reservoir. Generally about two-thirds of the energy is available for reuse after storage.

Potential energy can also be realized by compressing air, either directly or by isothermal hydraulic compression. Compressed-air methods are very inefficient.

4.6 LONG-TERM SOLAR ENERGY STORAGE

In most countries of the world, solar energy needs to be stored in the form of thermal energy for a period longer than two or three days. Several attempts were made to do so by using solar ponds, which rely on variable

TABLE 6

Comparison of Thermal Energy Storage Processes and Materials [a]

Process	Material	Temperature (°C)	Storage capacity (kcal/kg)	Storage capacity (kcal/liter)	Remarks
Phase-change					
1. Solid–solid	FeS	138		55	The latent heat is generally lower than that for solid–liquid phase changes, but solids may be used in the form of pebble beds with lower heat losses
	KHF_2	196		75	
	$LiSO_4$	575		140	
2. Solid–liquid					
(a) Pure	Water	0		72.9	
	Al_2Cl_6			80	Change of volume is very high
(b) Inorganic	$MgCl_2 \cdot 6H_2O$	117			Supercooling is a major problem in solid–liquid transitions. Nucleating agents are needed and they often become inoperative after several hundreds or thousands of cycles. Incongruent melting may be prevented by small amounts of additives
	$Ni(NO_3)_2 \cdot 6H_2O$	57			
	$Na_2S_2O_3 \cdot 5H_2O$	48–49		83	
	$Na_2SO_4 \cdot 10H_2O$ + 2.5% $Na_2B_4O_7 \cdot 10H_2O$	31.6		93	
	$Na_2SO_4 \cdot 10H_2O$	32		93	
(c) Organic	Naphthaline $C_{10}H_8$	80		41	
	Stearic acid	72		40	
	Beeswax	62		40	

(d) Eutectics	$CaCl_2$–$MgCl_2$–H_2O (41–10–49%)	25	41.7		Eutectics have sharp melting points, similar to the pure substances. Heat of fusion is equal to the weight average of heats of fusion of pure substances at the eutectic temperature plus the heat of mixing of the liquids
	$NaCl$–Na_3SO_4–H_2O (22.6–7.6–69.8%)	18			
	Urea–NH_4NO_3 (45.3–54.7%)	46	41		
3. Liquid–vapor					
(a) Aqueous solutions	$H_2SO_4 + H_2O$	40–60		150	Changes of volume is very large, unless the vapor is absorbed in a second container
	$Na(OH) + H_2O$	60–80		172	
(b) Solid hydrates	$LiBr \cdot H_2O$	103		530	To increase the vapor absorption rate, the liquid solution has to be stirred and the solids have to be made with sufficient contact area
	$LiSO_4 \cdot H_2O$	56			
	$NiCl_2 \cdot 6NH_3$	175	250		
	Silica gel + H_2O				
	Silica gel + NO_2				
	Silica gel + alcohol				
Heats of solutions	Ammonium Nitrate in water	80–100	27		Solubility of solids change with temperature. Heat is required to increase solubility of NH_4NO_3
Chemical reactions	$N_2O_4(l) \rightleftarrows 2NO_2(g)$	21	198		Large energy storage is possible in chemical reactions
	$NO_2(g) \rightleftarrows NO + \frac{1}{2}O_2$	>60	151		
	$Ca(OH)_2(S) \rightleftarrows CaO(S) + H_2O(l)$	580	200		

[a] From Bahadori (1976a).

salt concentration in a pool. The thermal energy can be trapped at the lower portion of the pool, which has the high density, and stay there for a long time. The heat loss is limited only to the floor of the pond. However, the cost of such a system and the fairly complex technology required have prevented many people from making use of it (Sayigh, 1977).

In most regions of the world, solar radiation is plentiful in the summer; therefore using large wells, cisterns, and caves during the summer to store thermal energy for use in the winter is a worthwhile and successful exercise. This practice has been used in the past to store ice from winter to summertime (Meyer and Todd, 1973; Bahadori, 1978).

4.7 CONCLUSIONS

Because of the infrequency of solar radiation in most parts of the world and the amount of thermal energy needed to heat or cool buildings, various storage systems have been used. The most widely used systems are those of sensible heat storage by means of either water or rock. Each system has advantages and disadvantages, is easy to construct and operate, and is very reliable. A combination of the two systems is highly recommended (Sayigh and El-Salam, 1975).

Phase change and chemical change heat storage have not yet been perfected, and more research is needed to make them durable, efficient, and safe. However, they have two major advantages over water and rock storage; namely, they require small volume, and they can store greater amounts of thermal energy. Table 6 shows most of the materials used in phase change and chemical change storage.

Solar pond technology could solve the problem of long-term thermal storage in the near future, but the use of already existing wells, cisterns, and caves will certainly give good results. This is mainly due to the low conductivity of the earth and to the temperature gradient over a fairly large thickness of the earth that will develop during the prolonged storage period (Yellott, 1973).

5

Solar Energy Utilization in Advanced Residential and Commercial Applications through Hydrogen Energy

T. NEJAT VEZIROGLU

CLEAN ENERGY RESEARCH INSTITUTE
UNIVERSITY OF MIAMI
CORAL GABLES, FLORIDA

WILLIAM J. D. ESCHER

ESCHER TECHNOLOGY ASSOCIATES
ST. JOHNS, MICHIGAN

5.1 INTRODUCTION

First-generation solar energy systems that attempt to address the needs of space heating and cooling and water heating are presently being deployed in many nations of the world in response to the increasing cost and scarcity of conventional fuels. These systems are based largely on flat plate collectors that limit working fluid temperatures to the order of 200°F. Evacuated tubular collectors of novel design offer promise of raising this limit substantially, but at additional costs for the collector. Increasing interest is being shown in concentrating collectors, despite the attendant increase in complexity as higher temperatures yet are sought.

In retrospect, this trend toward increasing temperatures in the working fluid is a direct result of the need to provide improved performance and increased flexibility within solar energy systems. Looking beyond today's systems, this trend toward expanded and more flexible service can be projected to "all-service" solar energy conversion systems of tomorrow. Not just space and hot water heating, and cooling (which re-

ISBN 0-12-620860-3

mains a marginal proposition with today's low-temperature systems) will be provided. High-temperature heat for cooking and special processes requiring high temperatures will be available. Electricity will be served on a demand basis. A fuel will be available for portable energy system needs, even for light transportation vehicles. And cooling needs will be met by total environmental control systems capable of cooling, heating, and humidity control.

But to accomplish these objectives, a considerably more sophisticated solar energy conversion system will be needed. Further, a new means of storing and distributing energy within the system will be required. Thermal means such as heated water can no longer do the total job.

It is proposed that "hydrogen energy" can serve as the new means of accomplishing these new systems objectives. Hydrogen energy refers to hydrogen as a fuel gas, or to the hydrogen–oxygen reactant combination produced when water is separated into its constituents using solar energy. How this can be achieved in context with advanced residential and commercial solar energy conversion systems, and the associated technical background, is related in this chapter.

The topics discussed are (1) solar production of hydrogen, (2) hydrogen storage and distribution in solar energy systems, (3) hydrogen as a fuel, (4) hydrogen heating units and appliances, (5) hydrogen air conditioning and refrigeration, and (6) electricity generation via hydrogen energy.

5.2 SOLAR PRODUCTION OF HYDROGEN

A. General Concept

Hydrogen, being a secondary fuel or synthetic energy form like electricity in most senses, must be produced by the application of a primary energy source. Usually this energy is applied in a water-splitting process, such as the well-proven water electrolysis, which yields both hydrogen and oxygen as free gases, the constitutents of water. More advanced water-splitting processes such as multistage thermochemical water-splitting processes are under research at present. Someday they may supplement and even supplant water electrolysis. However, it should be noted that electrolysis will be undergoing substantial advancements in the meantime.

Clearly, the sun is a principal, long-term, primary energy source candidate for the production of hydrogen energy. The solar production of hydrogen, reduced to its essentials, involves the capture and conversion of solar radiant energy for the purpose of dissociating water (as a "feed-stock'') into hydrogen and oxygen. This process can be envisioned as either a technological one, of one type or another, or a biological one. In the latter instance, it is well to observe that photosynthesis is fundamentally dependent on solar water splitting in a quite complex biocycle. Our presentation will be limited to ''technological'' systems, however.

Consideration has been given to the solar production of hydrogen on basically two different scales: large-scale ''central plant'' production in the tonnage hydrogen production range, and small-scale ''distributed'' production in the range of kilowatt equivalency, or less.

In the purview of the present chapter, we will be considering mainly the latter scale of operations. This would be compatible with residential and commercial usage, and even with the light-industrial usage situation being stressed in this volume. In summary, the use of the sun's energy to produce hydrogen (and oxygen) from water by means of water electrolysis (the only available method at present) will be considered in the context of an advanced solar energy conversion system associated with homes, institutions, commercial businesses, and light industrial applications.

B. Water Electrolysis

Water electrolysis is that electrochemical process performed in devices referred to as electrolyzers. Various designs of electrolyzers are sold all over the world by such firms as Lurgi, Brown Boveri, DeMag, The Electrolyser Corporation, Ltd., and DeNora and Teledyne Energy Systems. An electrolyzer system converts electrical energy into hydrogen energy at an efficiency dependent on the type and operating conditions of the unit, but today in the 65–75% range. This definition is based on the higher heating value of the hydrogen produced ratioed to the dc power applied to the electrolysis cells.

It is beyond the scope of this description to go into the theory of operation of water electrolyzers other than making brief mention of the basic processes involved. A direct current passing through water between two electrodes causes, under appropriate conditions, the water to dissociate into its elements. With the presence of an electrolyte, such as potassium hydroxide (water itself is a poor conductor of electricity), hydrogen and hydroxyl ions (H^+ and OH^-) are produced. The positive hydrogen ions migrate to the cathode or negative electrode where they combine with electrons to form hydrogen molecules (H_2). These build up until a bubble is formed, breaks away from the electrode, and rises to the top of the electrolyte where the gas can be collected. Similarly, the hydroxyl ions are attracted to the positive electrode, or anode, where an electron is given up and oxygen and water are formed. The oxygen goes to its molecular form (O_2) and is collected in a manner similar to the hydrogen gas.

There are two main approaches in the physical construction of an electrolyzer that are used today: unipolar cell and bipolar cell construction. The former is sometimes referred to as the "tank" design and the latter as the "filter press" design. Figure 1 shows schematically the layout of each of these. In either case, the voltage impressed across each individual cell, the repeating element involving the anode and cathode, a separator material (usually asbestos), and an electrolyte is of the order of 2.0 V. The lower the voltage, the higher the overall efficiency of the unit, with approximately 1.48 V equivalent to 100% conversion efficiency.

Electrolyzer units in the small size class are available from manufacturers. For example, the Electrolyser Corporation unit is rated at 20 ft^3/h of hydrogen produced. For comparison purposes, a standard "welding cylinder" of hydrogen at 2250 psi contains about 220 ft^3 of hydrogen or about 1 lb. Large industrial units are rated at input electrical power units of the range 1–2 MWe, with projections to the range 5–20 MWe per unit

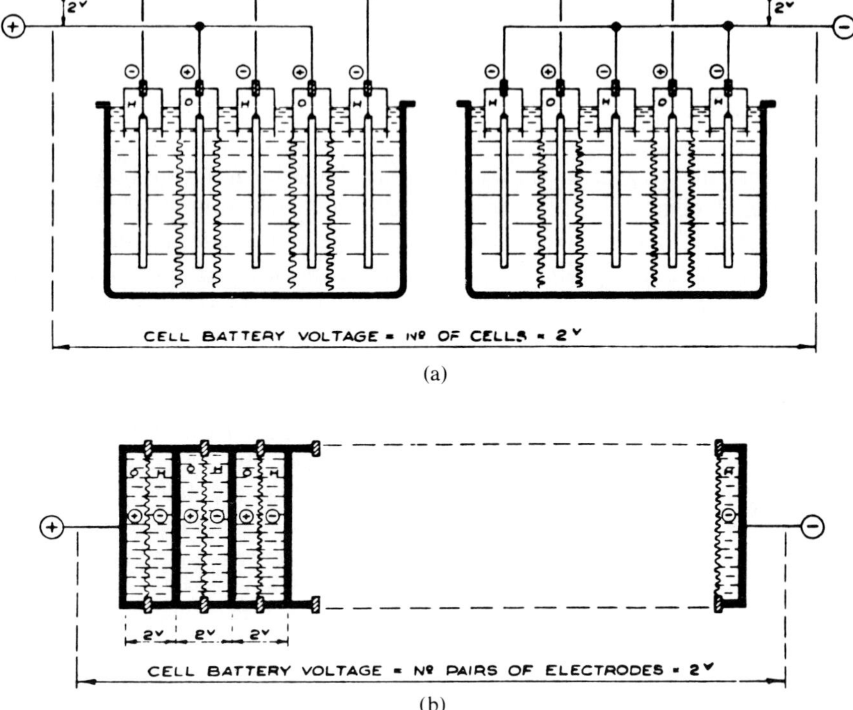

FIG. 1 Schematic diagrams of (a) unipolar and (b) bipolar electrolyzer construction. (Courtesy of The Electrolyser Corporation Ltd.)

in the future. The Aswan Dam installation of Brown Boveri electrolyzers in Egypt totals about 150 MWe using multiple units.

Not only are larger units projected for industrial production of hydrogen, but the efficiency and costs of tomorrow's electrolyzers are projected to improve considerably. Efficiencies above 90% are perceivable, with costs ranging downward from today's all-up system costs of $250/kWe to the order of $100/kWe. It should be noted that a considerable part of the cost is associated with "power conditioning" the delivered ac power to dc power for the cells. This involves rectifiers and often expensive transformers, as well as controls, switches, instrumentation, etc. If local electrical generation is to be used, it will pay to have the generated power available at conditions approaching those required by the cells. For example, studies have revealed that dc generation and direct coupling with the electrolyzer cells can cut costs in half, as compared to the use of ac power, which requires "conditioning." As will be discussed further, for the application under consideration here such dc generation may be a practical approach for reducing costs and raising overall conversion efficiencies.

C. Solar Generation of Electricity for Electrolysis

As stated, it is necessary to provide electricity for the water-splitting process. For solar energy systems there are various methods of generating electricity, some proved and in use and others that are speculative, though well enough founded technically to have credibility. We will concentrate on just two of these general approaches, those that are in demonstrated use: photovoltaic conversion (direct conversion) and heat engine/generator systems. With development, such conversion schemes as photoelectric, thermoelectric, thermionic, and various hybrid schemes may be applicable to solar hydrogen systems.

D. Photovoltaic Systems

The use of so-called solar cells for the direct generation of electricity through irradiation by sunlight is a proven and in-use scheme. Initially demonstrated as a means of providing earth satellite and spacecraft vehicles with electrical power with the launching of Vanguard 1 in 1958, many terrestrial uses have been developed for solar cell power systems, e.g., remote communication units, buoys at sea.

Photovoltaic devices are of several kinds: silicon cells, thin-film cells such as the cadmium sulfide type, and more complex ternary material cells. All are quite expensive and have somewhat limited conversion efficiencies of the order of 10 to 15%. However, prospects for new mass production techniques seem encouraging for reducing costs, and some increases in efficiency are offered in concept. Until such developments actually occur, the full deployment of solar cell conversion systems for residential and commercial applications will be limited. A positive aspect of photovoltaic conversion is that the dc electrical power produced may be directly connected to electrolysis systems without the need for intermediate energy conversion equipment and processes, which increase costs and reduce efficiency. Experiments in photovoltaic/electrolysis systems have already taken place at such institutions at the Jet Propulsion Laboratory in Pasadena, California, with encouraging results in terms of the matching up of components.

E. Heat Engine/Generator Systems

Provided medium- to high-temperature solar thermal collectors are employed, heat engines can be operated on solar energy to produce shaft power. This in turn can be used to drive conventional generators to produce electricity for electrolysis. Though generator efficiency is typically high, of the order of 90% and above, depending on type and size of the unit, the heat engine system efficiency is limited by temperature (Carnot) considerations, as well as by mechanical system efficiency. However, depending on the specific system selected, overall electricity generation efficiency levels significantly higher than comparable technology photovoltaics can likely be achieved. A key here is the achievement of high-temperature working fluid conditions at the outlet of the solar collector system. This implies concentrating collectors of the focusing mirror or lens type.

Thus focusing collectors capable of sun tracking are implied in any efficient solar thermal-to-shaft-power system. Once high temperatures are available, appropriate selection of a thermodynamic operating cycle and the associated mechanical equipment can be made.

In general, external-combustion-type power cycles are implied, e.g., Stirling, Rankine, and Brayton cycles. One configuration that comes to mind for this application is a tracking parabolic-dish-type mirror concentrator with a heat receiver for a Stirling cycle engine at its focus. Such a system can easily produce the 1500–1800°F conditions necessary to operate the engine at shaft efficiencies of the order of 40%.

Another application at the high-temperature end would be the closed cycle, Brayton cycle gas turbine using inert gases or even air as the working fluid. The heat exchanger again would be placed in the focus of a high-concentration-ratio concentrating collector of one type or another.

Requiring lower temperatures would be a Rankine cycle system, e.g., a steam turbine, which could be operated at working fluid temperature ranges of 500 to 1000°F to produce medium- to-high-efficiency shaft power. Depending on condensing or heat rejection conditions, efficiencies of the order of 25 to 35% would be possible with such systems. In the smaller sizes, reciprocating-type Rankine expanders might be used, e.g., conventional steam engines.

The selection of generators to best match the heat engine would be important. Similarly, as discussed in the preceding material on electrolyzers, a good electrical match to the cells will lead to significant cost reduction and gains in operating efficiency. For this reason it is expected that dc generation will play an important role in future solar hydrogen systems.

5.3 HYDROGEN STORAGE AND DISTRIBUTION IN SOLAR ENERGY SYSTEMS

A. Introduction

In solar energy systems utilizing hydrogen energy as an energy storage and carrier medium, it will be necessary to provide the means of storing the hydrogen (and probably the oxygen) and distributing it at required conditions to the points of use, e.g., heating appliances. This is analagous to natural gas supply systems that are in use today on a much larger scale. Underground gas storage fields are utilized to "load level" the production/transmission systems in view of the highly variable seasonal demand on gas by customers over the year. Once delivered to the "city gate," where pressure letdown and odorization are usually accomplished, a city distribution main system is needed to serve the gas to all customers. Finally, individual homes and institutions are required to be equipped with gas piping, valves, regulators, and meters. Most of these functions will have to be handled and the equipment supplied in a solar energy system providing the advantages of hydrogen energy.

Thus, hydrogen and oxygen storage and delivery systems must be considered in this context. Storage is required to provide an adequate match between the rate at which hydrogen energy can be produced (limited to maximum insolation periods), and the "peaky" demand made on hydrogen energy supplies by the user. Distribution is required to move the hydrogen and oxygen to the point of use at the proper pressure and flow rate, and to ensure safe operation at all times.

B. Storage Systems

Hydrogen and oxygen storage systems applicable to solar energy systems for commercial and residential applications are likely to be limited to systems capable of handling these constituents as gases. Cryogenic liquefaction can be considered in those cases where storage convenience and some unique users (such as certain kinds of transportation systems) may require the liquid form. Generally, this will prove to be uneconomical for systems of the scale under consideration here. Hence cryogenic hydrogen and oxygen will not be further discussed.

Hydrogen and oxygen gas can be stored in various kinds of holders or pressure vessels, and at various pressures. In addition, there is a special approach for storing hydrogen that provides significant advantages in terms of increased density of storage at low to moderate pressures. This is the area of metal hydride storage, which will be discussed below.

C. Pressure Vessel Storage of Hydrogen and Oxygen

Some electrolyzers produce hydrogen at elevated pressure, say 100 psi or so; however, most produce gases at only slightly above atmospheric pressure. This implies the need to compress the hydrogen to the level of maximum storage pressure. It is probable that this will be considerably less than the 2250-psi level used in commercial cylinder deliveries in view of the difficulty of compressing hydrogen over a very large pressure ratio. Oxygen, with a higher molecular weight, is a much denser gas and is more readily compressed, so that hydrogen will tend to "set the pace" on upper pressure level selection.

Pressure vessels for storing hydrogen and oxygen can be of metallic or nonmetallic construction. A conventional metal vessel design would consist of a length of welded or seamless pipe with pressure-dome-type head ends welded at each end. At higher pressures, forged "tube bank" units, such as those used in industrial gas deliveries, are applicable. Of the nonmetallic systems, the epoxy glass filament-wound structures have been used with successive, provided a sealing liner is installed within the container. One advantage of the filament-wound glass construction is that catastropic rupture failures do not occur, whereas they can do in the case of metal containers.

A special problem with hydrogen is referred to as hydrogen environment embrittlement, a phenomenon of direct hydrogen attack on metal, particularly where local yielding of the material is taking place. This leads to loss of ductility and strength in affected structures. With the exception of aluminum alloys and copper-based alloys, most other metals and alloys are susceptible to this mode of failure or attack. For this reason, recent experience with hydrogen pressure vessels should be consulted when hydrogen storage containers of this type are being designed.

D. Metal Hydride Storage Systems for Hydrogen

Certain metal alloys and intermetallic compounds provide another means of storing hydrogen gas that is of considerable interest, since it allows for a density of storage that is, on a volumetric basis, considerably more favorable than that of pressure vessel storage systems. Researchers at the Brookhaven National Laboratory, on Long Island, New York, have developed one such compound, a metal hydride combination, iron titanium hydride, to a level where useful applications are possible. Although the iron titanium hydride is marginally too heavy for transportation applications of most kinds (exceptions exist), it appears to be quite applicable to the stationary system storage of hydrogen under consideration here.

Brookhaven constructed an experimental iron titanium hydride storage unit for the Public Service Electric and Gas Company of New Jersey capable of holding about 13 lb of hydrogen. This unit (Figure 2) contains approximately 900 lb of the FeTi alloy in particulate form. Hydrogen is charged into the metal by being applied at pressure, and the heat of formation of the hydride is removed with cooling water (note the multiple conduits to the right side of the photograph). The hydrogen is released on demand by applying heated water, which is circulated in pas-

FIG. 2 Experimental iron titanium hydride hydrogen storage unit (Brookhaven). (Courtesy of Brookhaven National Laboratory.)

sages through the hydride bed. In many cases waste heat from an adjacent process can be used to supply the heated water, with the heat energy being basically conserved. The hydride bed has been operated over many cycles in an electrical energy storage experimental loop involving an electrolyzer and a fuel cell. Details of this experimental unit and its operation are available from Brookhaven National Laboratory.

E. Distribution Systems

Like a small-scale natural gas system, a solar energy system utilizing hydrogen energy will require piping, valves, fittings, regulators, and associated instrumentation to move the hydrogen, and oxygen if used, from the electrolyzer into storage and then out of storage to the using devices in the system, such as water heaters and space heaters.

In general, equipment for gas distribution is highly developed and readily available. However, fully qualified, low-cost equipment for home

FIG. 3 Experimental residential system hydrogen test loop (IGT). (Courtesy of the Institute of Gas Technology.)

and institutional applications, such as that in use for natural gas, is not presently available for hydrogen energy systems. The development of such equipment is likely when sufficient market forces are perceived by industry.

In the meanwhile, work is going on in the research laboratories in support of this eventuality. A most interesting program is being sponsored by the US Energy Research and Development Administration at the Institute of Gas Technology. This is a gas distribution system test-loop series for residential and commercial components currently used in natural gas service, but employing hydrogen gas. The basic objective is to evaluate the performance and operating reliability of production gas units under hydrogen flow conditions. Meters, regulators, valves, fittings, and pipes or various materials (metal and plastic), joined or combined by various techniques, are being exposed to circulating hydrogen at realistic operating pressures. The residential test loop is shown in Fig. 3. It is notable that all of the equipment being tested was donated by manufacturers and several natural gas companies. These contributors will be kept posted on the research data coming out of this operation, which commenced in spring 1977.

Outside of routine development and testing, there appears to be no major problems facing the distribution system requirements of advanced solar systems.

5.4 HYDROGEN AS A FUEL

A. General Background

The use of hydrogen as a gaseous fuel is not much practiced today even in view of its potential major role as a synthetic fuel or "energy carrier." Although, from a technical standpoint, hydrogen is a high-quality, clean fuel, its cost is substantially higher than naturally occurring fossil fuels such as natural gas (largely methane in its makeup). This stems from the fact that hydrogen does not occur in its free form on earth. Rather, the very large quantity of hydrogen to be found is chemically bound, e.g., water. As fluid fossil fuel availability becomes more limited, with consequent increases in cost, hydrogen's competitive position is likely to be improved. Also, the technology for efficient, large-scale production of hydrogen via one or more "water-splitting" processes will be developed to increasingly favorable levels, resulting in a reduction in hydrogen costs as compared to the situation today.

On the other hand, gaseous fuels with significant fractions of hydrogen content have seen, and continue to see, widespread application. Today's "low Btu gas," derived from coal gasification, and manufactured or town gas, used decades ago, represent cases in point. The latter was supplanted by natural gas, beginning in the thirties and forties in many industrial nations. Today, where impure hydrogen is available as an off-gas from processing or manufacturing facilities, it is sometimes used as a utility fuel, e.g., as a boiler fuel for steam raising.

One firm in the United States, North American Manufacturing Company, a builder of industrial "gas trains" and related combustion equipment, has designed and fabricated a number of hydrogen burner systems, including special controls. Specifically selected for hydrogen service, the components used in hydrogen gas trains, have worked quite satisfactorily.

Hydrogen has seen recent experimental use as an engine fuel for internal combustion piston- and rotary-type power plants, and as a gas turbine fuel. This work is in support of hydrogen's potential as a transportation fuel. Outside of its advantage of long-term availability, being producible from water using nonfossil, primary energy sources, such as solar energy, the subject to be addressed here, hydrogen has many advantageous technical characteristics. An example is the appeal of hydrogen as an aviation fuel. Hydrogen, in liquid cryogenic form, provides about 2.8 times the energy per unit mass as in-use hydrocarbon fuels. This is of special significance to aircraft where aircraft weight and performance are very sensitive to the gravimetric heating value of the fuel. On the other hand, hydrogen is characterized by compensating disadvantages and limitations, such as its very low volumetric heating value and its very low temperature and highly volatile nature (in cryogenic form).

At the present time, hydrogen's principal use is as a chemical intermediary and processing medium in the chemicals industry and in oil refining. The manufacture of ammonia, for example, requires the production of hydrogen from hydrocarbon feedstocks as a step in producing synthesis gas.

Returning to hydrogen's potential as a fuel gas for commercial and residential applications, a number of research organizations have operated hydrogen-fueled appliances and hydrogen delivery systems for experimental and demonstration purposes. Among these are the American Gas Association (AGA) Laboratories in Cleveland, Ohio, and the Institute of Gas Technology in Chicago, Illinois. Examples of the work going on in these organizations and in other locations will be cited in subsequent sections of this chapter.

B. Fuel-Significant Characteristics of Hydrogen

The lower (lean) flammability limit of hydrogen in air is about 4% (by volume) and the upper (rich) limit is approximately 75%. This is a much wider range of flammability than that of natural gas (about 5 and 15%, respectively). However, a number of in-use fuels—particularly liquid fuels—demonstrate substantially lower lean limits than hydrogen. In the case of a mishap due to a leak or spill, it is the lower-limit characteristic that signifies the possibility of a fire or explosion. In this sense, hydrogen may be a safer fuel than these liquid fuels, e.g., gasoline.

There are three governing technical characteristics that establish the behavior of a fuel gas in combustion systems such as an appliance burner. These are laminar flame speed, quench distance, and minimum ignition energy. These characteristics are portrayed for hydrogen and several comparison fuels in Figs. 4, 5, and 6, respectively. From these graphs it can be seen that hydrogen behaves in a markedly unique manner. Two general consequences of the behavior of hydrogen in terms of practical using devices, such as gas appliances, are

(1) Using devices designed for conventional fuels, e.g., natural gas, will not be optimum for hydrogen, and they may not, in fact, operate satisfactorily on hydrogen if at all.

(2) Using devices that are specially designed to take full advantage of hydrogen's special properties may prove to be quite superior with respect to performance and operating flexibility; furthermore, considerable opportunity for system simplification with cost savings may well exist in the use of hydrogen for such devices.

The following interpretation of Figs. 4–6 is offered to provide a practical insight into selecting and designing hydrogen-fueled devices from the standpoint of the preceding two generalizations.

Flame speed Hydrogen's very high laminar flame speed over a very wide range of stoichiometry (fuel/air ratio from lean to rich) is evident in Fig. 4 and is a direct consequence of its high reactivity as a fuel. As one ramification, where premixed fuel–air systems are employed, such

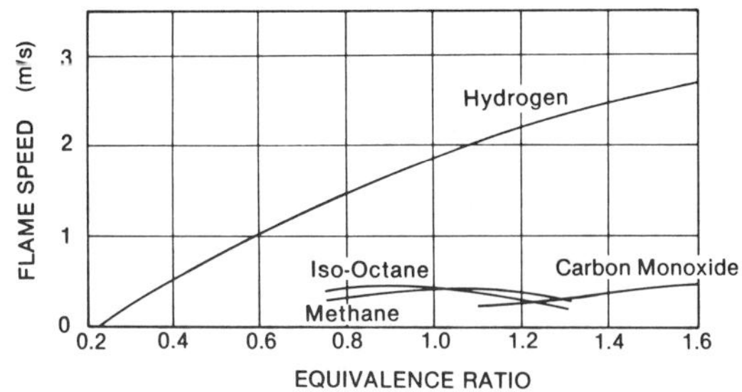

FIG. 4 Comparative fuel laminar flame speed characteristics. [From *J. Chem. Eng.* **4**, 226 (1959).]

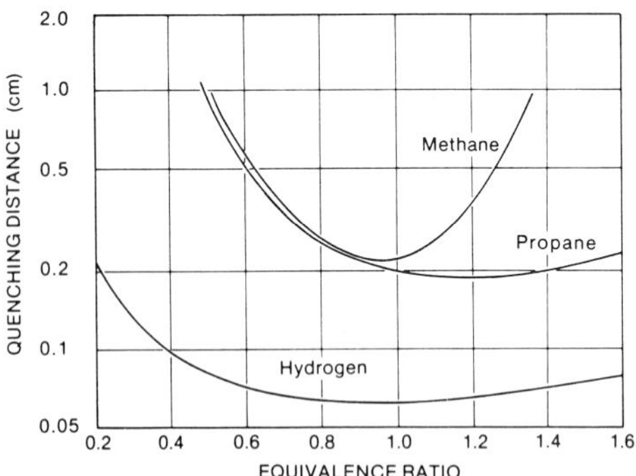

FIG. 5 Comparative fuel quench distance characteristics. [From *J. Chem. Phys.* **15,** 798 (1947).]

hydrogen–air mixtures will tend to burn extremely rapidly (short, intense flame patterns), and may burn upstream of the intended flame location; i.e., there may occur "backflashing" into the premix passage.

Quench distance Hydrogen flames can pass through much smaller openings or passages than other fuels without being extinguished owing to heat conduction and other surface effects. Flame arrestors and similar devices, which operate well with other fuels, may no longer be reliable or

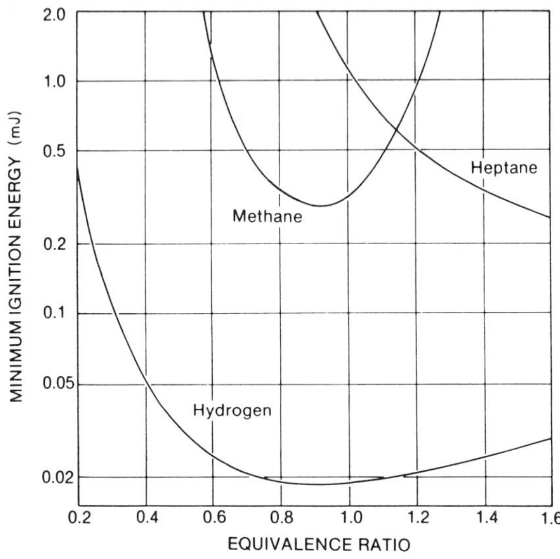

FIG. 6 Comparative fuel minimum ignition energy characteristics. [From *J. Chem. Phys.* **15,** 798 (1947).]

even function at all with hydrogen for this reason, since they function on a quenching principle. This also contributes to the backflashing tendency of hydrogen–air mixtures. (See Fig. 5.)

Ignition energy Hydrogen's minimum ignition energy level is more than an order of magnitude lower than that of other fuels: 0.02 versus 0.3 mJ (Fig. 6). On the one hand, this fact could lead to safety problems in the case of gas leakage in systems. A rule of thumb often stated regarding hydrogen leaks notes that one should automatically assume that one has a fire. On the other hand, this ready-ignitability aspect of hydrogen leads to certain ignition system design advantages, such as the elimination of pilot flames and other high-energy ignition means. Catalytic ignition systems become possible uniquely with hydrogen, as will be discussed.

C. Hydrogen Flame Combustion as Compared to Other Fuels

As will be covered below, there are two fundamental modes by which hydrogen can be combusted with oxygen in air or as a pure reactant material: flame combustion and catalytic combustion. It will be seen that hydrogen alone, among fuels, is well suited for catalytic combustion, and a number of interesting prototype appliances have been tested or proposed based on the catalytic combustion approach. However, since flame combustion is the more familiar general case and will likely be used in many applications, it is instructive to review hydrogen flame combustion fundamentals first.

Perhaps the most noticeable aspect of a hydrogen flame is its *transparency* and absence of color. In daylight conditions a hydrogen flame may be completely invisible, being detectable only by "heat striations," the consequence of density and index-of-refraction changes with the temperature of the combustion products and heated air. At night or in an indoor situation with low background illumination, the hydrogen flame takes on a very subdued luminosity, controlled in most cases by even trace contaminants or other gases in the hydrogen stream.

This nondectectable flame characteristic of hydrogen leads to the hazard of inadvertent exposure to flames, resulting in personal injury or property damage. For this reason, the inclusion of an added *illuminant* in the hydrogen fuel supply has been frequently suggested. Such an additive would provide a color or luminosity to the hydrogen flame. This would be analogous to the odorants added to natural gas. In fact, a combination odorant/illuminant might well be found for use in hydrogen systems to provide a means of leak detection and flame visualization. One precaution that must be observed in selecting such an additive is that catalytic devices may be quite sensitive to poisoning and other degradation effects due to specific compositions of the additive. This could negate the advantages of catalytic combustion (to be enumerated).

As an effect caused by hydrogen's high flame speed (see Fig. 4), the shape of the hydrogen flame may be quite different from that of other fuels of much lower flame speeds. For example, in a simple Bunsen burner normally operating on natural gas, hydrogen would lead to a much shorter and more intense flame with the same air port setting. There might also be backflashing problems, as mentioned. This will be the case with any air-premix-type burners, such as the conventional range-top element or atmospheric burner portrayed in Fig. 7. This is the type used with natural gas supplied to the gas orifice at low pressures, i.e., the order of several inches of water column.

As reported by the Institute of Gas Technology and Billings Energy Corporation, Provo, Utah, when conventional gas ranges are modified to operate on hydrogen, it is necessary to close off the primary air openings altogether and to effect a good seal to prevent air encroachment in order to prevent entry of the hydrogen flame into the mixing chamber (the central tubular part of the burner upstream of the burner ports). Even then ignition or start-up and shutdown (extinction) noises, or "pops," were experienced, as reported by IGT. Flame arrestors are probably not applicable because of the flow resistance offered in an intrinsically low-pressure system. IGT suggests minimizing the internal volume of the burner as the solution to these problems, implying a burner retrofit, in effect, for hydrogen operation.

However, in a premix burner (e.g., Bunsen burner), the problem of reduced flame length can probably be controlled by increasing gas supply pressures. IGT estimates that a change of a factor of 10 should return the original shape. Another important aspect of a hydrogen flame combustion system is the issue of exhaust products and emissions of consequence to the maintenance of air quality, particularly important in nonvented or open flame appliances. Having no carbon whatsoever in its makeup, hydrogen is uniquely free of the oxide of carbon and unburned hydrocarbon

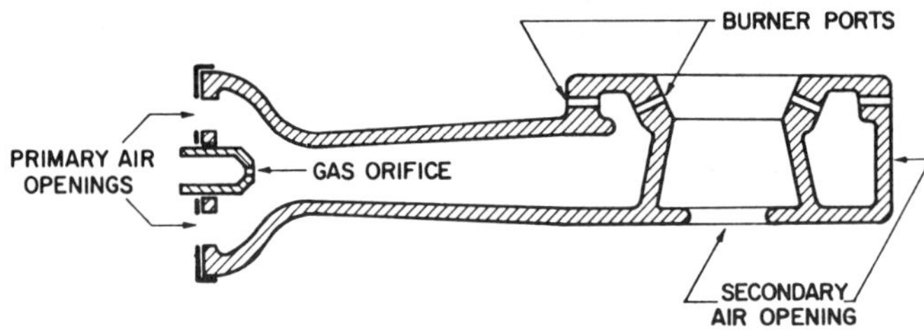

FIG. 7 Sectional view of a typical atmospheric range-top burner. (Courtesy of the Institute of Gas Technology.)

emissions that are of concern in all hydrocarbon flame burners. Water vapor and, in the case of inefficient combustion or failure to ignite, unburned hydrogen are the only direct effluents in a hydrogen-fueled device (unless there are complications, such as lubricant combustion as in a hydrogen-converted internal combustion engine; here oxides of carbon and hydrocarbon emissions may in fact occur). In view of the preceding, it can be seen that hydrogen flame combustion might well be carried on in ventless or flueless systems, since normally only water vapor exhaust will be present.

There is, unfortunately, one complicating factor: the issue of *oxides of nitrogen*. This is the only "pollutant" of primary concern in hydrogen-using devices. Further, with a hydrogen flame burning at somewhat higher temperatures than, say, methane (approximates natural gas), the emission level of oxides of nitrogen (NO_x) is significantly higher with hydrogen. By nature, flames that reach high local temperatures cause the nitrogen and oxygen constituents of atmospheric air to be combined into nitric oxide (NO), which subsequently can be oxidized to NO_2 (these are both covered in the "NO_x" designation). The higher the temperature, the stronger the NO_x emission tendency. The AGA Laboratory has determined the quantitative levels of NO_x emissions from natural gas appliances converted to operate on hydrogen, and compared the results with operation on natural gas. With a natural gas baseline level of 80 to 100 parts per million (ppm) being recorded for natural gas, Table 1 indicates the NO_x emissions level for the open flame hydrogen conversion for two settings of the primary air opening or air shutter (see Fig. 7). Thus, where it is mandatory to restrict oxides of nitrogen emissions to very low levels, there are definite problems with hydrogen flame combustion, at least that employing conventional gas burner devices. It may well be possible to design special hydrogen-specific flame combustion systems that will dramatically reduce NO_x emissions, but such work has not yet been reported.

TABLE 1

NO_x Emissions from Hydrogen Burner[a]

Primary air shutter	NO_x, (ppm)	% Increase compared to natural gas[b]	
		Maximum	Minimum
Full open	257	221	157
Full closed	335	319	235

[a] From AGA Laboratory.

[b] Referenced to 80–100 ppm with natural gas.

Otherwise, it can be seen that hydrogen flame combustion provides for an environmentally clean operation and an efficient conversion of chemical fuel energy to heat. The limited experimental and demonstration work to date has supported this judgment, but much more work remains to be done in deriving optimal hydrogen flame combustion devices.

D. Hydrogen Catalytic Combustion

It appears certain that a number of the problem areas discussed above for flame combustion devices using hydrogen can be avoided or ameliorated by employing *catalytic combustion* operating principles. As stated, hydrogen among all fuels is uniquely suitable for this approach. In this section catalytic combustion processes will be discussed, and advantages and limitations pointed out. In the following section a number of operating devices that have been developed to the prototype stage will be described to illustrate the potential of catalytic combustion.

By proper selection of a catalyst, such as platinum in rather small physical amounts, and the emplacement of the catalyst on a suitable substrate or holder, hydrogen fuel can be ignited and combusted in a controlled temperature fashion quite unlike flame combustion. This process, illustrated in Fig. 8 has been developed by the Institute of Gas Technology. Here it can be seen that with hydrogen admitted to the catalyzed area, spontaneous increase in activity of the catalyst results as the hydrogen and oxygen in air (or pure oxygen) are reacted together to form water vapor. A rise in temperature results (phase 1) until full activation is accomplished in a constant temperature, increased combustion rate step (phase 2). Beyond this point, a further rise in temperature occurs. If carried to the extreme, conventional flame combustion will be initiated (at approximately the autoignition point for the mixture). In general, this is to

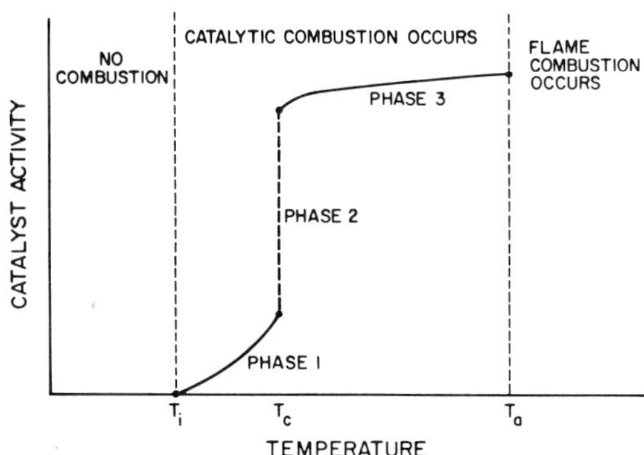

FIG. 8 Phases of catalytic combustion. (Courtesy of the Institute of Gas Technology.)

be avoided since it negates many of the advantages of catalytic combustion and may, in fact, permanently damage the burner, e.g., destroy the catalyst section.

The Institute of Gas Technology records several design and operating guidelines for hydrogen catalytic devices as follows:

> 1. Burners must be self-starting to be practical, but must not develop flame combustion. Standing pilot or electrical-ignition systems are undesirable.*
> 2. Catalyst configurations must be developed that will (essentially) completely combust the fuel. Many parameters are involved in this, and the physical configuration of the burners, catalyst activity and catalyst loading must be considered.
> 3. A system must be designed to allow not only for sufficient inducement of air into the burner, but also for diffusion of the fuel–air mixture to the catalyst surface and for diffusion of the products of combustion (water vapor) away from the catalyst surface.
> 4. Flashback cannot be tolerated. This occurs when the ignition temperature of the fuel gas is exceeded. A flame is initiated, and the flame propagates back to the orifice of the fuel-gas manifold.

In one of its patents covering a "Catalytic Fluid Heater" the Institute of Gas Technology researchers have described one specific formulation of a catalyzed surface (U.S. Patent No. 3.955,556) as follows:

> The procedure developed incorporates an anodization technique. This technique involves the electrolytic oxidation of aluminum to Al_2O_3 on the surface of the metal. Chloroplatinic acid can then be applied to this surface. The platinum can then be reduced with a soft flame to produce a high activity, self-igniting catalyst.

(Specific details of this process are listed in the patent).

Advantages of the catalytic combustion approach vis-à-vis flame combustion are several:

(1) Combustion is self-igniting leading to elimination of separate ignition systems that waste fuel, increase capital costs, decrease reliability, and/or increase maintenance costs.

(2) Combustion temperatures can be limited to the order of 500–600°F precluding the formation of NO_x, the only emission component of concern from the air quality standpoint (as discussed under flame combustion).

(3) Essentially complete combustion is obtained maximizing fuel energy utilization and obviating any hazards associated with free hydrogen in the effluent.

(4) Accurate and wide-range control can be effected over the heat release process to best match system energy demands.

* Their elimination is of considerable advantage in terms of energy conservation and equipment cost and reliability.

(5) Because of the moderate temperatures of the combustion products as controlled via catalytic combustion, it is relatively simple to arrange for water vapor condensation and attendant heat recovery in certain applications, e.g., water heaters.

Another form of catalytic combustion that has been demonstrated by the Billings Energy Corporation utilizes a modified flame combustion system. The company reported on its experimental range-top burners operated on hydrogen in a technical paper in 1974. Stainless steel thin-metal material similar to "steel wool" was placed around the gas outlet section of a conventional LPG burner. Once ignited, the hydrogen combusted in a catalytic mode over the stainless steel heating it to red-heat conditions without a prominent flame effect. Heat was transferred to cooking ware placed over the burner by both the hot gas effluent and infrared radiation from the glowing steel wool. This technique appears to be readily and cheaply implemented for conversion purposes.

5.5 HYDROGEN HEATING UNITS AND APPLIANCES

A. Hydrogen-Fueled Water Heaters, an Introduction

Residential and small commercial water heaters are of two fundamental types: storage and nonstorage systems. These are differentiated by the presence of a large-capacity water tank, normally holding heated water, in the storage systems. The nonstorage type provides for "instant heating" on demand of cold water to a preset temperature. There is no hot water storage. Each has its characteristic advantages and disadvantages, and both types are found in service.

Conceptually, in view of the similarities of hydrogen combustion to that of natural gas and other in-use fuels, hydrogen-fueled water heaters can be of both types. Further, as discussed in the preceding section, either flame or catalytic combustion could be employed in hydrogen units. Since flame combustion hydrogen water heaters would be quite similar to existing hydrocarbon units (with differences in detail in the burner designs), we will concentrate on the unique catalytic combustion water heater systems in this discussion. It will be recalled that these systems offer significant advantages: autoignition (no standing pilot flames or other ignition systems required; negligible NO_x emissions, and maximum combustion efficiency.

Nonstorage-type hydrogen catalytic combustion water heaters In this concept, recently carried to the experimental demonstration stage by researchers at the Institute of Gas Technology, flowing cold water is instantly heated in a combustion/heat exchange system to the temperature level indicated by an adjustable sensing element at the hot water outlet point. Combustion of hydrogen with air would occur on a catalyzed surface providing adequate heat transfer to the water circuit to provide rapid heating.

Figure 9 shows the IGT working model of a nonstorage hydrogen catalytic combustion water heater. This unit was designed and constructed under the sponsorship of the Southern California Gas Company in 1975. It is covered by U.S. Patent No. 3,910,255. The patent drawing of Fig. 10

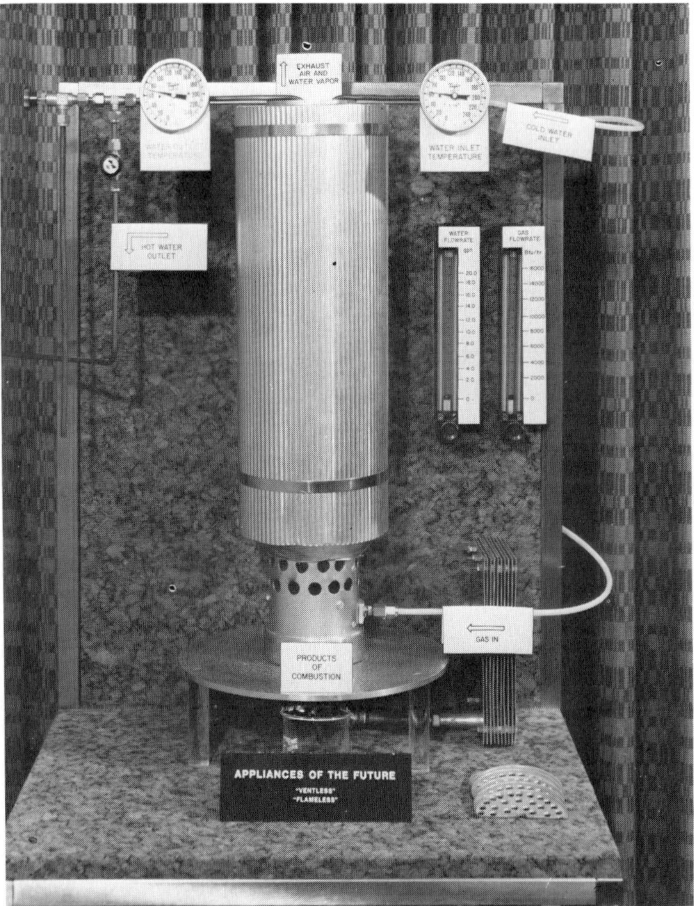

FIG. 9 Experimental working model of a catalytic combustion hydrogen-fueled nonstorage water heater (IGT/Southern California Gas Company). (Courtesy of the Institute of Gas Technology.)

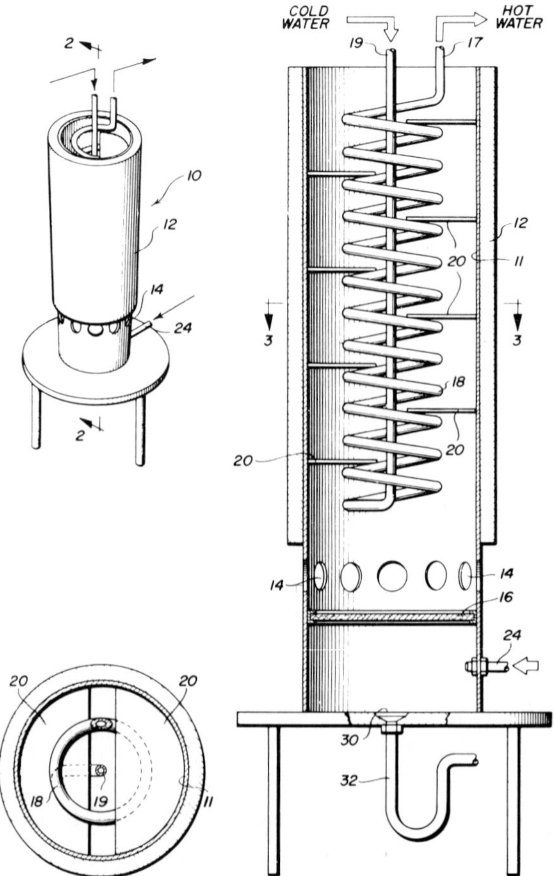

FIG. 10 Sectional view of a catalytic combustion Hydrogen-fueled nonstorage water heater design (IGT/Southern California Gas Company). (Courtesy of the Institute of Gas Technology.)

can be used to describe the operation of this unit. Hydrogen is admitted under controlled flow conditions at the bottom of a well-insulated cylindrical shell unit (24). It passes through a porous distribution plate (16) and rises upward mixing with air induced through the ports (14) by the chimney effect. Staggered half-circle metal baffle plates (20), which are catalyzed on their lower surfaces, are placed along the cylinder but are attached for heat transfer purposes to the water flow circuit (the coil-shaped unit). The baffle plates are perforated to permit water condensed from the exhaust to fall back to the bottom of the unit.

Catalytic combustion takes place progressively to completion as the hydrogen–air mixture rises through the unit. The heat released is transmitted to the water coil by convection and conduction. The heated airstream rises producing a chimney effect, inducing more air into the unit for combustion. As noted, some of the water vapor in the combustion products condenses into liquid on the cold water conduits. This water is eventually collected at the bottom of the unit and is conducted away (32) for possible use as a distilled water supply, e.g., for steam iron use.

Storage-type hydrogen catalytic combustion water heaters IGT also constructed and tested a working model, storage-type hydrogen catalytic combustion water heater in the program mentioned earlier. Views of this unit are provided in the patent drawing of Fig. 11 (from U.S. Patent No.

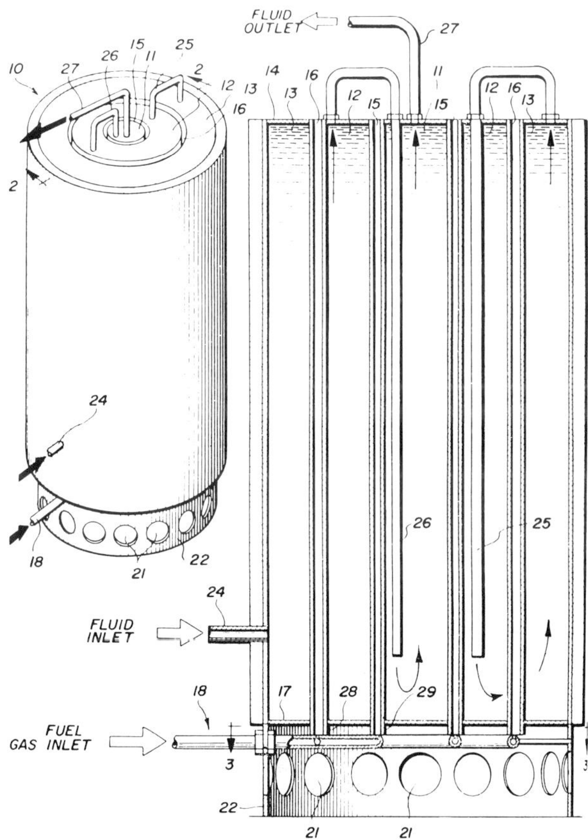

FIG. 11 Sectional view of a catalytic combustion, hydrogen-fueled, storage water heater design (IGT/Southern California Gas Company). (Courtesy of the Institute of Gas Technology.)

3,955,556). The specific configuration developed and tested by IGT is a series of concentric annular water container sections, with the water flow from the outside toward the center. Specifically, two annular containers (12 and 13) are concentrically mounted around a central cylindrical tank (11). Cold water flows in from the outside (24), and hot water from storage is extracted from the top center (27). The annular spaces formed between the water containers is where the catalytic combustion process takes place.

Hydrogen is admitted through appropriately located ring manifolds at the bottom of these spaces. It is mixed with air induced through ports (21) drawn in by the chimney effect. Only one surface of each annular combustion space is catalyzed; the other acts as a condensing surface for water vapor (which gives up its heat of vaporization to the water being heated). The catalyzed surface provides for autoignition and high-efficiency, moderate temperature combustion of the hydrogen as described earlier. Not shown in Fig. 11, the condensed water can be collected for use as in the nonstorage heater design. Both units would readily qualify as "ventless" appliances and would not normally require flues to exhaust to the outside, so long as circulation paths are provided such that fresh combustion air is admitted in sufficient supply.

Storage-type hydrogen–oxygen combustion water heater concept Were oxygen as well as hydrogen available for operating heating units, other concepts would become possible. The case of a storage-type water heater will be illustrated, although experimental work equivalent to that discussed above has not to our knowledge been reported. Hydrogen–oxygen combustion can be of the flame or catalytic combustion type analagous to the hydrogen-air systems discussed so far. A familiar example is the oxyhydrogen torch, such as that used for underwater cutting and welding. As early as 1924, Tucker patented an "Oxyhydrogen Steam Generator" in which water is added downstream of an oxygen–hydrogen flame operating at the stoichiometric point (i.e., at the chemically correct ratio for producing neither hydrogen nor oxygen in excess).

A concept for storage water heater might consist of a hot water tank with a submerged hydrogen–oxygen burner releasing heat directly within the water volume. Such a burner requires no air for combustion, and hence there would be no gaseous effluent at all. Heat transfer to the water would be direct and not through metal surfaces. The combustion water would all be recovered in a properly controlled system. Hence, the full higher heating value of the hydrogen would be converted to heated water, assuming complete combustion is achieved and a proper control of the hydrogen–oxygen mixture is obtained.

Such a unit would be extremely compact, have a very high response time, and, as already explained, would offer the maximum possible efficiency. Appliances based on hydrogen–oxygen combustion may have special significance to solar energy systems.

B. Hydrogen Catalytic Air Heaters for Space Heating

The use of catalytic combustion on surfaces, as described for water heaters in the preceding section, is a process that can be applied to direct air heating, as in space heating systems. The basic principle is illustrated in Fig. 12, a catalytic endurance test specimen. This unit has been built and extensively treated by the Institute of Gas Technology. As shown, hydrogen is admitted through orifices from a manifold between the two catalyzed surfaces where, mixed with air, it ignites and combusts catalytically on the surface. Air is heated directly by the flame and indirectly via

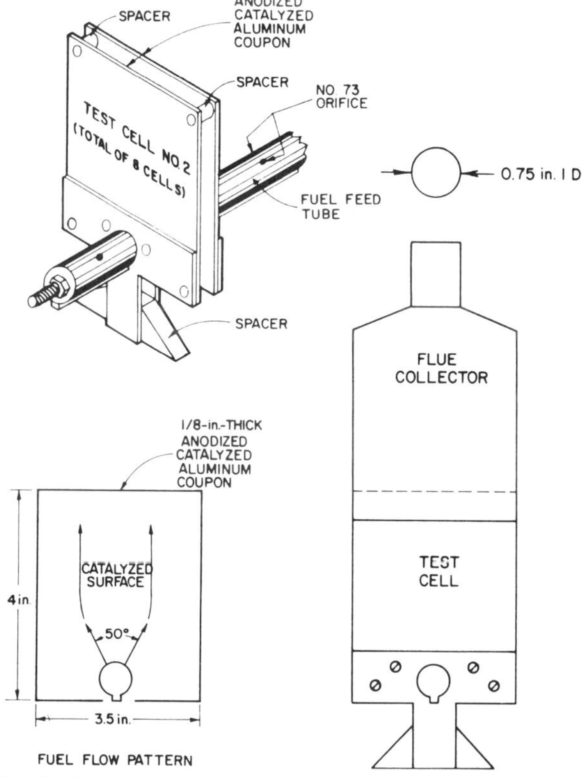

FIG. 12 Sketch of experimental catalytic combustion evaluation unit (IGT/Southern California Gas Company). (Courtesy of the Institute of Gas Technology.)

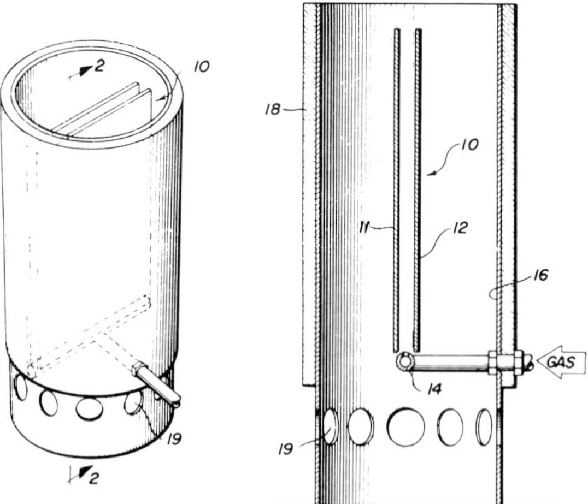

FIG. 13 Sectional view of a catalytic combustion, hydrogen-fueled, air heater design employing flueless direct space heating (IGT/Southern California Gas Company). (Courtesy of the Institute of Gas Technology.)

heat transfer from the opposite side of the metal coupon. Such a unit can be repeated indefinitely to form a "fin-tube"-configured heating strip for space-heating purposes.

The principle illustrated in Fig. 12 is shown in the various space heater designs from an associated IGT patent (US Patent No. 3,916,869). Figure 13 shows a cylindrical unit enclosing a two-surface catalytic combustor,

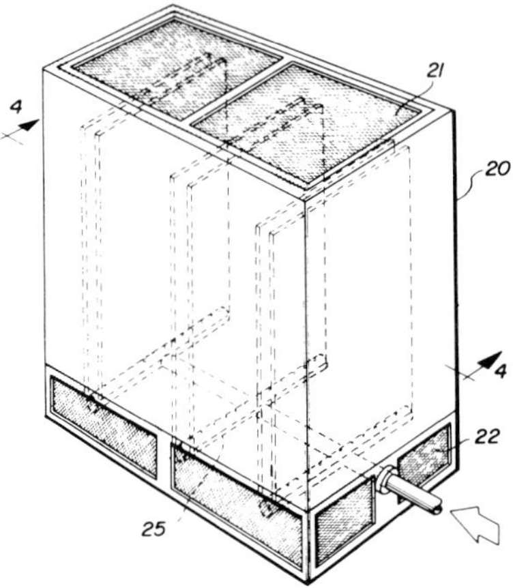

FIG. 14 Sectional view of a catalytic combustion, hydrogen-fueled, air heater design employing flueless direct space heating (IGT/Southern California Gas Company). (Courtesy of the Institute of Gas Technology.)

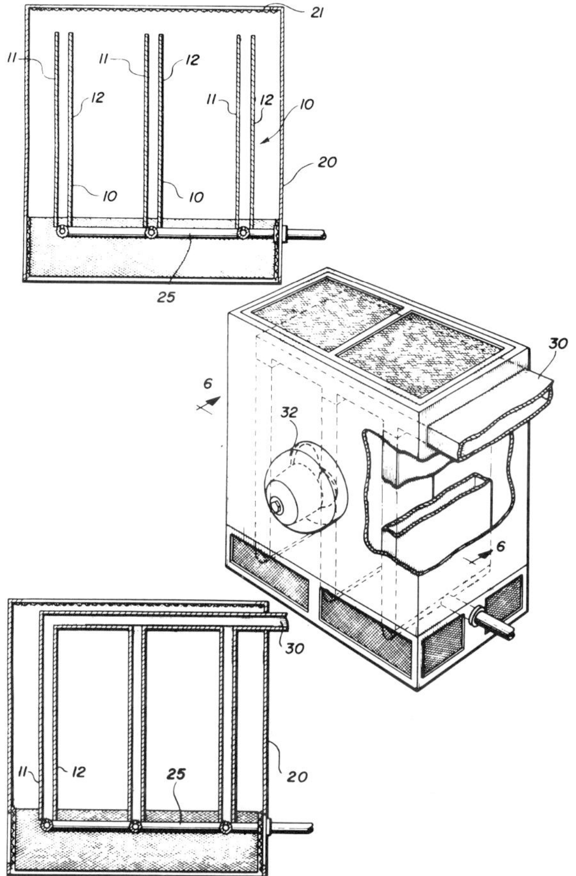

FIG. 15 Sectional view of a catalytic combustion, hydrogen-fueled, air heater design employing a flue and providing direct space heating with forced air (IGT/Southern California Gas Company). (Courtesy of the Institute of Gas Technology.)

similar to that of Fig. 12. Figure 14 shows a rectangular unit with several such combustion surfaces. In this case the effluent is circulated directly to the immediate space to be heated without a flue or other external conduit. In the design of Fig. 15, the effluent from the combustors is carried away in a flue duct (30), and an air blower unit is shown as an alternative to pure convection circulation. Such a design is entirely compatible with today's forced air-circulation central heating systems for residences. On the other hand, the use of catalytic hydrogen space heaters seems to lend itself to individual room-by-room heating, since such heaters can be unvented.

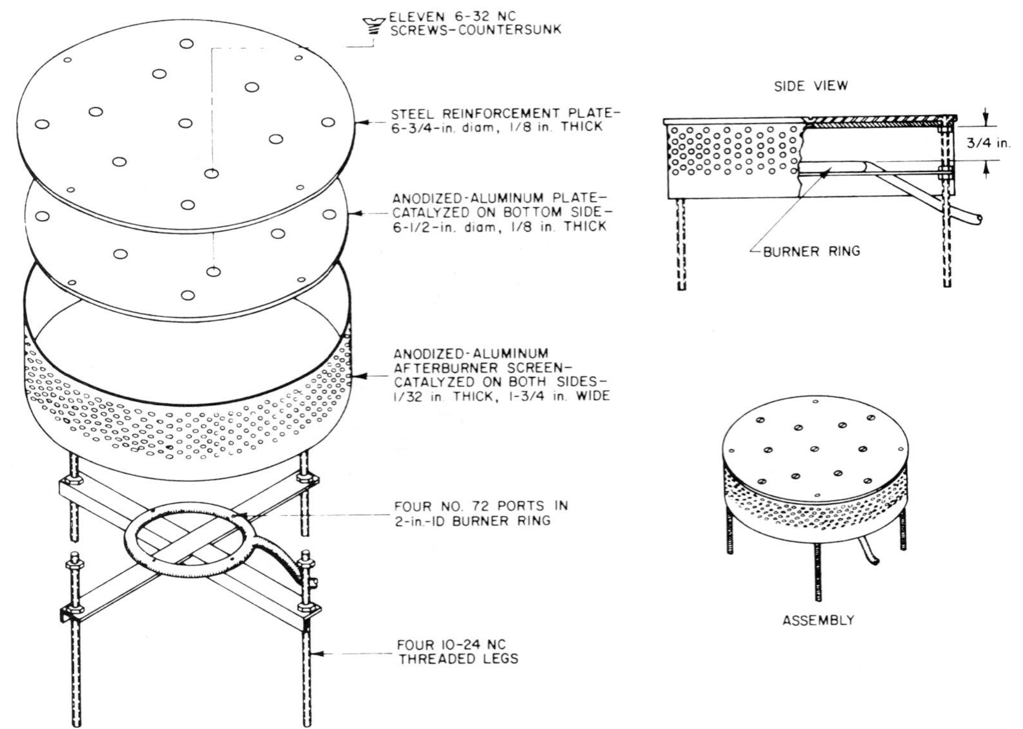

FIG. 16 Several views of a catalytic combustion, hydrogen-fueled, experimental range-top cooking unit (IGT/Southern California Gas Company). (Courtesy of the Institute of Gas Technology.)

C. Hydrogen Catalytic Combustion Range-Top Unit

As already mentioned, the Billings Energy Corporation has converted a hydrocarbon gas range element to hydrogen operation by placing a stainless steel mesh around the conventional burner. This arrangement has provided satisfactory catalytic combustion and appliance heating. The Institute of Gas Technology has constructed several catalytic combustion range-top elements employing the new principles demonstrated in water heaters, as described earlier.

One such unit is illustrated in Fig. 16. This unit uses a catalyzed aluminum surface directly connected to a steel reinforcement plate. The steel surface acts to support the cooking ware being heated. A large number of orifice patterns and flow circulations have been tried, and satisfactory ignition and operation have been demonstrated.

Another IGT unit demonstrated round 1970 in its "Home for Tomorrow" exhibit for the American Gas Association is a "clean-top" range. In this unit a catalyst was placed directly on the underside of a ceramic-type range top with appropriate hydrogen fuel distribution and control. The unit was demonstrated to the public along with other such hydrogen appliances as a radiant wall heater, a warming tray, and a residential environmental control system.

5.6 HYDROGEN AIR CONDITIONING AND REFRIGERATION

A. Introduction

The application of gas energy to the residential and commercial air-conditioning market has long been a recognized activity. In addition to the absorption-type "chiller" systems that are fuel-fired and/or utilize water heat recovery for air conditioning, there are new system concepts utilizing gas energy that provide for "total environmental control." In general, these concepts are geared to natural gas, which at present widely services residences and commercial establishments. Hydrogen, produced from solar energy conversion systems would appear to be quite applicable to such systems and may offer some unique advantages. The following discussion will concentrate on one such advanced system under development as a representative application.

B. Munters Environmental Control System

A system capable of providing air conditioning (cooling), heating, and humidity control and that uses gas energy is the Munters Environmental Control (MEC) system, known to many as the "Lizenzia" machine. The system is covered by numerous US and foreign patents owned by the Carl Munters Company of Sweden. The patent background has been licensed to the Gas Developments Corporation, a subsidiary of the Institute of Gas Technology, for further technical development and sublicensing for commercial development.

A complete description of the MEC system is beyond the scope of this chapter; further information is available from GDC and IGT. Figure 17 shows a schematic and a pictorial sketch of the system. It is predicated on the parallel counterflow of outside air and exhaust air from the living space, in which two rotating porous material wheels provide for humidity and heat exchange between the two airstreams. In the cooling mode the MEC unit provides for exhaust air heating via the heat exchange wheel, with further heating ahead of the drying wheel through a gas burner section. Where natural gas is used conventionally, it seems quite possible to substitute a hydrogen burner of either a flame-type or catalytic combustion design. Such burners have been described in previous sections.

C. "Servel"-Type Refrigeration

Natural-gas-operated refrigeration systems have been in widespread use for many years. Featuring gas flame heating of a refrigerant, the "Servel"-type units operate without moving parts thereby providing for

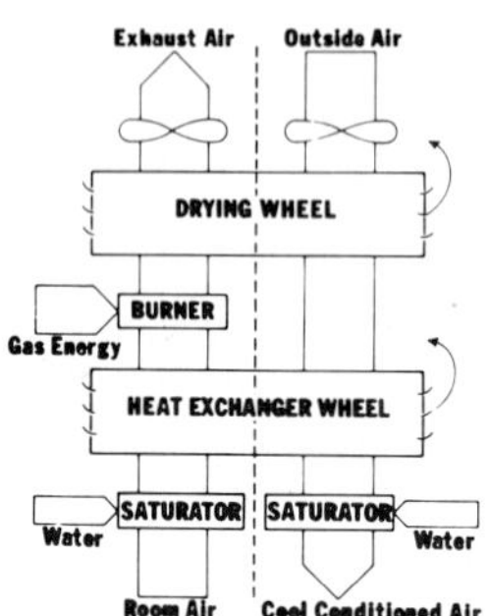

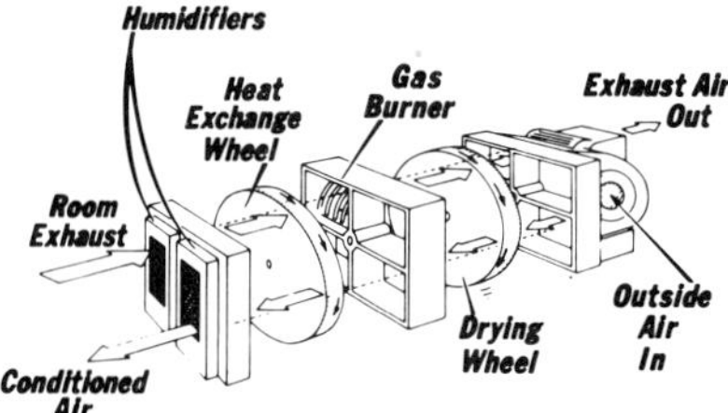

FIG. 17 Flow schematic and component layout of Munters Environmental Control (MEC) system. (Courtesy of the Institute of Gas Technology.)

long life and low maintenance. Again, a hydrogen-fueled refrigeration loop working quite analogously can be envisioned.

D. Mechanical Cycle Refrigeration via Heat Engines Operating on Hydrogen Energy

Since hydrogen is an excellent fuel for internal combustion engines, Rankine and Stirling external combustion engines, and gas turbine (Brayton cycle) power systems, it provides a means of supplying shaft power to mechanical air-conditioning systems of the conventional type. Heat pump systems providing heating and cooling can be similarly operated. In view of the complexity of a heat engine system, such applications would seem to be more suitable for larger commercial systems, especially for "total energy" systems where exhaust heat recovery can compensate for the relatively low efficiency of the heat engine.

5.7 ELECTRICITY GENERATION VIA HYDROGEN ENERGY

A. General Systems

A substantial number of studies have shown that where energy storage is needed in electrical utility systems, hydrogen and hydrogen–oxygen prove to be efficient and flexible means of storing chemical energy for subsequent conversion/reconversion to electricity. Similarly, in solar energy systems that are required to serve local electricity needs around the clock, hydrogen energy storage can be applied to meeting this need. For appropirate solar energy systems, presumably during periods of insolation, electricity can be directly generated up to the system capacity and placed on the line. However, for serving overcapacity peaks and—more significantly—for generating electricity in noninsolation periods, such energy storage and conversion will be necessary. Alternatively, nonchemical storage means can be used, e.g., flywheel (mechanical) energy storage systems. Perhaps, the baseline approach for comparison purposes is the conventional electric battery (electrochemical) system.

Two basic approaches for generating electricity from hydrogen will be discussed: fuel cell systems and heat engine/generator systems. It will be noted that either hydrogen–air or hydrogen–oxygen is applicable, with certain unique advantages accruing to the latter, though a less-developed option at present. Whether the hydrogen energy approach should, in actuality, be selected over the other alternatives cannot be answered on a general basis. Specific systems analyses are required for this determination. However, technically speaking, this option is an available one.

B. Fuel Cell Systems

Fuel cells directly convert chemical energy forms to electricity via electrochemical conversion. For this reason fuel cells are not subject to the usual "Carnot cyle restraints" of heat engine systems. Hence their efficiency can be quite high, approaching about 90% in ideal systems. Although fuel cells can be theoretically operated on a number of fuel types, hydrogen is by far and away the prime candidate as it is readily and directly reacted electrochemically with oxygen from the air or pure oxygen supplied. Illustrative of this, those fuel cells that have been developed for natural gas and light hydrocarbon fuels (e.g., kerosene and other middle distillates) actually use hydrogen produced from these hydrocarbons. The so-called TARGET natural gas fuel cell employs a methane steam-reforming step ahead of the fuel cell to produce a reformed gas that is largely hydrogen with some carbon dioxide in it.

Where hydrogen is available from the onset, such "pretreatment" of the fuel can be eliminated with gains in overall efficiency and reductions in cost for equipment and maintenance. Thus, hydrogen is the best fuel for fuel cells.

Fuel cells using air to support "combustion" of the hydrogen are usually based on a system using an acid electrolyte, such as phosphoric acid. Although alkali electrolyte fuel cells are significantly simpler and operate at a higher efficiency, with all things else being equal, they suffer from the problem of atmospheric carbon dioxide reacting in the fuel cell in such a way as to degrade its operation.

Thus nonalkali-type hydrogen–air fuel cells must be employed. If pure oxygen is available along with the hydrogen, then the advantageous alkali electrolyte cell can be employed with gains in efficiency and cost. This again is a significant point when the hydrogen energy is to be produced via water-splitting reactions, as in the case of water electrolysis. With oxygen being rather more readily processed and stored than its companion hydrogen, it seems only logical to consider hydrogen–oxygen fuel cells in preference to hydrogen–air systems. However, as of this time, the latter has undergone considerably more development.

Hydrogen–oxygen fuel cells have been developed for aerospace and military applications. For example, hydrogen–oxygen fuel cells supported the major electrical requirements of the Apollo space vehicle used in the US lunar missions. Although fuel cell units developed for these applications are very expensive and have limited operational lives, the fundamental technology is in all likelihood applicable to the ultimate development of commercial fuel cell systems such as might be integrated into advanced solar energy conversion systems. The target costs and efficiencies that have been cited for a developed hydrogen–oxygen fuel cell system are \$200/kWe and 70% HHV (Higher Heating Value).

C. Heat Engine/Generator Systems

The other alternative for generating electricity from hydrogen is the use of a heat engine to develop shaft power to operate a conventional generator. Hydrogen, being a flexible fuel, can be applied to essentially all heat engines, both the internal combustion and external combustion types. The selection of engines would depend on the tradeoff of efficiency and cost effects in view of the size, duty cycle, and other systems level stipulations to be placed on the electricity generation subsystem.

Hydrogen-fueled internal combustion engines have been operated over a range of sizes, generally with emphasis on use as vehicle prime movers. Because of its propensity for very lean burning where high efficiencies can be achieved, as well as other combustion-related factors, internal combustion engines using hydrogen may achieve substantially higher efficiencies than would be the case with hydrocarbon fuels. Similarly, hydrogen-fueled gas turbines, when especially designed to take full advantage of hydrogen's properties, e.g., its good cooling characteristics for turbine blade cooling, can probably reach higher efficiencies and be made more compact via increased turbine inlet temperatures than would be the case with conventional fuels.

Hydrogen–oxygen shaft power devices have been given limited development in connection with the space program as auxiliary power units. A system of particular interest because of its high-efficiency potential and its simplicity and compactness is the hydrogen–oxygen/steam combustion turbine system. Basically a steam turbine system (Rankine cycle) in which the steam is generated in a compact combustor, the system eliminates the conventional steam-raising boiler requirement. Further, it is able to go to much higher temperatures than a conventional system, leading to elevated efficiencies as high as 55% in large vacuum-condensing systems. Such systems are also pollution free and flexible in operation.

ACKNOWLEDGMENTS

The authors are pleased to acknowledge the support of the Institute of Gas Technology, Chicago, for information supplied in support of the preparation of this chapter. Particularly helpful were the advisements and contributions of Dr. Jon B. Pangborn, Walter J. Jasionowski, and Dale G. Johnson.

Photographs were supplied by the Institute of Gas Technology and Brookhaven National Laboratory.

6

Passive Solar Heating Systems

W. W. S. CHARTERS

DEPARTMENT OF MECHANICAL ENGINEERING
UNIVERSITY OF MELBOURNE
MELBOURNE, AUSTRALIA

6.1 INTRODUCTION

In the recent resurgence of interest in solar energy utilization in general and in the application of solar energy for building heating in particular, it is possible to discern a distinct division developing between active and passive solar heating systems. This division relies not so much on the type of solar equipment utilized in any one application but more on the type of operation of that equipment envisaged by the designer (Ballantyne, 1973; Olgyay, 1963).

It would seem reasonable at this stage to define an active system as one in which it is essential to supply input energy in the form of electrical power or mechanical work in order to transfer the collected energy from the point of collection directly to the interior of the building or to an assigned thermal storage. On the other hand, the basic feature of a passive system is that no input energy is required to perform this transfer/storage task. Because of their deceptive engineering simplicity, little attention has been paid to the potential of passive heating systems, and engineers have favored the more complex active systems that are inherently more capable of regulation and control.

Generally, passive systems have no separate collector elements, and every attempt is made to incorporate the collectors into the structure of the external envelope of the building. All thermal transfer processes then take place by natural convection and/or conduction and radiation. Evidently all buildings will perform as natural solar heating systems to some

ISBN 0-12-620860-3

degree, but it is only when the building structure is designed to maximize the solar energy collection to provide a significant proportion of the heating requirement that one would use the term passive solar building in practice.

In any particular design it is evidently relatively easy to vary the appropriate design parameters to significantly alter the solar contribution to the total building heating load (Szokolay, 1975; Shurcliff, 1976). These variables include, among others, the following factors;

(1) Type, area, and orientation of fenestration; and provision for shading devices, either fixed or variable.

(2) Ratio of window/wall area to control the amount of admission of incident solar radiation to the interior.

(3) Materials of construction (structural and insulating) and colors of the external wall and roof areas.

(4) Relativity of wall area facing the sun to all other wall areas of the structure controlled by site orientation.

By careful control of these design criteria it is possible to provide a substantial proportion of the total heating requirement (say 50–60%) on an annual basis. The basic problems to be overcome are those of thermal storage, which has to be designed into any such structure, and the associated fact of internal temperature control to achieve acceptable standards of temperature variation in the internal living quarters. However, the use of large areas of internal walls and/or ceilings at relatively low temperatures allows effective control and modulation of environmental temperature experienced in the living spaces. To assist with this control problem, one can use ventilation techniques in winter to reduce possibly unacceptable high temperatures and external shading devices in summer to prevent a large proportion of the solar gain due to the large variation in summer and winter solar angles (Ohanessian, 1976).

Storage of the trapped solar energy is normally built into the structure in the form of a concrete slab floor, a masonry or concrete wall, or a roof pond system (Fig. 1). The use of a vertical collector element such as a south-facing wall in the Northern Hemisphere or the opposite face in the Southern Hemisphere is recommended at locations of fairly high latitude (30° and above). For these locations the wall collector is ideally suited to collect well in the months of the year when the wall has to perform this radiation collection function, and it is simple to provide overhang shading of the wall in the summer months when this function is no longer required. The amount of storage required for the successful operation of such a passive system is critical and must be sized as a function of the solar collector area and the internal volume. It appears preferable from initial studies

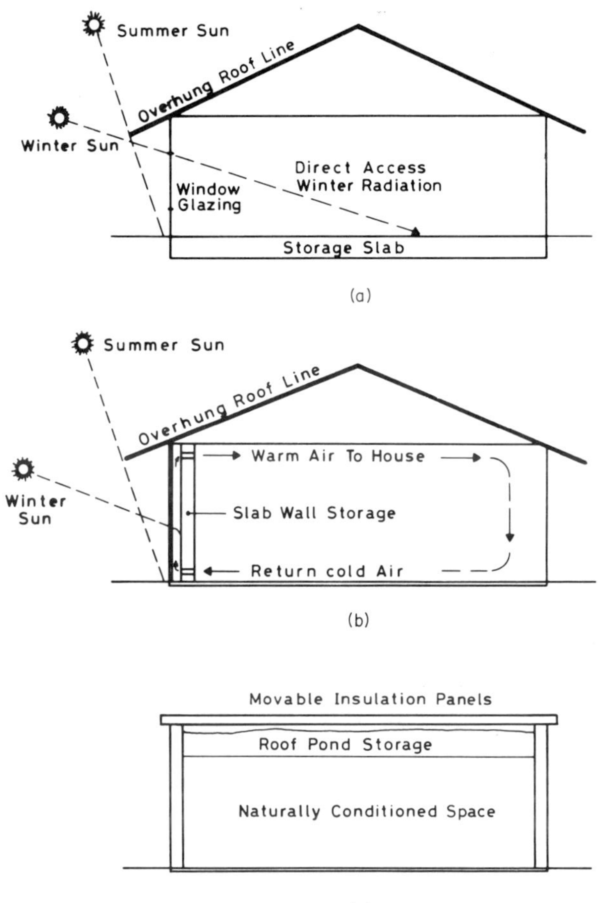

FIG. 1 Passive solar heating systems. Alternative methods of thermal storage: (a) floor storage, (b) wall storage, (c) roof and ceiling storage.

that the storage should be heated directly from incident radiation and be partially decoupled from the heated spaces. At present there is only very limited design data available to aid in the proper design of an efficient passive solar heating system, although many systems are now under construction. There is unfortunately also a lack of reliable performance data on the passive structures built to date with which one could validate the numerical models from the preliminary simulation studies reported in the literature. However, there is sufficient evidence to suggest that a correctly designed passive system will perform equally as well as the corresponding active system in terms of the amount of solar energy delivered to the heated space on an annual basis. In addition it has been shown in the correct climate areas that it is possible to maintain acceptable temperatures in the heated zones using such systems (Ohanessian and Charters, 1975).

It is equally difficult at this time to be definitive about the cost-effective aspects of passive systems, as these have not yet been built in sufficient numbers to allow an evaluation of the true extra costs of incorporating the solar collection elements into the basic building structure. Many of the passive houses proposed lend themselves readily to modular building techniques and factory prefabrication; and this aspect of the construction should help to substantially reduce the on-site costs. In addition the extra costs of the solar collection system appear to be small compared with those envisaged for active systems (Hay and Yellot, 1970; Hay, 1973a,b).

Some of the more famous attempts at designing effective natural house-heating systems are listed in the next section along with the relevant design data and any performance data available for the buildings involved.

There is no doubt in my mind that in the near future, with the ever escalating costs of conventional fuels and their scarcity in some regions, the introduction of solar heating systems in general and natural or passive heating systems in particular will receive urgent attention from practicing architects and engineers. The active liaison and cooperation of these two professional groups in the early design and planning stage will be essential for the effective incorporation and operation of passive heating as an integral part of the external envelope of the building (Hay, 1973c).

6.2 SOME EXAMPLES OF PASSIVE SOLAR HEATING SYSTEMS

The following examples are only a sample of the many passive solar heating systems proposed for buildings, but they summarize the thinking employed in the design and construction of such integrated structures.

A. St George's Secondary School

New wing construction (Szokolay, 1975; Shurcliff, 1976).

Wallasey (near Liverpool), England; latitude 53° N.

Outer glass wall area, 500 m^2.

Collector area (south wall), 500 m^2; 88% glass, 12% black painted masonry.

Spacing between inner and outer walls, 600 mm.

Note: 33% of the total area incorporates reversible aluminum panels that face bright side out from April to October and black side out from November to March.

Thermal Storage: 180-mm-thick concrete roof.
220-mm-thick brick walls on north, east, and west faces.
250-mm-thick reinforced concrete slab floor.
125-mm-thick polystyrene external insulation.

Typical winter performance monitored over a 10-year period indicates temperature control in the 16–20°C range with no auxiliary heating during the November-to-February periods.

B. Odeillo Houses

Trombe–Michel south-wall system; 1967–1974 (Szokolay, 1975; Shurcliff, 1976; Ohanessian, 1976).

Odeillo–Font Romeu, Eastern Pyrenees, France; latitude 43° N.

Double-glazed, vertical collector module elements.

Concrete thermal storage walls of thicknesses varying from 350 to 670 mm used in the various designs. Varying colors tested for the external face of the concrete range from dark blue or green, through dark red, to standard matte black finish (Trombe *et al.*, 1976).

Table 1 presents architectural and technical information on passive solar houses built by CNRS at Font-Romeu (Odeillo) in the French Pyrenees. Internal temperatures are controlled at 20°C with electrical boost elements. Solar contribution to heating load is estimated at 50 to 60%.

TABLE 1

Odeillo Solar Houses

Design elements	Date of construction: 1967	1974 No. 1	1974 No. 2	1974 No. 3
Number of levels	1	2	2	3
Number of rooms	4	6	6	7
Floor area (m^2)	76	210	180	250
Volume (m^3)	300	525	250	650
Mean wall conductivity (W/m^2 °C)	1.34	0.4	0.4	0.4
Collector area (m^2)	48	55	45	65
Collector area/volume (m^{-1})	0.16	0.10	0.10	0.10
Concrete slab thickness (m)	0.60	0.37	0.37	0.37
Glazing (mm)	2 × 3	2 × 4	2 × 4	2 × 4

The use of double glazing to minimize the heat loss to the ambient environment is probably necessary only in localities that experience low ambient temperatures. It is therefore doubtful that double glazing will prove cost effective in those areas where a moderate climate prevails. The effect of varying the thickness of the concrete slab wall is to alter the relative proportion of heat transfer to the internal environment by conduction through the slab and by free convection through the gap. Use of an excessively thick slab may well adversely affect the radiative heat transfer to the interior by lowering the internal surface temperature of the slab wall.

Use of numerical modeling techniques (Balcomb *et al.*, 1977; Ohanessian, 1976; Ohanessian and Charters, 1975) tends to support the thermal efficacy of a relatively thin slab of 150- to 250-mm-thick concrete rather than the much thicker slabs originally proposed in the Trombe–Michel wall system. Each of these numerical simulations also supports the more general contention that a well-designed passive system can prove effective in supplying a major proportion of the total building heating requirement under a variety of climatic conditions.

C. "Skytherm House"

Atrascadero, California; latitude 35° N; 1973 (Szokolay, 1975; Shurcliff, 1976).

Collector and roof pond storage formed using 8 bags (11.4 × 1.2 × 0.26 m) made from 0.5-mm-thick polyethylene.

Total water volume stored is 26.6 m^3 with total weight of 26,600 kg.

No antifreeze agent required in water stored.

Upward thermal losses inhibited by transparent plastic film suspended 50 mm above bags.

Roof pond supported by ribbed steel deck flat roof and covered by 9 panels of 50-mm-thick polyurethene-foam-insulated panels, which can be manually or automatically opened or closed.

Internal concrete block partition walls, sand filled. External concrete block walls, vermiculite filled. Designed to supply winter heating and summer cooling using heat rejection to the night sky.

D. "Zomeworks" Solar Houses

Corrales House 1 Corrales (near Albuquerque), New Mexico; latitude 35° N; 1970 (Szokolay, 1975; Shurcliff, 1976).

Single-story residence of interconnected polygonal units.

South-facing vertical collector wall, double glazed.

Thermal storage using 90 oil drums, each of 200-liter capacity, 95% filled with water, aligned north–south on their horizontal axes, and mounted 4–5 high on a metal support frame. South face of barrels painted with a nonselective matte black finish. Total weight of water in drum storage (20,160 kg) forms approximately half the total building thermal storage, the remainder being incorporated into a 125-mm-concrete slab floor and external adobe walls.

External, folddown, aluminum insulated panels for minimizing night heat loss through collector wall.

Internal curtain wall to reduce convective effects from drum storage to living space and variable ventilation.

90% solar heating contribution claimed if tolerance on temperature control is relaxed to allow a swing of 13 to 23°C.

Corrales House II 1974; One-story house of 100-m^2 floor plan.

Solar air heater used for collector situated on south-facing slope below house level. Collector area, 42 m^2. Collector operation by natural convection thermosiphon effect.

Thermal storage in rock bin built into house foundations: 40 m^3, 45,900 kg.

Designed for 75% solar contribution.

7

Passive Cooling of Buildings

A. A. M. SAYIGH

DEPARTMENT OF MECHANICAL ENGINEERING
COLLEGE OF ENGINEERING
UNIVERSITY OF RIYADH
RIYADH, SAUDI ARABIA

7.1 INTRODUCTION

Solar energy has been used adequately in many ways, but its most widespread and profitable use has been in the heating and cooling of buildings. Obviously, the hotter the climate, invariably the greater the availability of solar radiation and hence the greater the need for cooling. Thus, the supply of energy and the demand for it are closely matched, the greatest need occurring during periods of greatest insolation; this matching extends even to variations within the day.

In this chapter emphasis is on the conservation of supplied energy. Therefore passive cooling is better than active cooling for which a refrigerant and its accessories are used.

In fact, solar insolation does not come into this application directly. However, because of the solar radiation dissipated during the day and the high environmental temperature rise, a need for cooling arises.

7.2 GENERAL REMARKS

Air conditioning, in general, is the provision of a comfortable environment by changing, heating, cooling, humidifying, or dehumidifying the surrounding air. In a real situation the given surrounding atmosphere is a mixture. This mixture consists of a vapor, air, steam, and a noncondensible gas (Sayigh, 1977b). One of the most important properties of such a

ISBN 0-12-620860-3

mixture is the moisture content. This is defined numerically as the mass of steam per unit mass of air. This is conveniently represented by the relative humidity (RH), which is the ratio of the partial pressure of steam to the corresponding value for saturated air at the same temperature. Because of the effect of barometric pressure, the RH does not exactly equal the ratio of the moisture content to its value for saturated air. The difference is very small. Another way of expressing the moisture content is in terms of the wet-bulb temperature (WBT). This is the temperature indicated by a thermometer whose bulb is kept covered with a film of water. If the air is made to move relative to the bulb, an equilibrium temperature that results from the air temperature and the relative humidity will be reached. Therefore, the difference between the temperature and the wet-bulb temperature (WBT) is related to the RH.

Figure 1 shows a psychrometric chart for moist air. Another expression of the moisture content is the adiabatic saturation temperature. This is the temperature that would be reached if the air were maintained in contact with free water at the same temperature. It happens that the WBT and the adiabatic saturation temperature are nearly equal (Jones, 1973).

Comfort is a relative term, and the relationship between comfort and the thermodynamic state of the environment cannot be fully defined.

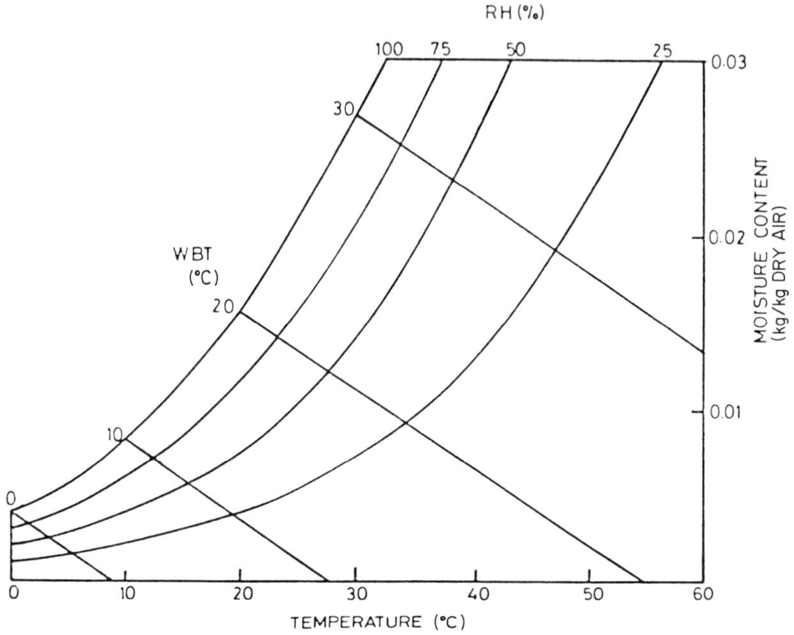

FIG. 1 A psychrometric chart for moist air.

Human beings maintain a surface temperature of about 33°C by heat and mass transfer with the surroundings, and comfort is evidently related to the ease with which this can be accomplished. As a result, only a narrow range of environmental temperatures, say 18–27°C, is considered comfortable. The humidity tolerance is much wider, about 25–60%, with a tendency for lower RH to be preferred at higher temperatures to assist heat loss by evaporation.

The temperature and RH limits, within which most people would feel comfortable are drawn differently by different authorities. A consensus of the requirements for continuous occupation and light activity is shown roughly in Fig. 2.

Conditions well outside these boundaries can, and often must, be tolerated. Outdoor air temperatures in excess of 50°C occur and 100% RH is common in coastal regions. For example, in Kuwait (Sayigh 1975a) the maximum temperature varies from 20 to 45°C, while the RH varies from 40 to 90%. In central Saudi Arabia, e.g., Riyadh, the maximum temperature varies from 30 to 45°C, while the RH varies from 20 to 60% (Sayigh 1975b). However, there are no universal rules about acceptable conditions that can be held to apply in every case. As a rule of thumb, the in-

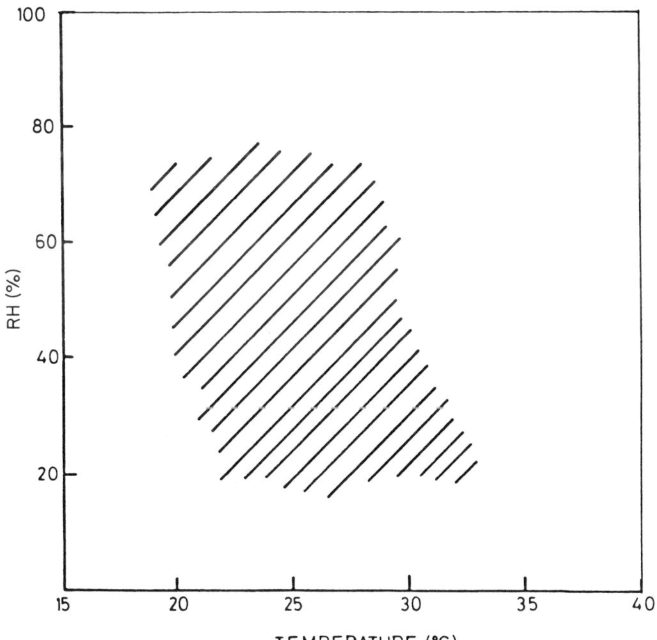

FIG. 2 Region of human comfort.

door temperature should be between 4 and 11°C lower than the outdoor temperature, with RH of about 50% (Sayigh and El-Salam, 1975).

7.3 PSYCHROMETRY

Let us consider the desired conditions in a given room and that these conditions are shown by point *R* in Fig. 3. In order to maintain such conditions, the supplied air should be at some state, say *S*, from which the change in its temperature and its moisture content, on diffusing into the room, will take it to the room condition *R*. The supplied air is thus absorbing the sensible heat and moisture (or latent heat) released into the room. The line containing *S* and *R* is called the *room ratio line*. The slope of this line is determined only by the ratio of the rates of release of sensible heat and latent heat into the room. Thus, the state of the supplied air must always lie on the room ratio line passing through the desired condition *R* if this condition is to be maintained. The required mass flow rate of the supplied air will depend on how far the entry point *S* is from *R*. Therefore if air is supplied from outside, whether or not it is partially mixed with recirculated air, it will normally need to be further conditioned in some degree

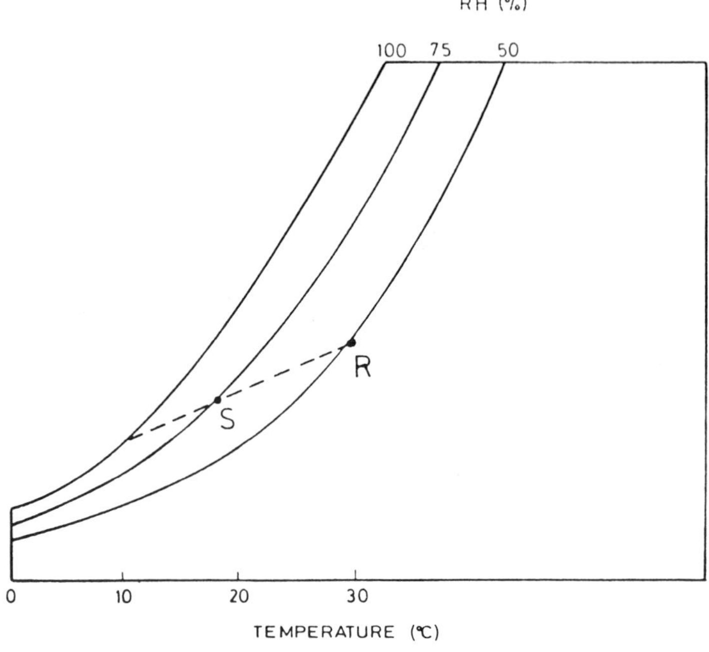

FIG. 3 The room ratio line.

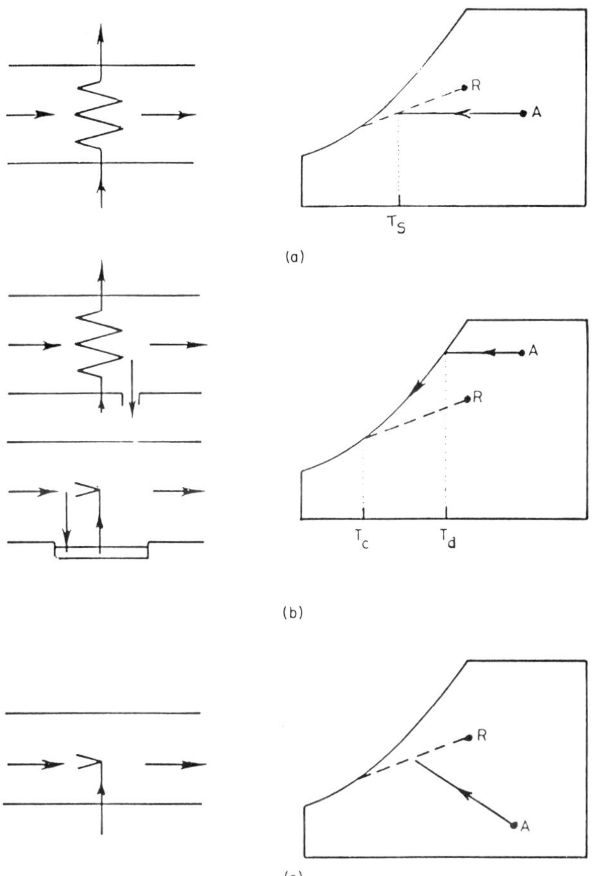

FIG. 4 Air-conditioning processes: (a) sensible cooling, (b) cooling and dehumidification, (c) cooling and humidification.

so as to lie on the room ratio line. Many processes, of which some basic ones are outlined in this chapter, enable this conditioning to occur. Figure 4 outlines such processes.

The state of the available air is represented by point A. This point can be moved to a suitable position on the room ratio line by three means.

Sensible cooling (Fig. 4a) Basically, this process consists of a passage over a cooling coil that carries a chilled liquid. In this method the coil reduces the temperature at constant moisture content. The coil temperature is required to be somewhat lower than T_s.

Cooling and dehumidification (Fig. 4b) If the coil temperature is below T_d, which is the dew point corresponding to state A, the air becomes saturated and liquid water condenses on the coil. Cooling may continue down to T_c, with dehumidification; the air exit temperature T_s is somewhat higher. A similar effect can be obtained with a recirculated spray of chilled water in place of the coil.

Cooling and humidification (Fig. 4c) In this case, we assume that the sprayed water is all evaporated into the air; then cooling is achieved by the reduction of sensible heat at the expense of an increase in the latent heat. In the limit, the process approaches adiabatic saturation. Whatever the initial water temperature, this process takes place approximately along a line of constant WBT. If regenerative cooling is employed, the temperature may be reduced a few degrees below the WBT.

7.4 EVAPORATIVE AIR COOLERS

Evaporative air cooling is achieved by evaporating water at ambient temperature into the airstream, so that the dry-bulb temperature of the air is reduced along a line of constant wet-bulb temperature (WBT), with a consequent increase in latent heat and moisture content. Such systems have their applications mostly in hot and dry areas such as desert environments, where it is possible to obtain some measure of relief by removing sensible heat only.

Figure 5 consists of three types of evaporative air coolers. The *drip- (or desert-) type cooler* is a fan unit with evaporative pads over which water is passed by a small circulating pump. The pads are normally made of aspen wood, glass fiber, metal wire, or expanded paper. The *spray-type*

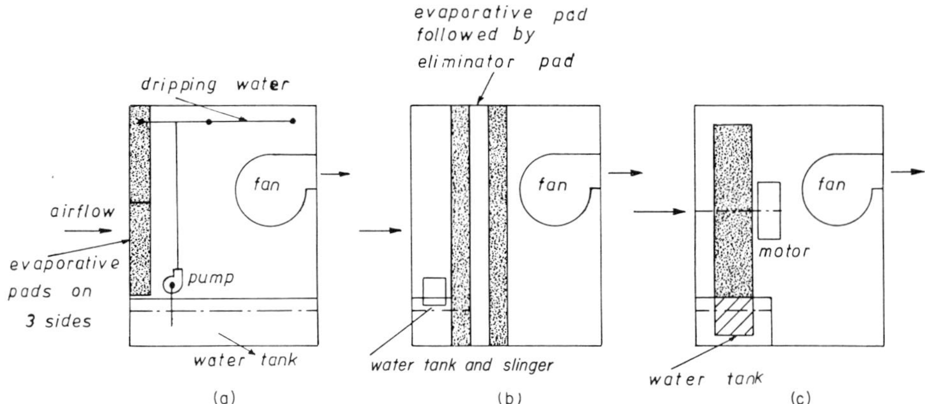

FIG. 5 Evaporative air coolers: (a) drip type, (b) spray type, (c) rotary type.

cooler has evaporative pads, and a fine water spray that is thrown into the air and onto the pads by a water slinger, a centrifugal vapourizer, or spray nozzles. This type can be manufactured with or without an integral fan unit. The *rotary-type cooler* is a device that continuously wets and washes the evaporative pad by rotating it through a water bath and presenting it to the airstream (Faber and Kell, 1958).

7.5 EVAPORATIVE WATER COOLERS

In air-conditioning work, water is frequently used to absorb the heat rejected by the refrigeration condenser. For economy of operation the water is normally used again and again, and it must be continually re-cooled. Figure 6 shows some of the evaporative methods available (Croome and Roberts, 1975).

The *cooling pond* is the simplest and cheapest form of water cooling and relies on natural wind effect to evaporate water from a large surface and hence cool the main body of water. Performance can be improved by the use of a *spray pond* where the water is sprayed several meters above the pond surface, thereby increasing the effective transfer area. *Natural draught (or atmospheric) towers* rely mainly on wind effect to circulate air through frames made mostly of wood and wetable material called fillers.

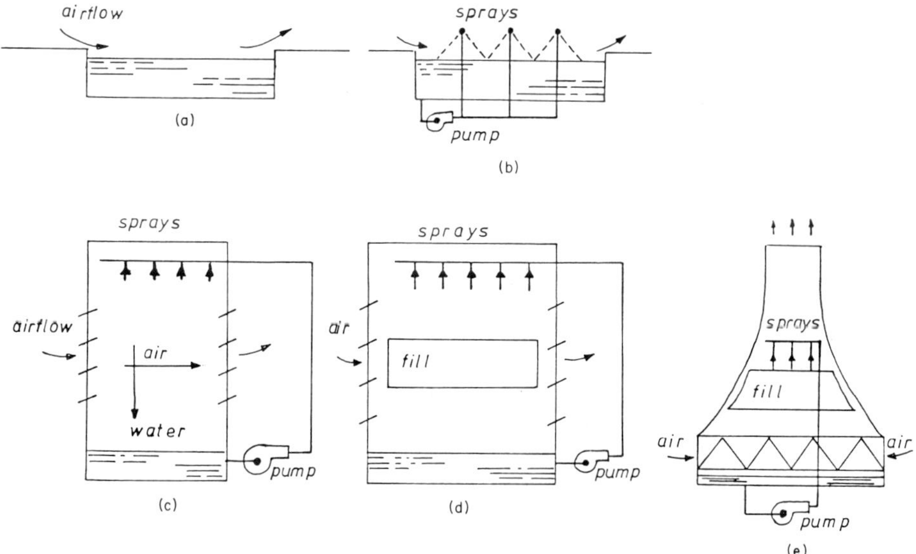

FIG. 6 Evaporative water equipment: (a) cooling pond, (b) spray pond, (c) spray-filled tower, (d) wood-filled tower, (e) hyperbolic tower.

This system has spray nozzles on top with wooden sides and fillers. The object here is to increase the wetted surface area and allow time for air/water contact. The *hyperbolic tower* uses the stack effect of a chimney above the packing to induce airflow up the tower in counterflow to the descending water droplets. However, in air-conditioning work the majority of refrigeration plants that use recirculated water for condenser cooling employ a mechanical draught cooling tower, but this proves to be fairly expensive.

7.6 RADIATIVE COOLING OF SELECTIVE SURFACES

The radiative cooling technique is fully explained in a paper by Catalanotti *et al.* (1975) and is based on the Stefan–Boltzmann principle of energy emitted from a special radiator. This energy per unit area is

$$E = \sigma T^4 \tag{1}$$

If the area of an ideal radiator is 1 m^2 and its temperature is 30°C, then for ideal conditions

$$T = 273 + 30 = 303 \quad {}^\circ\mathrm{K}$$
$$\sigma = 5.7 \times 10^{-8} \quad \mathrm{W/{}^\circ K/m^2}$$

Therefore,

$$E = 480.447 \quad \mathrm{W/m^2} \tag{2}$$

But because of the availability of heat from diffuse radiation and the excess heat that is provided by the surroundings, mostly by convection and conduction (the convection–conduction effect is proportional to $T - T_0$, where T is the temperature of the radiator and T_0 the temperature of the surroundings), and the thermal radiation emitted by the atmosphere and absorbed by the radiator (proportional to $\bar{\epsilon}\sigma T_0^4$, where $\bar{\epsilon}$ is the average atmospheric emissivity), the final temperature of the radiator will be somewhat less than T; i.e.,

$$\bar{\epsilon}\sigma T_0^4 - \sigma T_E^4 = 0 \tag{3}$$

where T_E is the equilibrium temperature. Therefore,

$$T_E = T_0(\bar{\epsilon})^{1/4} \tag{4}$$

$\bar{\epsilon}$ can be determined by dividing the incident flux on the radiator by the black body emissive power at the ambient temperature.

Therefore $\bar{\epsilon}$ is a function of location and atmospheric conditions:

for a cloudy sky $\quad \bar{\epsilon} = 1$

for a clear sky $\bar{\epsilon} = 0.8 - 0.9$ (at sea level)
$\bar{\epsilon} = 0.5 - 0.6$ (at arid zones with high elevation)

Let us consider a black radiator placed horizontally to face the sky. The radiation emitted by the sky and absorbed by the radiator is clearly shown in Fig. 7a as a full line. The spectrum consists of two parts, one due to diffuse radiation ($\lambda < 4$ μm) and one due to long-wave radiation, which is the black body radiation of the atmosphere ($\lambda > 5$ μm). Therefore, the black body radiation of the atmosphere can be represented by

$$S = \epsilon(\lambda)W(\lambda, T_0) \tag{5}$$

This in fact follows the perfect black body spectrum at temperature T_0, except for a few dips (upper dotted line); i.e.,

$$S_1 = W(\lambda, T_0) \tag{6}$$

The dips represent the *transparency windows of the atmosphere* through which the cold space can be seen. The window considered here is the 8- to 13-μm transparency window. Another dotted line shown in Fig. 7a is that of a black radiator at a temperature T ($T < T_0$) and its spectrum:

$$S_2 = W(\lambda, T) \tag{7}$$

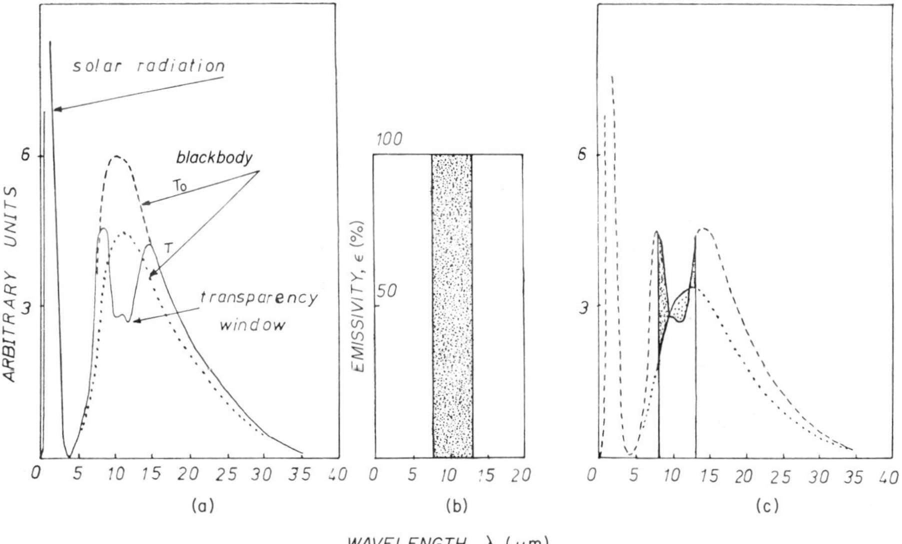

FIG. 7 (a) Radiative cooling of selective surfaces. (b) Emissivity of 8- to 13-μm selective radiator. (c) Equilibrium between absorbed and emitted radiation for selective radiator ($T_E \ll T_0$).

Therefore, the temperature of an ideal radiator, i.e., one that does not interact with its surroundings, can be determined by equating the area of the spectra; i.e.,

$$\int_0^{\infty} [S \text{ (solar radiation } \lambda < 5) + S(\lambda)]\, d\lambda = S_2 \int_0^{\infty} W(\lambda, T)\, d\lambda \quad (8)$$

If a radiator has optical properties matched to the atmospheric window, where it is completely black, then the state of equilibrium is limited to the 8–13 μm of the atmospheric window. Figure 7b shows the properties of such a radiator, while Fig. 7c shows the interaction between the radiator and the atmospheric window.

If T is the radiator temperature, then it will emit only the part of the black body spectrum $W(\lambda, T)$ that corresponds to the temperature T. Therefore, the equilibrium temperature can be determined from the equation

$$\int_8^{13} \epsilon(\lambda) W(\lambda, T_0)\, d\lambda = \int_8^{13} W(\lambda, T)\, d\lambda \quad (9)$$

This result leads to a much lower temperature than T_0 since in the region of 8 to 13 μm, $\epsilon(\lambda)$ has a minimum value. Also, such a radiator will be effective during the daytime as well, since the diffuse radiation has a minimum effect on it.

7.7 NIGHT COOLING CONCEPT

The ground cools as a result of the radiative effect of long-wave energy to space through the atmospheric window at night. The low surface temperature is propagated to the atmosphere by sensible heat exchange. In other words, a downward flux cools the atmospheric layers near the ground so that atmospheric temperature is reduced below wet-bulb temperature with the consequent formation of liquid water drops and fog. This phenomenon is certainly true for a metallic surface, such as a car body or metallic roof, although in the case of soil, fog formation and supersaturation are not so frequent. This is because of the hygroscopic nature of the ground (the ocean is also hygroscopic) and the downward flux of sensible heat, which has a divergence nature and tends to heat the lowest layers and maintain dry-bulb temperature above wet-bulb temperature (Paltridge and Platt, 1976).

The emissive power of a surface, with an area A_1 to the whole hemisphere it faces, is

$$E_1 = 0.1713 A_1 \epsilon_1 (T_1/100)^4 \quad \text{Btu/h} \quad (10)$$

Opposing this is the incoming long-wave atmospheric radiation from the whole sky, provided the reflected rays from the surface do not vary with the wavelength. For simplicity let us assume that the sky has a temperature of T_s; therefore the thermal radiation exchange would be

$$E_1/A_1 = 0.171\epsilon_1[(T_1/100)^4 - (T_s/100)^4] \quad \text{Btu/h/ft}^2 \tag{11}$$

Factoring the terms in the brackets, we have

$$E_1/A_1 = 0.171\epsilon_1[(T_1 - T_s)(T_1 + T_s)(T_1^2 + T_s^2)] \times 10^{-8} \tag{12}$$

Another factor called the geometric view factor (F) can be introduced, so that the simplest case for direct radiation interchange between two planes or convex surfaces is

$$E_1 = \frac{A_1F_{12}\epsilon_1\epsilon_2[0.171(T_1^4 - T_2^4) \times 10^{-8}]}{1 - F_{12}F_{21}(1 - \epsilon_1)(1 - \epsilon_2)} \quad \text{Btu/h} \tag{13}$$

where F_{12} is the fraction of radiation leaving a black surface A_1 in all directions (hemispherical) and intercepted by surface A_2 with view factor F_{21}.

For the simple case of exchange with the sky,

$$\frac{E_1}{A_1} = \frac{0.1713}{(1/\epsilon_1) + (1/\epsilon_s) - 1}\left[\left(\frac{T_1}{100}\right)^4 - \left(\frac{T_s}{100}\right)^4\right] \quad \text{Btu/h/ft}^2 \tag{14}$$

Several empirical formulas have been derived, which relate the atmospheric emission E_1 to the emission power of a black body at T_1 and vapor pressure e. Brunt (1932) proposed a formula

$$E_1/T_1^4 = a + be^{1/2} \tag{15}$$

where a and b are constants that depend on the locality. Brunt (1932) proposed $a = 0.52$ and $b = 0.065$, while Boldrine *et al.* (1974) proposed $a = 0.637$ and $b = 0.084$. Swinbank (1963) proposed a more universal formula that is a function only of the temperature:

$$E_1 = 5.31 \times 10^{-14}T_1^6 \quad \text{mW/cm}^2 \tag{16}$$

Equations (15) and (16) can be true only outside the atmospheric window (8–13 μm). Equation (16) does not apply in desert areas or on the tops of mountains where conditions are fairly dry.

A simplified analysis is to assume the atmospheric emissivity as viewed from the ground (i.e., $E_1/\sigma T_0^4$) to be a constant 0.7 independent of temperature. Outside the window net exchange is always small and hence can be ignored. Net exchange occurs only in the window region and in clear sky; it consists only of the upward component from the ground; i.e.,

$$\overline{E}_1 \simeq (1 - 0.7)\epsilon_0\sigma T_0^4 \tag{17}$$

If a cloud cover exists, it affects the net flux only in the window region. A cloud with a temperature T_c, emissivity ϵ_c, and cloud cover amount CC increases the downcoming long-wave flux from E_1 to

$$E_2 = E_1 + (1 - 0.7)\epsilon_c \sigma T_c^4 \cdot \mathrm{CC} \tag{18}$$

Again for a black body at temperature T, the energy emitted is

$$E = \sigma T^4 \tag{19}$$

If E is in watts per square meter (W/m^2) and T in degrees Kelvin (°K), then $E = 5.7 \times 10^{-8}$ W/°K/m^2.

In reality a lot of thermal energy exists in the atmosphere as a result of diffuse radiation and buildings and surroundings that retain high temperatures during the daytime. Therefore because of the convection–conduction effect that is a function of $(T - T_0)$, where T_0 is the surrounding temperature, a black radiator can absorb only a fraction of the radiation emitted by the atmosphere; i.e.,

$$E = \bar{\epsilon}\sigma T_0^4 \tag{20}$$

Several models have been used to predict the average emissivity of the atmosphere with turbidity and, in particular with water vapor content. A

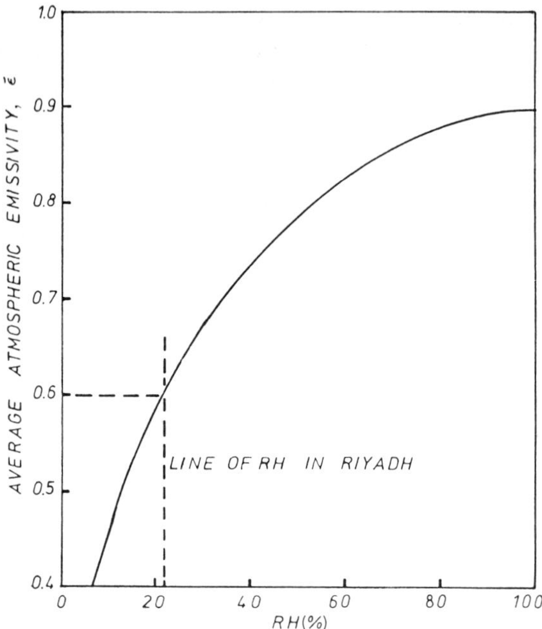

FIG. 8 Variation of $\bar{\epsilon}$ with RH. For low cloud $\bar{\epsilon} = 1.0$.

typical model used by Sayigh (1976c) is based on practical results and is a function of locality and turbidity; for example, in an arid zone $\bar{\epsilon} = 0.5$, while in a coastal area with high humidity $\bar{\epsilon} = 0.9$. For a low cloud area $\bar{\epsilon} = 1.0$ (see Fig. 8).

The final temperature of the radiator T_E is less than T, but at equilibrium T_E is equal to T and equal to $T_0(\bar{\epsilon})^{1/4}$. Therefore the atmospheric downward radiation is

$$q = \sigma T_s^4 = \bar{\epsilon}\sigma T_0^4 \tag{21}$$

where T_s is the equivalent temperature of the sky and the net radiative energy is

$$q = \epsilon\sigma(T^4 - T_s^4) \tag{22}$$

The efficiency of a radiator η can be defined as the ratio of the actual radiation exchange to the ideal radiation exchange between the sky and a per-

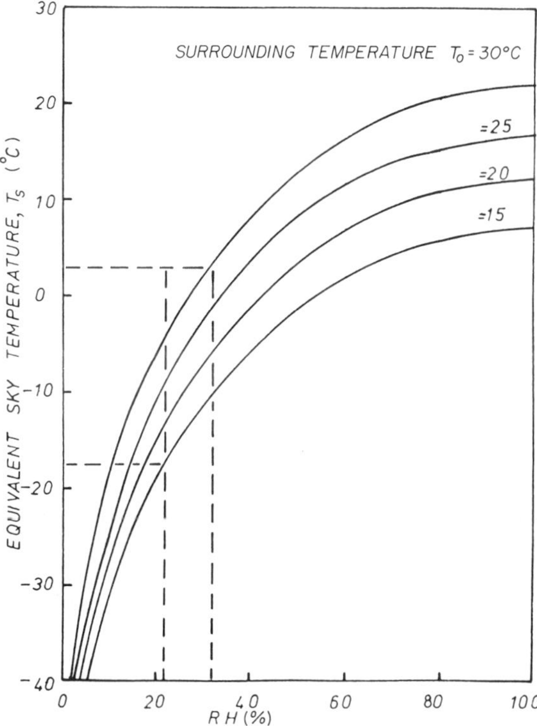

FIG. 9 Equivalent sky temperature in Saudi Arabia. Sky temperature in Saudi Arabia is within the range 3 to −17.5°C.

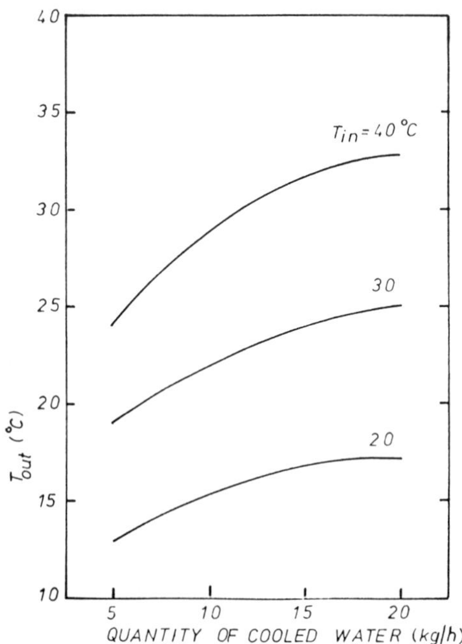

FIG. 10 Outlet temperature for a given flow in a flat plate collector. RH = 30%, surrounding temperature = 30°C.

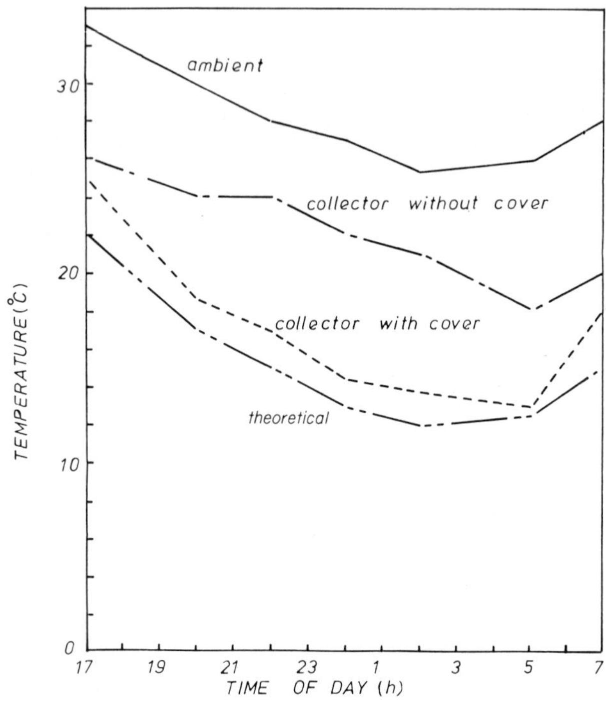

FIG. 11 Use of polyethylene in night cooling.

fectly insulated black body:

$$\eta = \bar{\epsilon}(T_0^4 - T_s^4)/(T^4 - T_s^4) \tag{23}$$

Using the model proposed by Sayigh (1976c), the equivalent sky temperature can be calculated; i.e.,

$$T_s = (\bar{\epsilon})^{1/4} T_0 \tag{24}$$

for various relative humidity ratios (see Fig. 9).

Using a flat plate collector with a 0.15-mm-thick polyethlene cover that is transparent in the infrared region, we can derive Fig. 10 (Sayigh, 1976c).

In this method, the low temperature of the cool night during the summertime provides cooled air that can be stored and used during the hot daytime. The extent of such coolness is clearly shown in various experimental tests carried out in Saudi Arabia. The results of tests conducted at

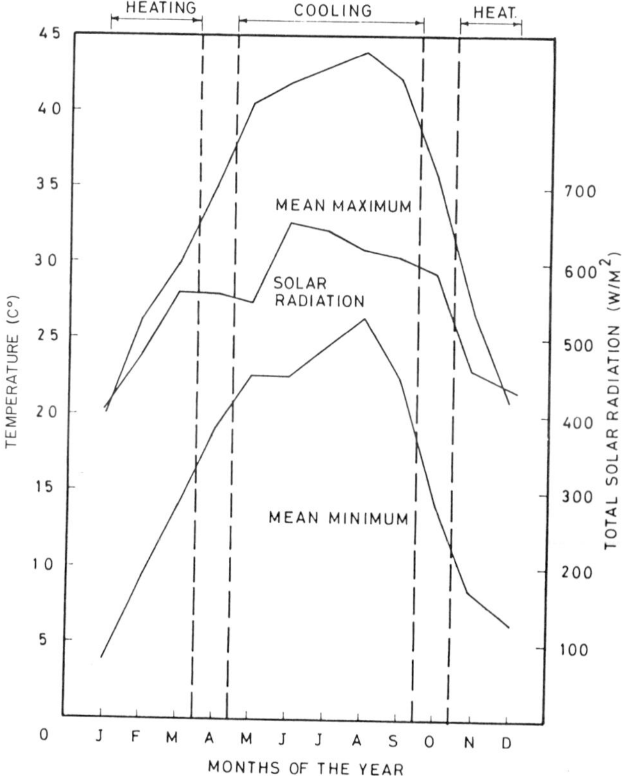

FIG. 12 Requirements for heating or cooling during the year.

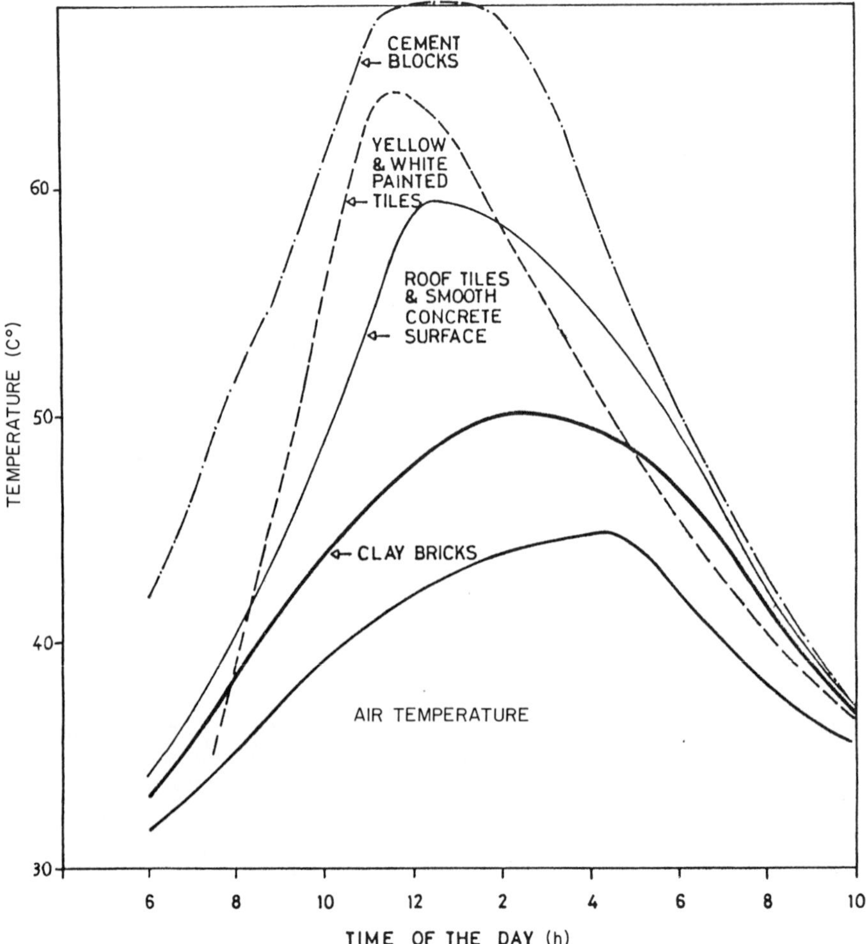

FIG. 13 Surface temperatures for various building materials.

the College of Engineering, University of Riyadh are indicated in Figs. 11 and 12. Figure 12 shows the insolation, the mean maximum temperature, and the mean minimum temperature in Riyadh during the year. This Figure also shows the obvious need for cooling during five months of the year, since in these five months the temperature during the day ranges from 33°C in the morning, say about 8 o'clock, to about 45°C at 2 o'clock in the afternoon. Various building materials were exposed to the sun in the same manner during the day in June 1975, and Fig. 13 shows how their

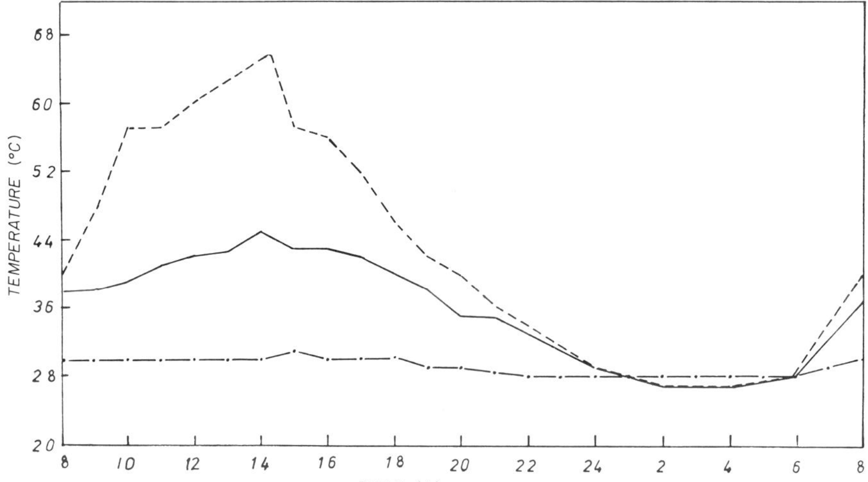

FIG. 14 Swimming pool temperatures: ---, surrounding floor; —ambient temperature; ·–·, water temperature.

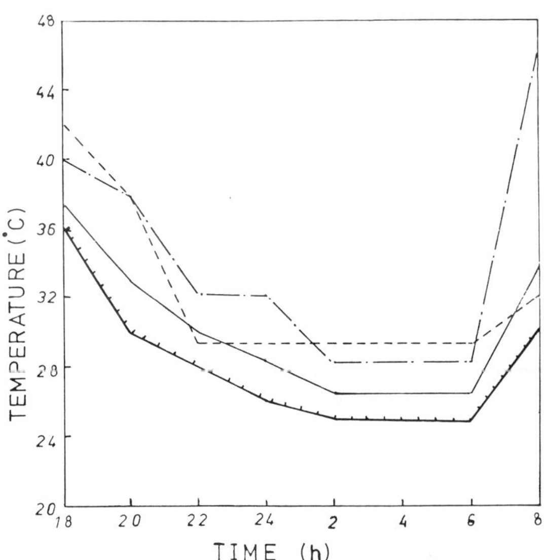

FIG. 15 Roof temperature at nighttime on July 1 and 2. ——, spray; ——, ambient; ----, lower surface flowing west; —·— upper surface flowing west.

temperatures vary with time. The best building material was found to be clay bricks.

During July 1976, several tests were carried out to measure the temperature of a 9- × 18- × 2-m swimming pool, the temperature of the floor surrounding the pool, the temperature of the specially water-cooled corrugated roof, and the ambient temperature. Figures 14 and 15 show the results of these tests.

7.8 CALCULATIONS AND COST ANALYSIS

Let us assume we are dealing with the air conditioning in Saudi Arabia, which has a hot desert climate during the summer and a cold dry climate during the winter.

The shortage of potable water is another factor (Sayigh 1976a) that must be taken into consideration in choosing suitable air-conditioning equipment. Figure 16 shows the electrical consumption during the summer season, while Fig. 17 shows the evaporation rate per annum (in

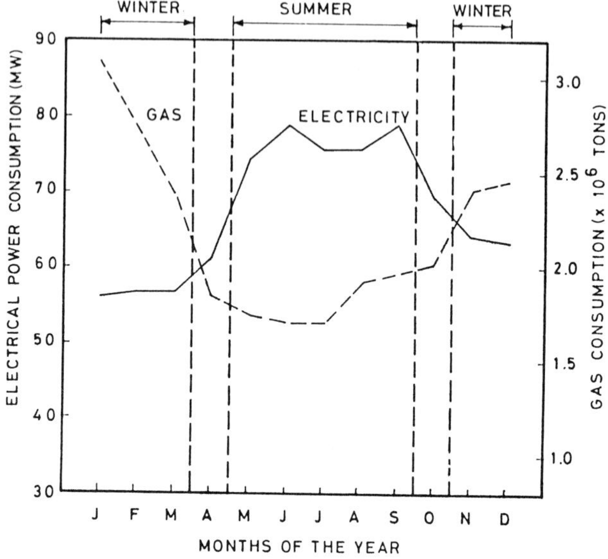

FIG. 16 Average electrical power consumption per day in Riyadh for the year 1974. Average daily electrical power per year = 66.77 MW; average daily gas consumption = 0.15×10^6 tons.

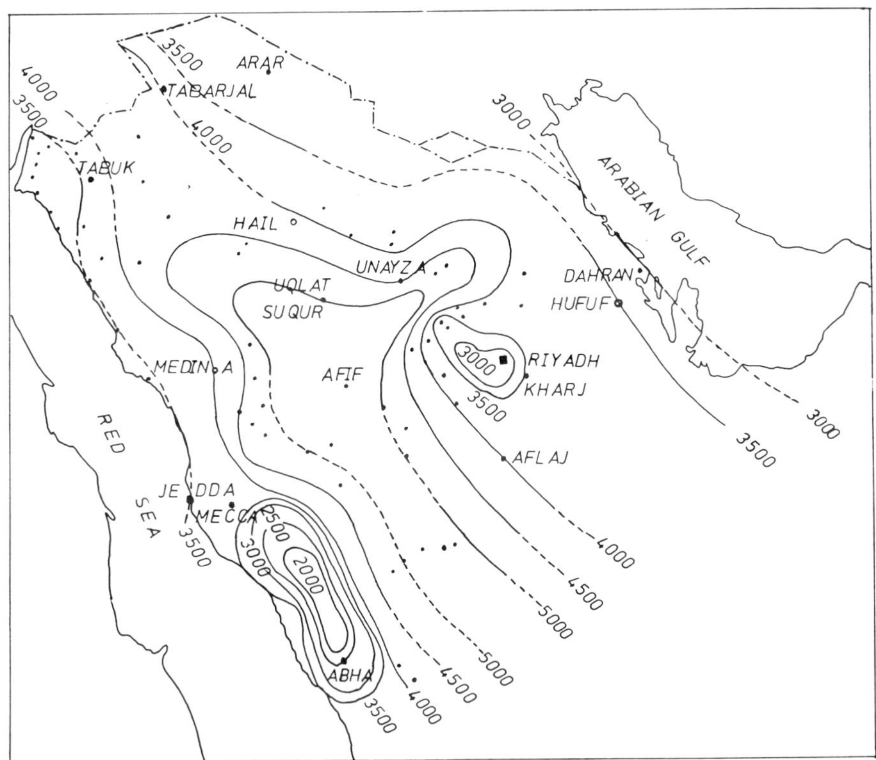

FIG. 17 Evaporation from "A" pan, 1969–1974.

millimeters) from "A" pan in the Arabian Peninsula. Obviously oil is cheap in Saudi Arabia and hence the electricity is fairly inexpensive, about 3¢/kWh (Sayigh 1976b). Electrical equipment, say freon window-type air conditioners, will cost about \$60 per month per family, while a mechanical draught cooling tower or desert cooler will cost about \$30 per month per family. An average apartment or home requires 1.5 m^3 of water per day. Using an open pond or a tray of water with a surface area of 50 m^2, and since the average rate of evaporation during the summer months in Riyadh is 300 mm per month or, say, 10 mm per day, the amount of evaporated water will be 0.5 m^3 per day. During the November meeting of COMPLES, 1975, in Dhahran, the author proposed a house with night cooling that uses 160 m^3 of rock and 30 m^3 of water as storage and has a ground area of 100 m^2; other details are given by Sayigh and El-Salam (1975). By means of such cooling, the rise in temperature with regard to

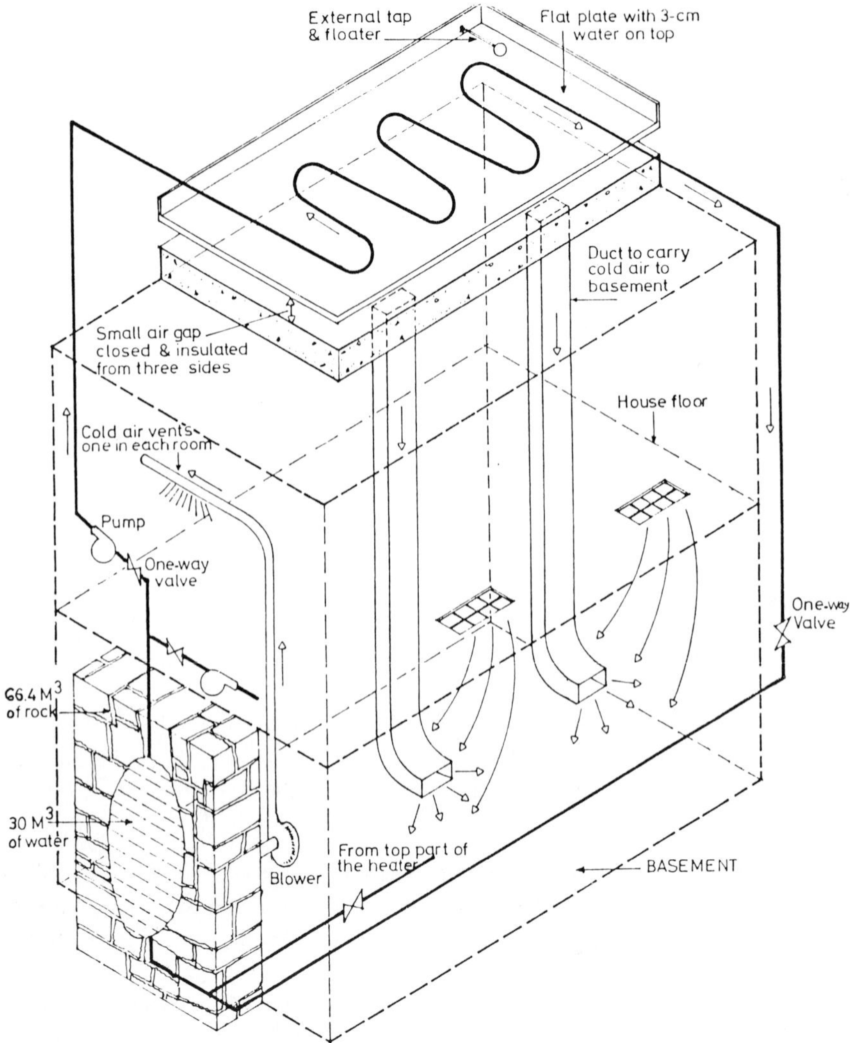

FIG. 18 Night cooling system.

the night temperature is only 1.56°C, which can be tolerated. Figures 18 and 19 show this house.

7.9 CONCLUSIONS

Air conditioning is a process that depends on locality, environment, and the community. The human body has a surface temperature of 33°C,

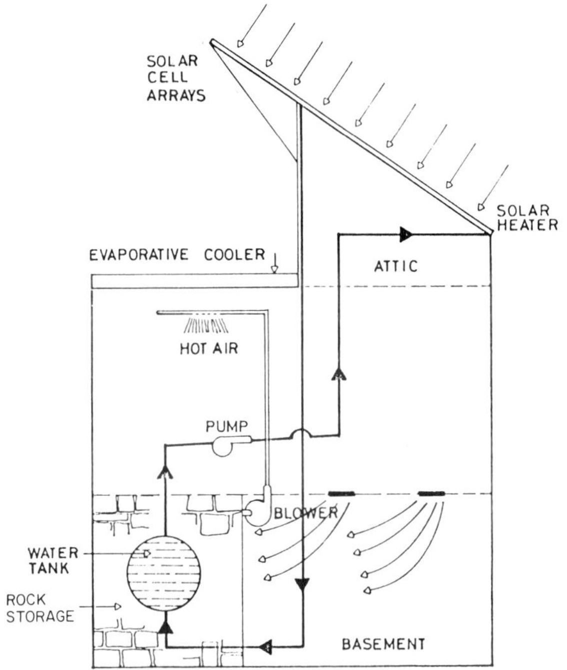

FIG. 19 Solar heating system.

and due to mass and heat transfer with the surroundings, this temperature can be as high as 37°C in a hot and dry climate.

When choosing a suitable system to cool a hot room or a dwelling during the summertime, one should bear in mind the cost of such a system, its reliability, and its continuity. A 3-ton refrigerant-type air conditioner, for example, will cost about $5000 to purchase and install. The electrical power required to run an air conditioner, say for 10 hours a day, is 105 kWh. This is obviously very expensive. An alternative is to use a solar air conditioner. This type (Swartman *et al.*, 1974) will save electricity, as it requires about one-fifth the electrical power used by the preceding example. The initial cost is still high, and unless the operating temperature is not above 93°C, the cooling is not very efficient.

An attractive alternative is to use night cooling. In most hot countries of the world, the temperature of the atmosphere drops considerably during the night; for example, in Saudi Arabia this drop is 20°C. Therefore if some way is devised to harness the coolness and store it for use during the daytime, then cooling is achieved. The important criterion for such cooling is the balance between the temperature and humidity in order to get the *room ratio line*.

This can be achieved by several methods. Figure 20 shows some of these methods as if applied in Riyadh, which has an indoor temperature of 38°C and relative humidity of 27% during the five summer months. If the relative humidity is maintained, then air can be circulated in a swimming pool with a temperature of 30°C, or the same air can be circulated in an evaporative water cooler, resulting in a temperature of 28°C. The use of an artificial roof with night water spray can result in a temperature of 25°C; using a 5-cm-deep water tray on the roof of a building and circulating the air through it will lead to a temperature of about 23°C. If now the humidity is changed, then an evaporative air cooler giving a temperature of 28°C with a RH of 60% can be used.

A very attractive method is the use of selective cooling material. This method is still in the experimental stage and is likely to cost just as much as the initial cost of a freon air conditioner. If a special material (yet to be discovered) is used (Sayigh (1976d), then air conditioning can be achieved with minimum running expense. The increase in humidity can be effected by the use of an aquarium inside the house or with some indoor plants.

From Fig. 20 it is clear that a certain degree of comfort will result if the

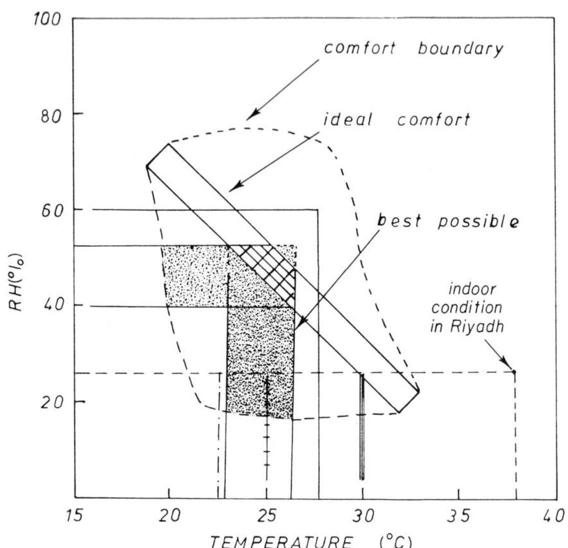

FIG. 20 Degree of comfort in high cooling. ≡, swimming pool; ——, spray cooling; —·—, tray of water; ——, desert cooler. Temperature = 38°C, RH = 27%.

room ratio line falls within the comfort boundary. For ideal comfort, this ratio must fall within the triangle; for the best possible cooling results, the ratio line must fall somewhere within the shaded area. Such optimum air conditioning can certainly be achieved, for example, in Riyadh, with minimum cost by using a swimming pool, a shallow roof pond, or a night roof spray. Coupling these effects with some indoor increase in humidity, excellent summer cooling is achieved.

8

Solar Cooling for Buildings

ROBERT K. SWARTMAN

FACULTY OF ENGINEERING SCIENCE
THE UNIVERSITY OF WESTERN ONTARIO
LONDON, ONTARIO, CANADA

8.1 INTRODUCTION

Of all the possible applications for solar energy, the solar cooling of buildings seems the most attractive. This is a use in which the demand for energy closely matches the supply of solar energy, even to the variation within the day. Solar energy is an appropriate source for the heating and cooling of buildings because the temperatures required are usually moderate and high-temperature sources are not required. A combination of solar heating and cooling, which results in higher use factors on the solar energy equipment, is, in most places, more economical than heating or cooling alone (Duffie and Beckman, 1976).

In the developed countries the refrigeration of food is normal, and the air conditioning of buildings fairly common. In the southern climates of the United States, most new houses are air conditioned. Even in the cooler climates, office and institutional buildings are air conditioned. Cooling is a significant element in the energy economy of the developed countries, with air conditioning causing peaks in electrical demand on summer afternoons and evenings.

If solar energy can be utilized in the cooling of buildings in the developed countries, the application will soon spread to the developing countries. The same basic ideas that are used in solar cooling of buildings can be extended to solar refrigeration for food preservation. The benefits would be substantial if solar cooling can be successfully developed.

ISBN 0-12-620860-3

8.2 CONDITIONING SPACE FOR COMFORT

Normally the space in a building is conditioned to provide a comfortable environment for the occupants. The conditioning involves heating, cooling, humidifying, dehumidifying, or ventilation. The atmosphere is a mixture of gases, mostly nitrogen and oxygen, and water vapor. The water vapor is really steam, at very low pressure, with a saturation pressure equivalent to the dew point. Water vapor is always present in atmospheric air; it is one of the most important factors in human comfort and has a significant effect on many materials. Although it is only about 1% of the weight of atmospheric air, its impact on human activities is disproportionate to its relative weight.

The art of measuring the moisture content of air is called *psychrometry*. The science that investigates the thermal properties of moist air, considers the measurement and control of the moisture content of air, and studies the effects of moisture on material and human comfort may be called *psychrometrics*.

The following definitions are commonly used in psychrometrics:

Dry-bulb temperature, DB Temperature of air indicated by an ordinary thermometer.

Wet-bulb temperature, WB Equilibrium temperature of a thermometer when a water film is evaporating off the bulb.

Adiabatic saturation temperature Temperature at which water will saturate air by evaporating adiabatically into it.

Dew point temperature, DP Temperature to which air must be cooled to cause condensation of any of its water vapor. It is the saturation temperature corresponding to the actual partial pressure of the water vapor in the air.

Specific humidity or humidity ratio, W The ratio of the weight of water vapor to the weight of dry air. The specific humidity can be evaluated from $W = 0.622[p_v/(p - p_v)]$ where p is the total pressure and p_v the partial pressure of water vapor.

Absolute humidity The weight of water vapor per unit volume of air.

Relative humidity, RH The ratio of the partial pressure of the water vapor in the air to the saturated partial pressure at the DB temperature.

Sensible heat Heat that changes the temperature of a substance when added to or rejected from it.

Latent heat Heat that does not affect the temperature but changes the state of a substance when added to or rejected from it.

Degree of saturation, ϕ The actual W over the W if saturated. So $\phi = W_{act}/W_{sat} = \mathrm{RH}(p - p_s)/(p - p_v)$ where p_s is the saturated pressure for DB.

The properties of moist air are readily depicted in a psychrometric chart as in Fig. 1. The psychrometric chart shows the interrelationships among DB, WB, and RH. *W* is along the RH ordinate, DP across the upper left side, and RH diagonally across. A value for heat content is assigned to each WB line. The psychrometric chart is usually constructed for the standard sea level barometric pressure of 29.92 in. Hg.

Comfort is subjective, but there is a range of temperatures and moisture content for air in which most people feel comfortable. The physiological behavior of the human body demands an equality between the rate of internal chemical heat production and the rate of external physical heat loss. This loss occurs by radiation, convection, and evaporation. The body maintains a remarkable system of temperature control. The proportion of loss by each method depends on the total heat production due to activity, amount of clothing, temperature of surrounding walls, and properties of the ambient air.

Figure 2 shows an envelope for humidity ratio and temperature that describes the comfort zone. The human body maintains a surface temperature of around 33°C, giving up heat and moisture to the surroundings. The interior body temperature is 37°C; so a narrow range of environ-

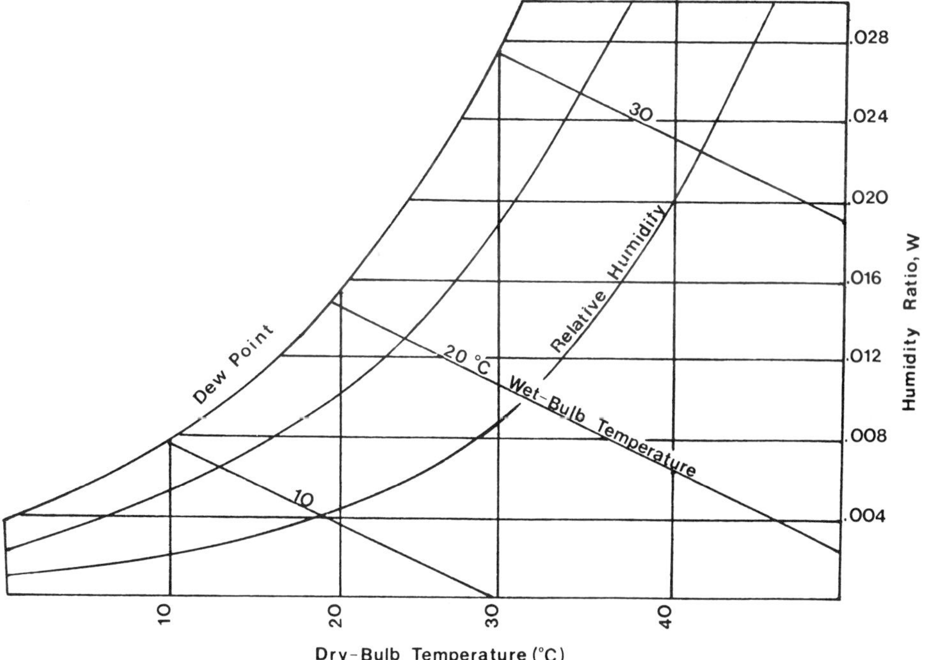

FIG. 1 Psychrometric chart.

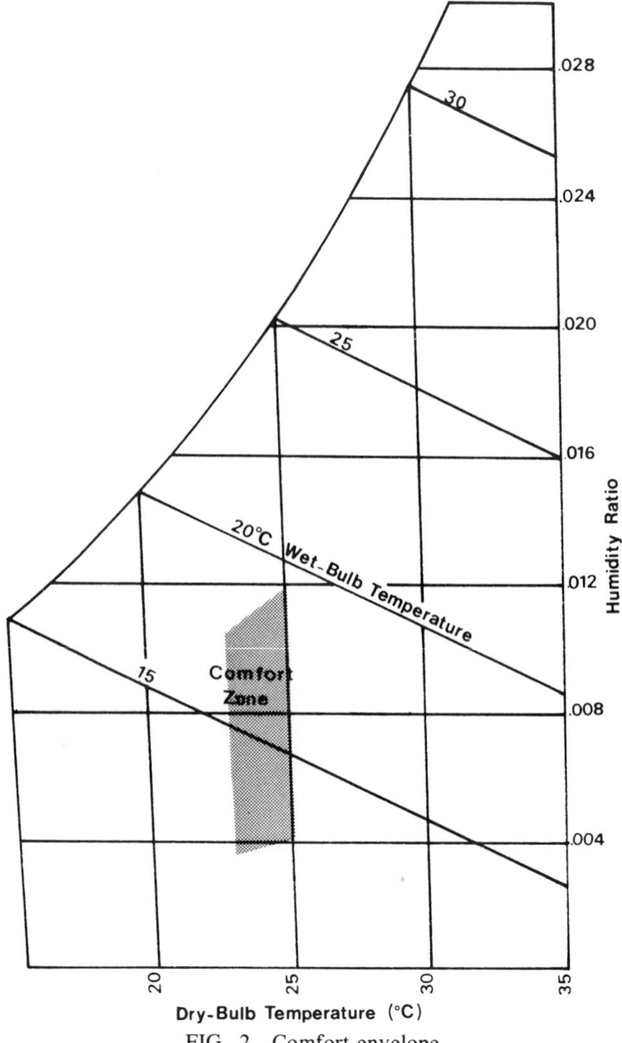

FIG. 2 Comfort envelope.

mental temperatures, from about 18°C to 27°C, is considered comfortable. That range of temperature allows the heat and mass transfer that is perceived as comfortable. The tolerance of humidity is much wider, between 25 and 60% RH, with lower humidity preferred at higher temperatures, presumably to assist heat loss by evaporation.

Conditions often occur outside the envelope describing comfort in terms of relative humidity and temperature, as in Fig. 2. Increased propensity to accidents and decreased industrial productivity occur above about 20°C. The drop in industrial output is in the order of a few percent per degree Celsius, reaching about 50% at 43°C and 50% relative humidity (Brinkworth, 1977). Jones (1973) has suggested that indoor temperatures should be between 4 and 11°C lower than the outdoor temperature in hot weather, with a relative humidity of about 50%. Even if this difference cannot be achieved, it is certain that any reduction in temperature gives some improvement in comfort on entering the cooler space.

The heat gain of a building is primarily due to solar energy but not entirely. The sensible heat gain is by solar radiation through the windows, conduction through the walls from solar radiation incident on the outside, electric lights and motors, people, and cooking. Other sources such as internal combustion engines may also add to the sensible heat gain. Sensible heat can be removed by sensible cooling or refrigeration, and by evaporative cooling. Latent heat and moisture content are gained from people, cooking, and the combustion of a hydrocarbon fuel. Latent heat can be removed by sensible cooling below the dew point and by adsorption with a desiccant.

The conditions in a room are maintained by adjusting the temperature and humidity of the supply air. The supply must be cool enough and dry enough so that it absorbs the sensible heat and moisture and the latent heat released in the room. The typical cooling needs of a building in Miami, Florida, are shown in Fig. 3. It shows clearly the buildup of latent heat early in the day and the gradual leveling off of sensible heat.

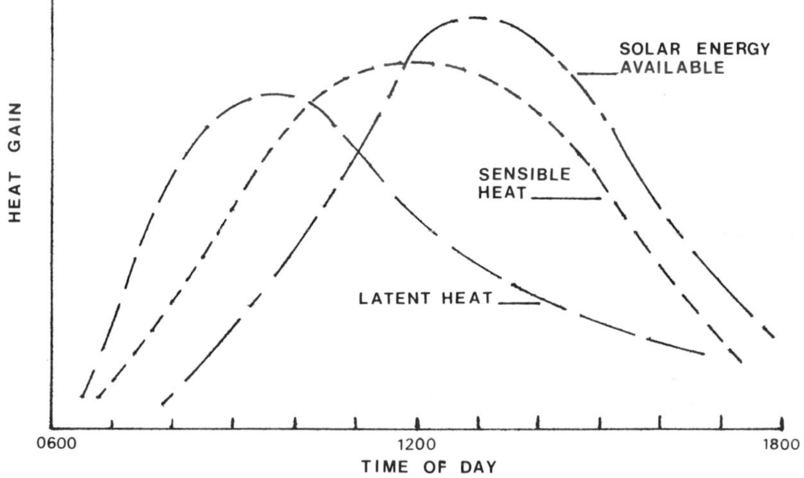

FIG. 3 Cooling needs in Miami, Florida.

A. Air Conditioning

The processes for conditioning the supply air can be shown on a psychrometric chart, as illustrated in Fig. 4. They are

(1) *Sensible cooling* Passage of air over a cooling coil, which carries a chilled liquid, reduces the temperature at constant moisture content.

(2) *Cooling and dehumidification* Air passing over a cooling coil is cooled to the dew point corresponding to A, and liquid water condenses. If cooling continues, the air is dehumidified as well. A spray of chilled water has a similar effect as a cooling coil. If air is passed over a surface or through a spray of water that is at a temperature less than the dew point temperature of the air, condensation of some of the water vapor in the air will occur simultaneously with the sensible cooling process. The air that does not contact the surface will be finally cooled by mixing with the portion that did.

(3) *Cooling and humidification* If sprayed water is evaporated into the air, cooling is obtained by the reduction of sensible heat at the expense of an increase in the latent heat. The process is along a line of constant wet-bulb temperature.

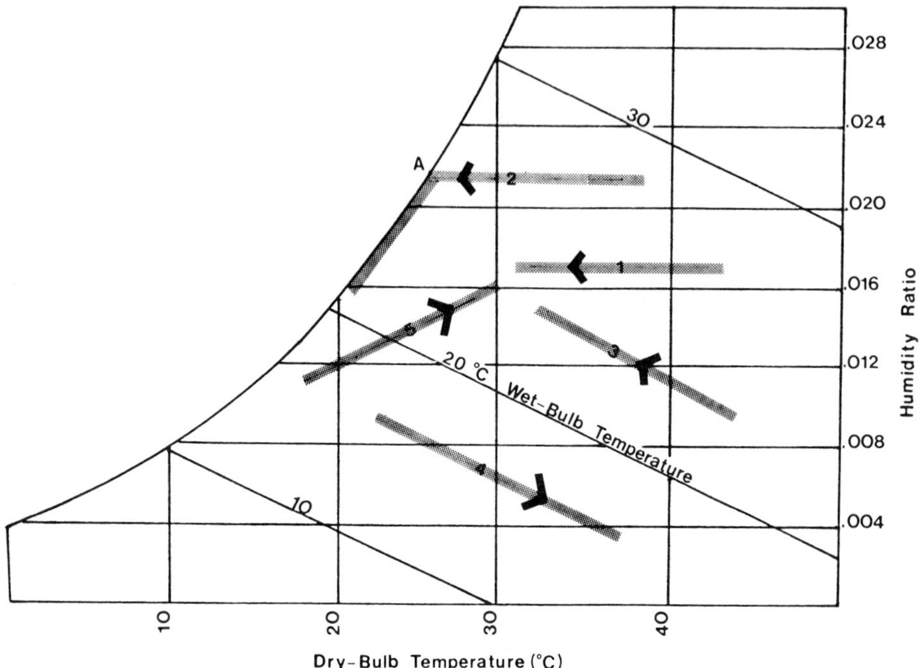

FIG. 4 Psychrometric processes.

When unsaturated air is passed through a spray of continuously recirculated water, the specific humidity increases while the dry-bulb temperature decreases. This is adiabatic saturation or evaporative cooling.

If the wet-bulb temperature of the supply air is sufficiently below that of the desired room condition, evaporative cooling may be sufficient. At other times chilled water may be required in the temperature range 5–10°C, which could be obtained from solar cooling.

(4) *Heating and dehumidification* Simultaneous heating and dehumidification can be accomplished by passing air over a solid adsorbent surface or through a liquid absorbent spray. Heat of condensation is liberated as is the heat of adsorption or absorption. Common adsorbents are silica gel and activated alumina. A common absorbent is ethylene glycol.

(5) *Heating and humidification* When air is passed through a humidifier that has heated instead of simply recirculated spray water, the air is humidified and may be heated, cooled, or unchanged in temperature. The air increases in specific humidity and enthalpy, and DB changes according to the initial temperature of the air and of the spray. Heating and humidification can also be accomplished by evaporation from an open pan of heated water or by direct injection of hot water or steam.

B. Calculating the Building Cooling Load

The cooling load is the rate at which heat must be removed from the building to maintain a constant inside temperature. The cooling load differs from the building heat gain because of heat storage in the structure. Heat is gained from

(1) solar radiation falling on the outside walls and entering the building through windows or other transparent surfaces;
(2) heat conduction through the skin of the building because of the temperature difference between inside and outside;
(3) internal sources including lighting, occupants, and electric motors;
(4) ventilating air that uses outside air either partially or completely;
(5) latent heat from processes such as cooking.

The following information is needed to calculate the building cooling load:

(1) The characteristics of the building structure such as size, shape, and materials in the skin.
(2) The building location and orientation so that shading of the structure can be determined.
(3) The outdoor weather conditions including dry-bulb and wet-bulb

temperatures. Information on solar heat gain is given in the ASHRAE "Handbook on Fundamentals" (ASHRAE, 1972) for various latitudes, at various times, and on various dates. The latitudes vary from 24 to 56°, the times vary hourly from 6 A.M. to 6 P.M., and the tables provide solar heat gain factors for the twenty-first of each month. The solar heat gain factors are in Btu h/ft^2 for various orientations 45° apart starting at north.

(4) The indoor design conditions such as dry-bulb and wet-bulb temperatures and ventilation rate.

(5) The internal heat gain of the building from lighting, occupants, electrical equipment and appliances, and processes such as cooking.

The instantaneous heat gain must be calculated for a specific time of the day and day of the month, and should include the following heat gain sources:

(1) solar heat gain through glass surfaces, roof, and exterior walls;
(2) heat flow through exterior and interior walls, ceilings, and floors;
(3) internal heat gain;
(4) infiltration and ventilation;
(5) latent heat gain.

It is assumed that the energy gained by radiation is averaged over several hours by the thermal storage of the building. For light construction the average is taken over 2 to 3 h and for heavy construction is taken over 6 to 8 h. The total instantaneous cooling load is the sum of the convection portion of the heat gain and the time-integrated average of the radiant portion. It is necessary to separate the heat gain into convective and radiant components. This can be done using Table 1.

TABLE 1

Convective and Radiant Heat Gain to Cooling Load[a]

Heat gain source	Radiant heat (%)	Convective heat (%)
Solar, without inside blinds	100	—
Solar, with inside blinds	58	42
Fluorescent lights	50	50
Incandescent lights	80	20
People	40	20
Transmission through walls and roof	60	40
Infiltration and ventilation	—	100
Machinery or appliances	20–80	80–20

[a] Anonymous (1970, Chapter 22, Table 34).

8.3 SOLAR COOLING SYSTEMS

The three classes of solar cooling systems are solar sorption cooling, solar–mechanical systems, and solar-related systems that are not solar operated, but use some components of the solar heating system for cooling. There are variations within each class, such as the use of continuous or intermittent cycles, hot or cold side energy storage, different control strategies, and various ranges of operating temperatures and thus different collectors. Now a closer look at each of the three classes.

A. Solar Sorption Cooling

Sorption cooling includes absorption and adsorption cycles. In an absorption system thermal energy is applied directly to produce cooling. Absorption involves mechanical retention as in a sponge. An adsorption system is suitable when a large portion of the cooling load is used in latent heat removal. Moisture is adsorbed on a drying agent; then the agent can be regenerated using solar energy.

Absorption systems In an absorption system a thermal energy input produces a cooling effect. In general, an evaporating refrigerant is absorbed by an absorbent on the low-pressure side, the absorbed refrigerant is generated by direct thermal energy input on the high-pressure side, the generated refrigerant is liquefied in the condenser, and the liquid refrigerant evaporates in the evaporator. In a continuous cycle, the generator and condenser are contained in one reservoir with the evaporator and absorber in another reservoir, as in Fig. 5. Refrigeration is accomplished as the liquid refrigerant evaporates, and heat is rejected as the refrigerant liquefies in the condenser. The performance of the system depends on the temperatures in the generator, absorber, condenser, and evaporator, and the capacity depends on the generator and cooling water temperatures.

The behavior of absorption refrigerators depends on the thermodynamic characteristics of the refrigerant/absorbent mixture. There is a threshold value of generator temperature that must be exceeded if the machine is to function. If the evaporator temperature is specified for a particular application, this determines the pressure in the evaporator and absorber. When the temperature at which heat is rejected is specified, this temperature and the absorber pressure determine the concentration of the refrigerant in the absorber. Specifying the heat rejection temperature also determines the pressure in the condenser and generator. For a given pressure there is a unique relationship between temperature and refrigerant concentration in the generator. The system cannot function, however, unless the concentration of refrigerant is lower in the generator than in the

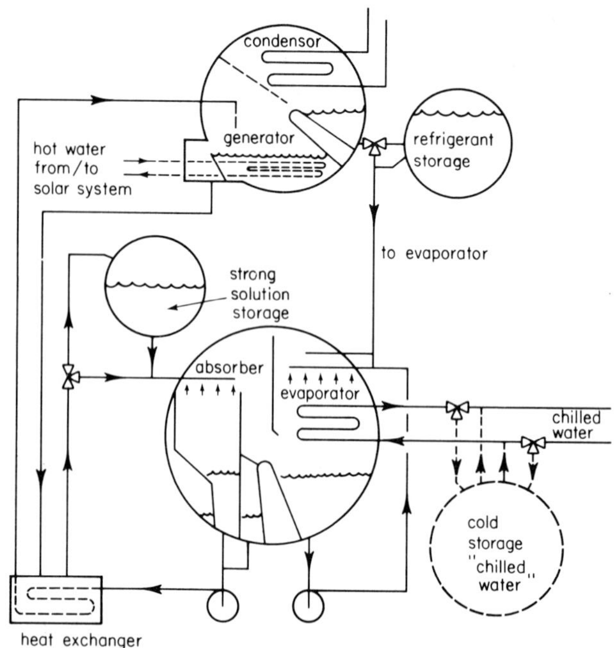

FIG. 5 Absorption cooling system; refrigerant storage within and outside the absorption cycle.

absorber. This determines the lower limit for the generator temperature above which operation is possible.

The cooling effect is approximately equal to the enthalpy of evaporation at the evaporator temperature, and the heat supplied is approximately equal to the enthalpy of evaporation at the generator temperature. The ratio of the cooling effect to the heat supplied is the coefficient of performance (COP). For an ammonia/water system just above the threshold temperature, the COP is about 0.7, falling to about 0.6 when the generator temperature is at 130°C.

The water/lithium bromide system operates at around 10°C and therefore is suitable for air conditioning. Most machines using $LiBr/H_2O$ have a water-cooled absorber and condenser, which in turn requires a cooling tower. The pressure differences between the high- and low-pressure sides are low enough that these systems can use a vapor lift pump and gravity return from absorber to generator, instead of a mechanical pump. The COP of $LiBr/H_2O$ systems is usually in the range 0.6–0.8. If water is used to cool the absorber and condenser, the generator temperature is in the range 75–95°C. Variations in the generator temperature with changes in solar flux vary the capacity of the cooler. The operating temperatures required of the solar collector because of the high generator temperatures make flat plate collectors marginal in this application. For this reason focusing solar collectors are being considered for many solar cooling projects.

The ammonia/water system is similar to the water/lithium bromide system except that a rectifying section must be added to the top of the generator to remove water vapor from the ammonia vapor going to the condenser. The pressures and pressure differences of the NH_3/H_2O system are much higher; so mechanical pumps are required to return solutions from the absorber to the generator. The condenser and absorber are often air cooled with generator temperatures in the range 125–170°C. When water cooling is used, generator temperatures are in the range 95–120°C. The generator temperatures are usually too high for flat plate solar collectors. Again, focusing solar collectors are a possibility but are not yet well developed. Most work has been directed at development of cycles using higher concentrations of NH_3 to lower the generator temperatures.

Adsorption cooling Another class of sorption air conditioners utilizes an adsorbent for adsorbing moisture from the air and then evaporative cooling of the air. The drying agents can be regenerated using solar energy. Rotary dehumidifiers can be used as in the Munters Environmental Control (MEC) system (Rush *et al.*, 1975). Silica gel and active charcoal are two solid adsorbents in common use. However, the solid adsorbents are usually poor heat conductors; so the amount of refrigerant adsorbed deteriorates with time. Liquid desiccants can also be used, such as triethylene glycol that requires generation at about 90°C, which is feasible with flat plate collectors.

Dunkle (1965) describes a rotary dehumidification system containing solid adsorbents, such as silica gel or activated alumina, with generator temperatures between 80 and 105°C. Dannies (1959) describes a passive system for adsorbent cooling that uses solar-driven convection of air.

Adsorption cooling is in commercial use now, but not with solar operation. Dehumidifier/regenerator combinations are manufactured in various sizes handling from 1000 to 40,000 ft^3/min of air. The liquid desiccant system uses a hygroscopic liquid so that any air passing through a spray of the desiccant leaves at a low relative humidity. The spray temperature must be kept high for regeneration so that the dew point of the air leaving is higher than the air entering. There is often a heat exchanger to cool the air as it exits from the unit, thereby increasing the relative humidity and reclaiming most of the adsorbent (Anonymous, 1968). An interesting proposal was made a few years ago to incorporate a solar adsorption cooling system on the Citicorp Building, then under construction in New York. It was not included, as the necessary design could not be completed in time.

When the outside design conditions and the desired inside conditions are specified, the total sensible heat load and the total moisture load will determine the condition and quantity of the entering air. The type and nature of the coolant available will influence the condition of the air. It is best to assume a dew point temperature for the entering air that can be obtained by the most economical coolant: city water, well water, or cooling tower water. When the dew point is very low, it may be necessary to cool the cooling water by refrigeration.

After moisture has been removed from the air, it may be necessary to cool it further, possibly using evaporative cooling. Evaporative cooling is essentially an adiabatic process, with negligible heat flow, as air is passed through a wetted media or water spray. Some of the water evaporates, increasing the latent heat content of the air, but lowering the sensible heat in the air. The air leaving an evaporative cooler has a lower dry-bulb temperature and a higher relative humidity than when it entered. The process can be approximated by a constant wet-bulb-temperature line on a psychrometric chart.

There is no system in existence yet that utilizes solar energy to regenerate an adsorbent, with additional sensible cooling of the air accomplished by evaporative cooling. The closest thing at this time is the Solar MEC system being developed by the Institute of Gas Technology in Chicago, Illinois (Rush *et al.*, 1975). A schematic of a possible solar-powered adsorption system is illustrated in Fig. 6.

B. Solar–Mechanical Systems

A solar cooling system that has received some attention in recent years couples a conventional air-conditioning system, involving a vapor compression refrigerator, with a solar-powered prime mover. This may be done by converting solar energy into electricity by means of photovoltaic devices, and then using the electricity in an electric motor to drive the vapor compressor. The conversion efficiency of solar energy into electricity by photovoltaic means is in the order of 10%. This low efficiency, coupled with the efficiency of the mechanical system, results in low overall efficiencies.

The solar-powered prime mover can also be a Rankine engine. The combined system would have an overall efficiency of 17 to 23%, which makes this system more attractive than the photovoltaic/electric

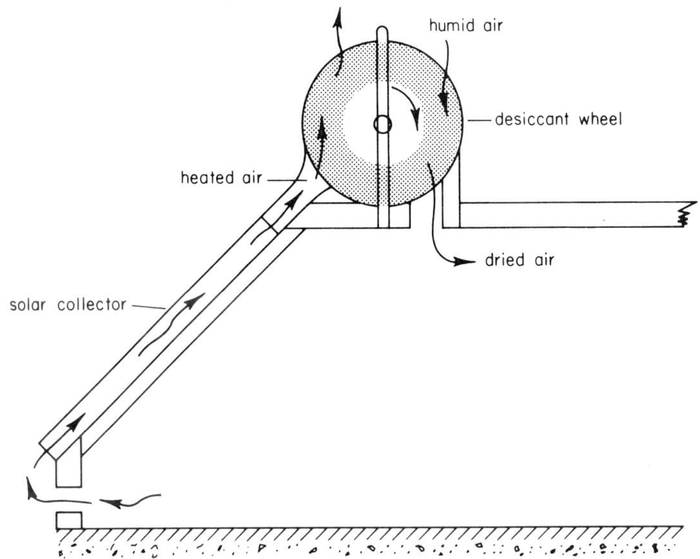

FIG. 6 Solar adsorption cooling system.

motor/air conditioner system (Sargent and Teagan, 1973). The problems associated with both systems are basically the problems associated with converting mechanical energy from solar energy to drive a compressor. Even though the design of conventional air-conditioning systems is well established, the secondary problem of adapting air-conditioning equipment to the solar energy source is difficult, particularly for part load operation.

So far, most studies on Rankine cycle solar air conditioning have been theoretical. However, an experimental system was installed in the Honeywell mobile laboratory more than two years ago (Prigmore and Barber, 1975) although no long-term operating data are yet available. A schematic of a simple Rankine cycle cooling system is shown in Fig. 7. Energy from the collector is stored, then transferred to a heat exchanger which, in turn, transfers energy to the heat engine. The heat engine drives a vapor compressor, finally producing a cooling effect at the evaporator. Figure 8 shows that the efficiency of the solar collector decreases as the operating temperature increases, whereas the efficiency of the heat engine for the same system increases as the operating temperature increases. It also shows how the efficiencies of collector and engine combine giving an optimum operating temperature for steady state operation. The thermodynamic analysis of such a cycle is straightforward, but not for unsteady operation. The prediction of component performance at off-design condi-

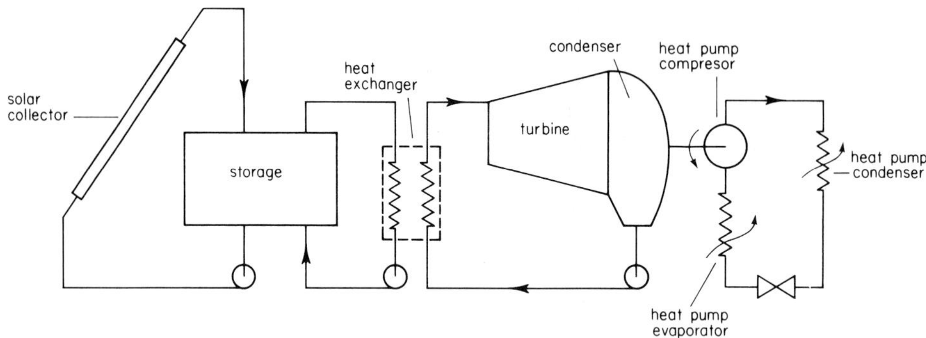

FIG. 7 Rankine cycle cooling system.

tions is very difficult, and the matching of all components into a complete system that optimizes the overall performance is currently being researched. The reader can appreciate the difficulties in designing a system in which the storage tank temperature changes through the day, and hence in which the temperature to the boiler will change. Less energy will be added to the working fluid as a result. To ensure that only vapor enters the turbine, either auxiliary energy must be added at the boiler or the circulation rate of the fluid must be reduced. Meanwhile, the cooling load varies. When a Rankine heat engine is coupled with a constant speed air conditioner, the output of the engine seldom matches the input required by the air conditioner. When the engine output is greater than needed, the matching can be accomplished by having the engine operate at an off-design condition and wasting the available energy. The excess energy

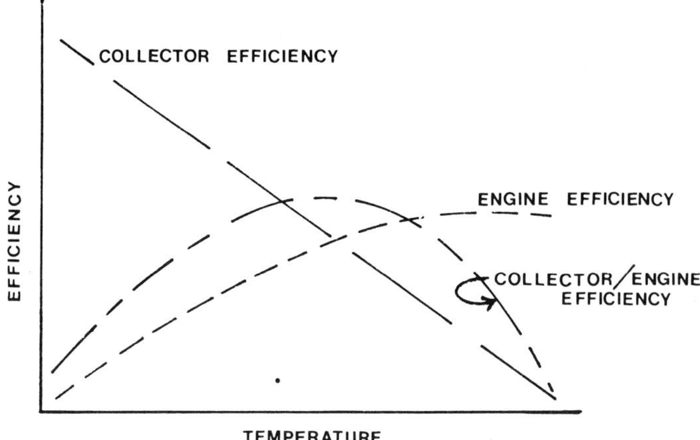

FIG. 8 Efficiencies of collector, engine, and collector/engine.

could be used to produce electricity for other purposes. When the engine output is less than that required by the air conditioner, auxiliary energy must be supplied. The system can be designed to operate at variable speed. Then the air conditioner will be operating off-design with a lesser output.

C. Solar-Related Cooling Systems

When solar heating equipment is installed in a building, some of it can be used to cool the building but without the direct use of solar energy. Some examples of this combined application include *heat pump* systems, *sky radiation* systems, *rock bed regenerator* systems, and *passive* systems.

Heat pump Lord Kelvin proposed in the 1850s that refrigeration equipment could be used for heating by utilizing the heat rejected. But the heat pump remained a curiosity for many decades.

The heat pump is a device that pumps heat from a low temperature to a higher temperature, appearing to violate the normal flow direction of heat. Usually vapor compression refrigeration machines are used as heat pumps; so the evaporator can take heat into the system at low temperature and the consenser can reject heat from the system at high temperature. In the heating mode a heat pump delivers thermal energy from the condenser for space heating and can be combined with solar heating. In the cooling mode the evaporator extracts heat from the air to be conditioned and rejects heat from the condenser to the atmosphere. In this mode, however, solar energy does not contribute to the energy for cooling.

The *coefficient of performance* (COP) indicates the efficiency of the cycle. It is defined as the useful effect divided by the work done. The COP typically ranges in value from 2 to 5. For refrigeration $COP_{ref} = Q_A/W$, and for heating $COP_{htg} = Q_R/W$;

$$COP_{htg} = Q_R/W = (Q_A + W)/W = COP_{ref} + 1.$$

Both the condenser and the evaporator can be either air cooled or liquid cooled (Gilman, 1976). The heat pumps installed in residences today are usually air–air; that is, they use air as the heat source and give up heat to the inside air. Heat pumps are used for cooling as well as heating. In the heating mode electricity is the supplementary energy form. Some heat pumps are liquid–liquid when a convenient heat source such as underground water is available.

A heat pump can be combined with solar heating and a large energy storage in a hybrid system known as the annual cycle energy system

(ACES). ACES operates as an air source heat pump down to an atmospheric temperature of 4°C. Below that temperature the system becomes a water source heat pump, which uses the latent heat of fusion given up by water in its change of phase to ice. The ice that is produced while supplying the winter heating is stored and used during the summer months to satisfy the cooling requirements of the building. Feasibility studies show that the system will provide from 35 to 70% of the annual cooling requirements of a building from the stored ice.

Heat pumps are becoming more popular. Since 1952 there have been over 1.8 million unitary heat pumps installed. They increase the effectiveness of electricity by a factor of 2 or 3 over electric heating. Also, the combination of heat pumps with solar energy in hybrid systems looks very promising.

Sky radiation Another method of cooling that uses solar equipment already installed relies on dissipating energy to the night sky by radiation. In extensive experiments in Arizona, Bliss (1964) found that on a monthly average his uncovered collectors of 93 m^2 could dissipate approximately 30 ton-h of energy. This would provide over 1 ton of cooling for each hour of the day. Bartoli *et al.* (1976) report experimental results that are better than the theoretical predictions of natural radiative cooling. The method uses the characteristic windows in the atmosphere, which are transparent to certain bands of radiation. Radiation losses to the sky in these bands can be increased by employing selective surfaces on the cooling surfaces. If this method is utilized with flat plate collectors, they must have properties opposite to those needed for efficient collection. Hence, a compromise is necessary. An alternative is to use movable insulation as used in the Skytherm concept developed by Hay (1975). This system combines collector, radiator, and storage capabilities in the basin of water on the horizontal roof of a building. The movable insulation allows the system to provide heating in the winter and cooling in the summer. A house incorporating the concept in a clear, mild California climate keeps conditions inside the house within acceptable limits over the year. Figure 9 shows the two modes of the Skytherm house.

Rock bed regenerator Another method that uses solar energy equipment but is not a solar system is the rock bed regenerator discussed by Close (1965). For an air heating system that has a rock bed for thermal energy storage, the rock bed can be used to store coolness as well as heat. During the summer, when the rock bed is not used for storing solar heat, cool night air can be passed through an evaporative cooler and then used to cool down the rocks. During the day ventilating air can be drawn into the rock pile and cooled. As the air leaves the rocks, its temperature can be further reduced in an evaporative cooler.

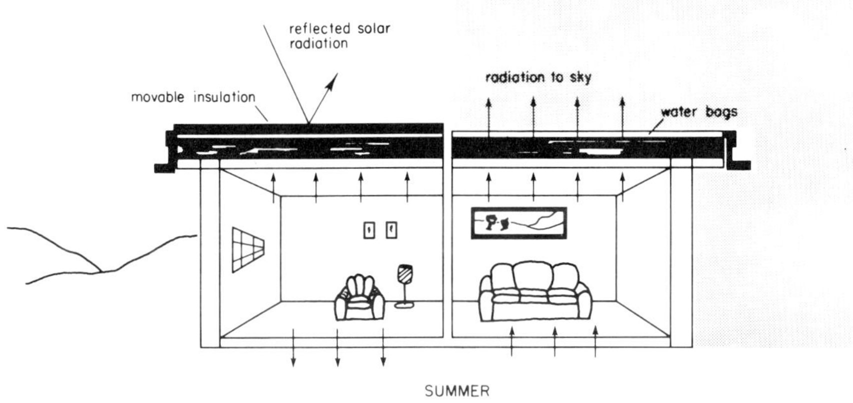

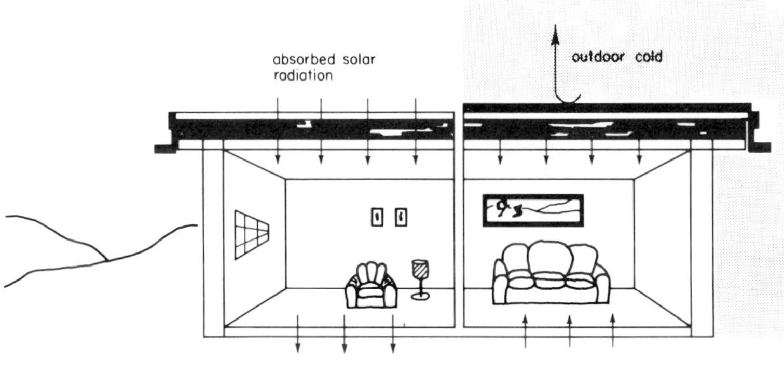

FIG. 9 Skytherm house.

Passive systems Natural air conditioning can be accomplished by a passive solar heating/cooling system. The solar collector and heat storage are combined in a large thermal mass that does not include any machinery for pumping or blowing the working fluid. An application of this method is incorporated, as an example, in the CNRS solar houses in France. A south-facing, vertical, double-glazed window with a blackened concrete wall behind comprises the collector–heat storage, as in Fig. 10. In summer, cool air from the shaded north side of the building is drawn into the living space by thermosyphon action through the space between the collector window and the concrete wall and exits at the top of the collector.

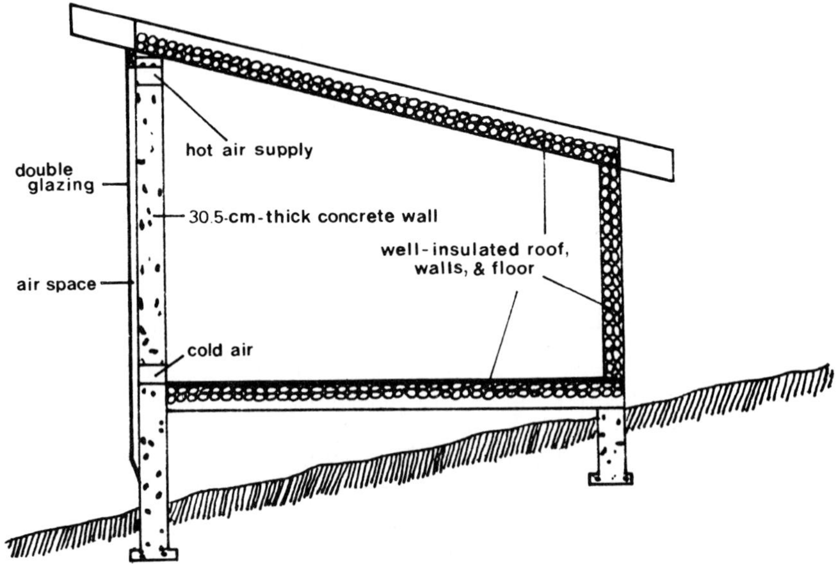

FIG. 10 CNRS house.

8.4 STORAGE AND PERFORMANCE

A. Storage

Systems for producing cooling, as well as for space heating and service hot water, utilize the flat plate collector. This unique heat exchanger uses a black absorber plate to absorb solar energy. Transparent covers over the absorbing plate reduce thermal losses from the front face, and insulation at the back and sides reduce losses in those directions. Current research on flat plate collectors is aimed at reducing the losses and increasing the energy absorbed. The investigation of selective surfaces, for example, is attempting to enhance the absorptivity while reducing the emissivity of absorber surfaces. It is possible to achieve higher temperatures with concentrators, but generally the extra energy required to track the sun is excessive for the benefit received. Stationary concentrators, oriented east–west, would require less energy than concentrators oriented north–south. The north–south concentrators would be adjusted continuously to match the diurnal variation of the sun, whereas the east–west concentrators would be adjusted only every few days as the season changed. Current research is investigating the application of solar concentrators, as well as evacuated tube collectors, to solar cooling.

The temperature limitations of the collector provide a practical limit on what can be expected of solar cooling. As collector temperatures are pushed upward, storage may then become a critical problem. Energy storage is the other major component in a solar energy system. The storage accumulates solar energy when it is available and makes it available to meet energy demands at other times. Liquid systems usually use insulated water tanks for storage, and air systems usually use insulated rock beds. A third method of storage uses the latent heat of a phase change. Many phase change materials suitable for storing coolness or "coolth" have been reported (Anonymous, 1975a).

B. Solar System Performance

One method of calculating the performance of a solar heating system is to use the "F-chart" correlation developed at the University of Wisconsin. Two quantities D_1 and D_2 are correlated by F-lines that represent the fraction of the heating load that will be carried by a solar heating system of given characteristics:

$$D_1 = \frac{\text{energy absorbed by collector}}{\text{total heating load}}, \qquad D_2 = \frac{\text{collector energy loss}}{\text{total heating load}}$$

$$\begin{aligned}\text{total heating load} = {} & \text{energy absorbed by collector} \\ & - \text{collector energy loss} + \text{auxiliary energy}.\end{aligned}$$

Dividing by "total heating load," we obtain

$$\begin{aligned}1 = {} & \frac{\text{energy absorbed by collector}}{\text{total heating load}} - \frac{\text{collector energy loss}}{\text{total heating load}} \\ & + \frac{\text{auxiliary energy}}{\text{total heating load}}\end{aligned}$$

or

$$1 = D_1 - D_2 + \frac{\text{auxiliary energy}}{\text{total heating load}}.$$

The "F-chart" correlates the first and second terms with lines that characterize the fraction of the heating load that can be supplied by solar heating.

Worksheets are available (Anonymous, 1977d) to provide a straightforward method of calculating the performance of a solar heating system using the F-Chart correlation.

D_1 and D_2 can be expressed mathematically as

$$D_1 = \frac{A_c F'_R(\tau\bar{\alpha}) S K_4}{L}$$

$$D_2 = \frac{A_c F'_R U_L (t_{ref} - t_0) \Delta_{time} K_1 K_2 K_3}{L}$$

where A_c is the collector area, S the total incident solar radiation normal to the collector for an average month, L the total heating load including the hot water load, $F'_R(\tau\bar{\alpha})$ and $F'_R U_L$ the collector performance characteristics, t_{ref} the 212°F arbitrary reference temperature, t_0 the monthly average daily temperature, Δ_{time} the number of hours in month, K_1 the air collector flow capacitance rate factor, K_2 the storage mass capacitance, K_3 the hot water factor, and K_4 the heat exchange factor. $F'_R(\tau\bar{\alpha})$ and $F'_R U_L$ can be found from the graph of collector thermal efficiency against $\Delta T/I$ shown as Fig. 8. The Y intercept is $F_R(\tau\bar{\alpha})$ and the slope is $F_R \cdot U_L$. $F'_R(\tau\bar{\alpha})$ and $F'_R U_L$ are calculated from $F_R(\tau\bar{\alpha})$ and $F_R U_L$.

From a conventional point of view there is a unique index of performance for rating cooling processes, the coefficient of performance or COP, which is the amount of energy required to produce a given amount of cooling. For solar operation there is an additional index, the temperature level required to drive the process.

For air conditioning,

$$\text{total cooling load} = (\text{energy required to produce cooling}) \times \text{COP},$$

or, if solar energy is used to provide energy for cooling,

$$\text{total cooling load} = (\text{energy absorbed by collector} - \text{collector energy loss} + \text{auxiliary energy}) \times \text{COP}.$$

This is similar in format to the analysis for solar heating shown previously. Dividing by "total cooling load" gives

$$1 = \left(\frac{\text{energy absorbed by collector}}{\text{total cooling load}} - \frac{\text{collector energy loss}}{\text{total cooling load}} + \frac{\text{auxiliary energy}}{\text{total cooling load}}\right) \times \text{COP}.$$

The correlation has not yet been developed, but a design procedure similar to solar heating appears likely for solar cooling.

8.5 ECONOMIC CONSIDERATIONS

The usual basis for economic evaluation of solar cooling and heating systems is to compare the yearly costs of the solar system with those of a conventional system. The yearly costs include amortized capital costs, operating costs, and maintenance costs. It is usual in designing a solar system to assume the collector area and, hence, calculate the capital costs, which, amortized over a period of time, would be a certain amount per year. The system performance is calculated on the basis of the assumed collector area. The calculated performance gives the yearly output in energy, which can be assigned a value based on the price of conventional energy sources. If the total yearly costs are greater than the value of the energy saved, then the system is not economically justified and a different collector area must be assumed.

Löf and Tybout (1974) developed a series of cost analyses for solar heating and cooling. They showed that the combination of solar heating and cooling, which results in higher use factors on the solar energy equipment, is generally more economical than heating or cooling alone.

An interesting analysis was reported by Bilgen (1976) on a hypothetical single-story building of 5500-m² area and 17,000-m³ volume. The specific heating requirement of the building was 80 kJ/m³ degree day, an acceptable value for a well-insulated building. The calculations were made for Montreal, which has 4388 heating degree days annually (based on 18°C), and approximately 800 h of cooling operation. The calculations were also made for 2000- and 4000-h operation. The results of these calculations expressed as savings in dollars per cubic meter space per year are shown in Fig. 11 along with annual capital costs. It is apparent that solar

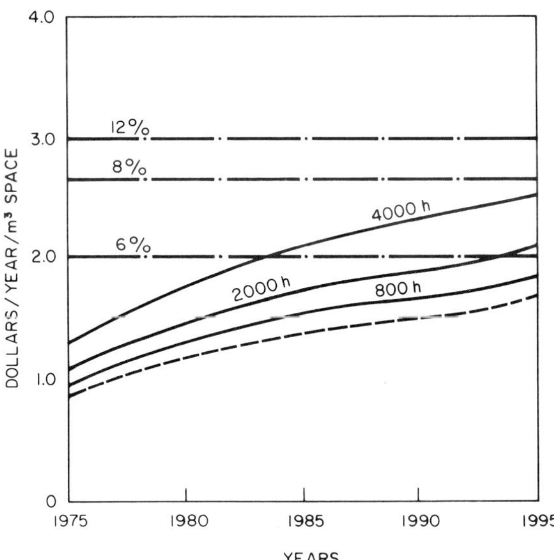

FIG. 11 Savings for solar heating/cooling: ---, with solar heating; —, with solar heating and cooling; –·–, additional annual capital cost for solar air conditioning.

cooling would not be economical for 800-h operation. For a longer period of cooling (4000 h) or a low capital cost (such as 6% interest), solar air conditioning would be economical.

Solar energy processes are usually capital intensive; large investments are made in equipment to save operating costs. The essential economic problem is balancing the annual cost of the extra investment against annual energy savings. As fuel costs rise and the supplies of low-cost fuels are depleted, solar energy will become more competitive and the fraction of annual load to be carried by solar energy will increase. The many heating/cooling demonstrations now being funded by the U.S. government will undoubtedly lead to more economical hardware and provide more realistic assessment of the economics of solar cooling.

8.6 CURRENT RESEARCH AND DEVELOPMENT

The relationship of an air-conditioning system with the building it heats or cools, the occupants, and the outdoor ambient temperature to which the building is exposed has many variables (Newton, 1976). Combining solar energy with such systems adds the variables of diurnal changes in solar radiation which influence the quantity and availability of energy. There are four conditions varying to give a dynamic interaction:

(1) the cooling load to maintain a building at a specified temperature and humidity,
(2) the temperature at which the cooling equipment receives heat from the building,
(3) the temperature at which spent heat is rejected,
(4) the temperature at which heat is supplied to energize the system.

The maxima and minima of the different conditions are usually out of phase and not under the control of the designer. The designer, however, must use the out-of-phase relationships effectively to improve operating efficiency and sizing of solar components.

The technology of solar cooling is not so advanced as that for solar heating. The Energy Research and Development Administration in the United States, now part of the Department of Energy, initiated considerable work in solar cooling with a large infusion of money since 1975. Figure 12 shows the paths that have been identified by ERDA for solar cooling of buildings. The American program is overwhelming! However, there may be approaches being overlooked in the U.S. program that could be attractive in other applications in other parts of the world.

Two recent ERDA Request for Proposals based on Fig. 12, have objectives significant for the development of solar cooling. One, the Solar

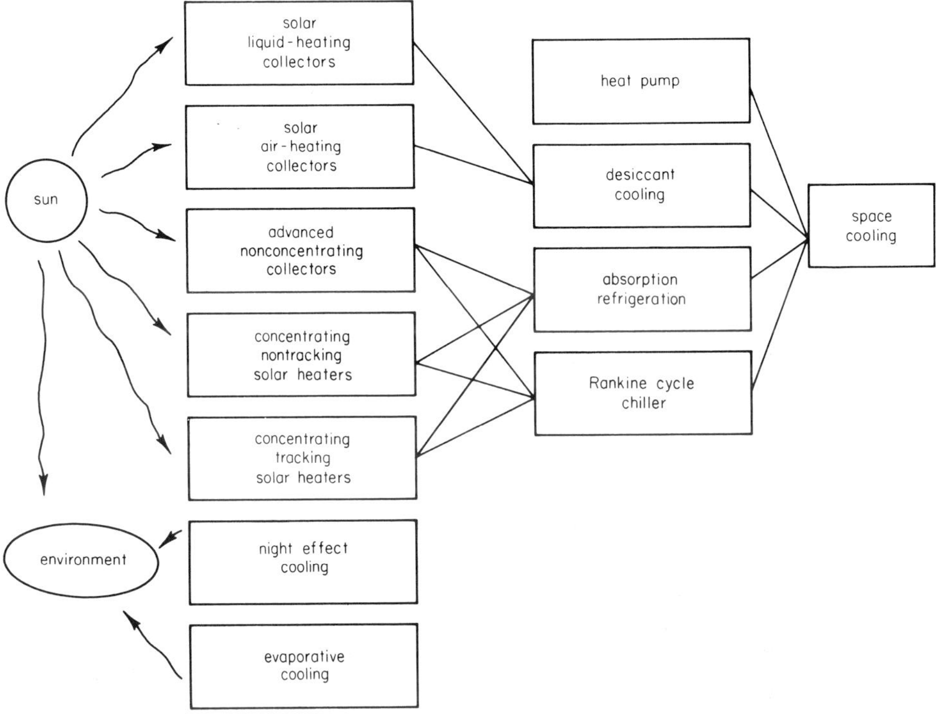

FIG. 12 ERDA paths to solar space cooling.

Activated Cooling Projects (Anonymous, 1977a) has the following objectives:

(1) To develop and analyze new refrigerant pairs, or additives to presently used absorbent/refrigerant pairs, or other cycle modifications that can potentially improve the performance or reduce the cost of an absorption chiller in the size range of 3 tons.

(2) To procure the design, fabrication, test, and analysis of a single-family-size residential solar-powered absorption chiller having improved performance and reduced cost.

(3) To procure the design, fabrication, test, and analysis of solar-powered heat engine/vapor compression cycle chiller sized for multifamily residential and commercial applications, normally of 15- to 25-ton capacity.

(4) To procure the design, fabrication, test, and associated analysis of a solar desiccant dehumidification cooling subsystem sized for multifamily residential or commercial applications.

The other, requesting proposals in Controls and Passive Systems (Anonymous, 1977b), has these objectives.

(1) To conduct studies that will provide information required for the determination of the economic characteristics of passive systems and/or meaningful quantitative evaluation of their thermal performance.

(2) To develop control strategies and control subsystems that will lead to improved performance and cost effectiveness of solar heating and cooling systems.

8.7 CONCLUSIONS

The practical implementation of solar cooling will depend on social acceptability, economic competitiveness, political decisions, and technical suitability for local conditions. Local factors include the availability of electric power and cooling water. It is difficult for solar cooling to compete with vapor compression refrigeration where electric power is plentiful. Political decisions that will affect the competitive position of solar energy include deregulation of natural gas prices, costs of petroleum, and tax incentives. Solar cooling is still in the experimental stage, but improvements in this application of solar energy should result in substantial markets. New buildings and retrofitting existing buildings will make significant contributions in saving money and energy.

9

Natural Cooling in Hot Arid Regions

MEHDI N. BAHADORI

DEPARTMENT OF MECHANICAL ENGINEERING
SCHOOL OF ENGINEERING
PAHLAVI UNIVERSITY
SHIRAZ, IRAN

9.1 INTRODUCTION

Before the discovery of oil and its abundance, people made use of solar and other types of energy to meet some of the energy needs of their dwellings. It was only after oil and natural gas became inexpensive and readily available that building designs and technology changed. The designer, for example, allowed more heat into the building in summer and then tried to remove it by active cooling systems, using electricity or other forms of energy. With the help of active heating and cooling systems, it was possible to design a dwelling irrespective of the environment and maintain any desired environment within the structure throughout the year. While this provided a great deal of freedom for the designer, it was, nevertheless, an energy-wasting system.

With man's increasing awareness of the true value of liquid and gaseous fuels and his wastefulness in using these valuable commodities, he is now becoming more interested in eliminating waste and exercising greater conservation of energy.* Conservation-minded man has to learn the old techniques employed by people who provided comfort in their dwellings by making use of such natural sources as solar energy, sky radiation, wind, and daily temperature range.

* Thermodynamically speaking, energy is conserved and not wasted; but the expressions employed here are in accordance with their common usage, with "conservation" implying, for example, less utilization of the prime source of energy and less loss of energy to the surroundings.

ISBN 0-12-620860-3

9.2 CHARACTERISTICS OF HOT ARID REGIONS

Arid regions are characterized by very low annual rainfall, low relative humidity, a high degree of solar radiation and hot days, a high rate of radiation loss to the sky during the night and cool nights, and rather fixed patterns of diurnal and seasonal winds. People who have lived in these regions have learned to cope with and make use of these severe environmental contrasts and to create comfortable dwellings for themselves. They have also adopted a way of life highly influenced by the climate. For example, in the arid regions of Iran (shown in Fig. 1), the people have accommodated their dwellings to the climate in the following ways:

(1) They reduced the exposed surfaces of the buildings and the heat transfer with the outside air by constructing buildings attached to each other with common walls, in a cluster form. This type of town development served security and defense purposes as well. Figure 2 shows the fortress in Bam that once housed the entire city population and was being occupied up to 150 years ago. Figure 3 shows the city of Yazd with its attached buildings.

FIG. 1 Map of Iran showing the desert regions and their surrounding towns, in which passive cooling systems are employed. The cities of Tehran, Esfahan, and Shiraz are shown for reference only.

FIG. 2 The fortress of Bam that was being used up to 150 years ago.

FIG. 3 The city of Yazd, showing the attached buildings.

(2) They made use of the high daily temperature range in summer and employed thick adobe walls and a small number of doors and windows for their rooms. By means of the latter they also reduced the energy exchange with the outside air, as well as the infiltration and collection of dust in the building.

(3) They made use of the radiation losses to the sky by producing ice in winter and storing it for summer, and sleeping in the open air, mostly on the roof, during the summer nights.

(4) They built large, deep cisterns, filled them with cold water in winter, and kept them cool through the summer by providing a continuous and natural evaporation of water from their surfaces.

(5) They reduced solar heat gain on the walls by building deep courtyards surrounded by rooms, planting trees and shrubs in them, and entrapping the cool night air for several hours of the next morning (Tavassoli, 1975). They further reduced the effects of dusty winds by employing tall parapets, which also provided the privacy needed for summer night use of the roofs. The tall walls and the narrow streets provided good shade for pedestrains and reduced solar heat gain. Figure 4 shows such a narrow street in Yazd.

(6) They made use of the prevailing summer winds, developed wind towers, and created excellent natural circulation and cooling of the outside air through the building.

FIG. 4 A narrow street in Yazd. Note the shaded walk.

(7) They made use of the low temperature of the ground in summer and lived in the basements, especially in the summer afternoons.

(8) They employed dome-type roofs for many rooms, especially the kitchen, in order to increase the heat transfer coefficient and area, to create a suction on top of the dome, and to vent the hot air underneath the dome. In locations where dusty winds predominate, one finds these natural air vents, instead of wind towers, employed for air circulation.

(9) They used solar energy in the rooms designed for winter occupancy, and stored the energy in the heavy walls and roofs of the building.

(10) They did not heat all the rooms in winter, but only those that were utilized and even then only at the time of occupancy.

Of course, technological progress has necessitated many changes in the life-style of the people. It is a challenge for architects and engineers to utilize natural phenomena as "primitive" people have in the past, incorporate the changes in life-style into them, and create comfortable dwellings.

9.3 WIND TOWERS

Wind towers or "Baud-Geers," are masonry structures designed to provide natural circulation and cooling of the ambient air through the building. Figure 5 shows the cross section of a typical wind tower. The openings (1) on top of the tower may face in all directions or only in the direction from which the wind is predominant. In locations where dusty and dust-free winds blow from opposite directions, the openings face in the direction to intercept only the dust-free and favorable winds.

A. Operation and Design

There is always a circulation of air through wind towers (Bahadori, 1976b, 1978).

Night operation When there is no wind blowing at night, the wind tower (Fig. 5) acts as a chimney. The tower walls A, B, C, and D that have been heated during the day transfer heat to the cool night ambient air. The heated air is then exhausted at the openings (1). This chimney action of the tower maintains a circulation of the ambient air through the building and cools the structure of the building as well as that of the wind tower. The natural circulation of air through the building and the wind tower continues all through the night. Further night cooling of the roof and external walls of the building is accomplished by thermal radiation to the sky, which is extremely clear in the desert regions.

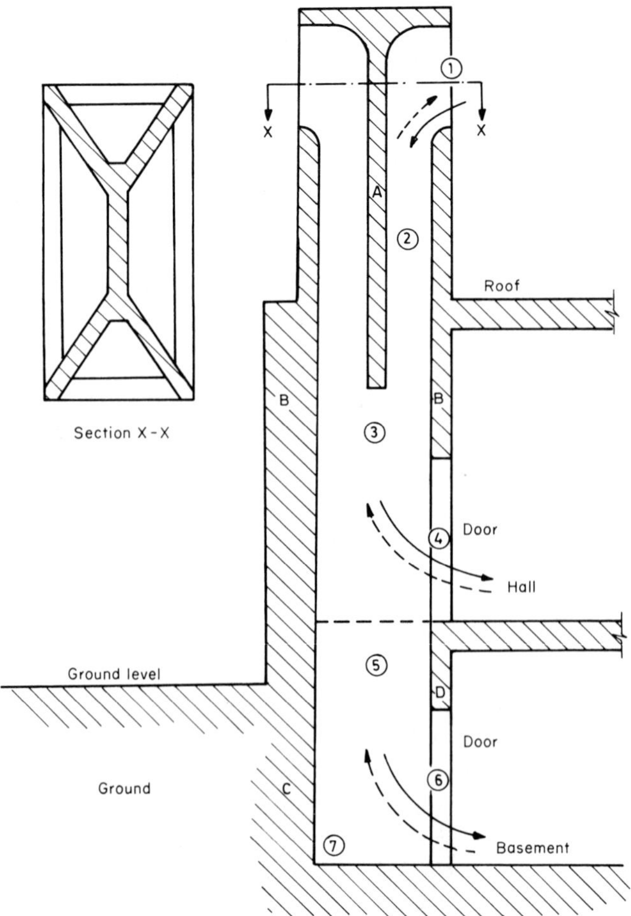

FIG. 5 Operation of a wind tower in summer. Air flow during the day, →; air flow during the night with no wind, ←--.

When there is a wind blowing during the night, the air circulation is in the direction opposite to that described, but the walls of the tower are cooled, and some cooling of the rooms may also result. The configuration of the wind tower cross sections at points (1) and (2) and the thicknesses of walls A, B, etc., are such as to provide sufficient heat transfer area and thermal energy storage capacity.

Day operation When there is no wind blowing during the day, the tower operates as the reverse of a chimney. The hot outside air in contact with cold walls A and B (cooled during the previous night) is cooled and pulled down through passages (2) and (3). This air may be let out through door (4) or door (6). When there is a wind, the air circulation and the rate of cooling are increased, and the cold air can reach far distances. The air going through door (4) or (6) enters a central hallway that contains openings to adjacent rooms. By opening the door of any one of these rooms and closing the others, the air may be circulated through the room. By opening or closing door (4) or (6), airflow through each floor may be controlled. When it goes through section (5) and when wall C is moist, the air is further cooled evaporatively and sensibly. Before household refrigerators became popular, area (7) was used as a cold storage.

The psychrometric chart presentation (Severns and Fellows, 1964) of sensible and evaporative cooling processes through wind towers is given in Fig. 6. The cooling process if an evaporative cooler were employed is shown by line 1–8. It is clear that wind towers, in addition to not requiring active energy, are in many cases more effective than evaporative or desert coolers (which have become very popular in Iran recently).

Design There are numerous wind tower designs. They include various tower heights and openings, and different cross sections for the airflow passages. All of the designs attempt to provide desired airflow

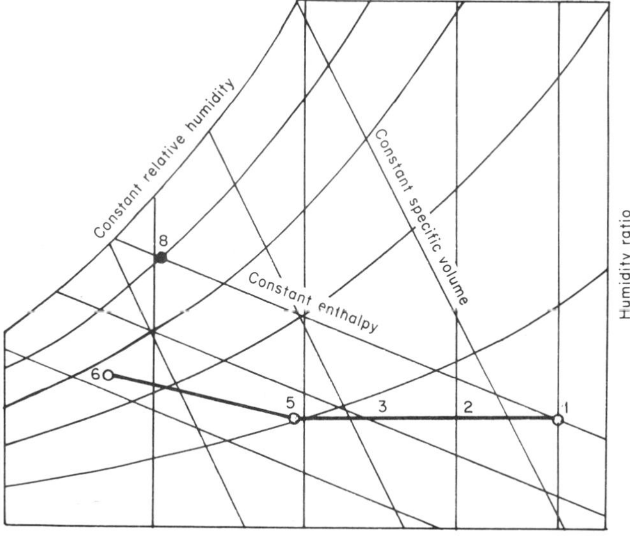

FIG. 6 The psychrometric presentation of the cooling process through the wind tower.

rate, heat transfer area, sensible heat storage capacity, and evaporative cooling surfaces. Some of these designs are shown in Figs. 7–16. The cross section of the double-story wind tower (Fig. 13) at the roof level is shown in Fig. 17. Figure 18 is the bottom view of a corner of this wind tower. The tall towers are wood reinforced, and the ends of the wooden beams extend out to be used for scaffolding. Figure 19 shows the cross section of the wind tower in Dowlat-Abad Garden, Yazd (see Figs. 8–10), where a small pool with a fountain is situated directly under the tower. High-velocity air coming down the tower is first sensibly cooled and then further cooled evaporatively by the pool and the fountain, and enters a central hallway, shown in Fig. 20. There is another larger pool in this central hallway, as shown in the figure. The conditioned air from the hall may be directed through one or more rooms adjacent to the hallway by

FIG. 7 The wind tower of a residence in Yazd. The openings have a height of about 3 m, and the tower is about 13 m tall.

FIG. 8 The wind tower in the Dowlat-Abad Garden in Yazd. The openings have a height of about 11 m, and the tower is about 34 m tall.

opening their doors to the outside and closing those of the others. Figure 21 shows the cross section of a wind tower in Bam (see Fig. 14), where the tower is placed about 50 m away from the basement and connected to it by an underground masonry tunnel. Water used for the lawn, trees, and shrubs in the yard diffuses through the ground and keeps the walls of the tunnel moist. Air is further cooled evaporatively when pool D contains water and its fountain is operating. Figure 22 shows the air-conditioning processes through this wind tower in a psychrometric chart.

Figure 23 shows the cross section of another special wind tower, which has an access to an underground water stream. The underground water is generally cold, and the scheme employed in this figure can pro-

FIG. 9 The wind tower of Dowlat-Abad Garden, Yazd, showing the relative height of the tower with respect to the building. Note the dome and its cap next to the tower.

duce a cooling rate more effectively than many other wind towers. The house has access to the underground stream by shaft D, and the wind tower is built favorably with respect to this shaft. The high-velocity air coming down through the tower and entering the basement causes an entrainment at point c. The entrained air enters the underground waterway through openings like a, which may serve several wind towers of similar design in the neighborhood. Figure 24 represents approximately the air-conditioning processes of the wind tower shown in Fig. 23. The air at point c is very cold and moist, and this location was used as a natural refrigerator for food preservation. The windless night operation of this wind tower is similar to that explained earlier, and the tower acts as a chimney. The outside air still flows over the underground stream and up

FIG. 10 The openings of the wind tower in Dowlat-Abad Garden, Yazd. Note the screens added to prevent birds from entering the tower.

through well D, is mixed with the room or basement air, and is finally vented from the tower top.

The wind tower operation just described, although applicable only in special places, seems to produce the best cooling effect in the basement.

B. Special Features

(1) When all the house doors are closed or the wind is prevented from entering the house, the high-velocity air coming down the tower returns through the opposite pathways and back to the atmosphere from the leeward openings. In fact the suction created in the leeward helps this short-circuiting of air flow. In normal wind tower operation there is

FIG. 11 The wind tower of a new house in Yazd. The openings have a height of about 1.25 m, and the tower is about 5.5 m above the ground.

always an entrainment of room air through the leeward openings. The suction created at this point and the entrainment of air through the building and up through the opposite portion of the tower have been used for maintaining a constant evaporation of water surfaces in cisterns, which will be discussed in Section 9.5.

(2) In locations where dusty winds always blow in one direction and the more pleasant winds in the opposite direction, wind towers are built with openings to intercept the pleasant winds only. Since the openings are in the leeward, during the periods of dusty wind, air is entrained up through the tower.

(3) Airflow rates through the wind towers may be controlled by doors (4) and (6), shown in Fig. 5. To prevent any air leakage through

FIG. 12 The wind tower openings of the Ebrahim-Khan School in Kerman. The openings have a height of about 6 m, and the tower is about 18 m tall.

these doors in winter, the practice in Bam, with low and one-sided wind towers (see, e.g., Fig. 15), has been to partially or fully block the openings by a thin masonry wall during this season.

(4) The sensible cooling effect of the wind towers (process 1–5 in Fig. 6) is limited by the thermal energy storage capacity of walls A and B (Fig. 5). With a small energy storage capacity the sensible cooling may stop after several hours of operation in the hot summer days. After this time, and except for the evaporative–sensible cooling of the air flowing over the wall C (Fig. 5), the wind tower acts only as an air circulator.

(5) In designs where the rooms open into a large courtyard, e.g.,

FIG. 13 A double-story wind tower in Kerman. The top openings circulate the air through the central core of the tower.

Dowlat-Abad Garden in Yazd (Fig. 9), and when the wind is blowing from the yard toward the wind tower the air may enter the building from the yard and leave through the leeward openings of the wind tower. In this case the air entering the tower also leaves at the leeward openings.

C. Disadvantages

(1) Dust, insects, and birds coming down through the wind towers are the main disadvantages of this system. To prevent birds and insects

FIG. 14 The wind tower of a house in Bam. This tower is connected to the basement by a 50-m-long underground masonry duct. Note the dome roof and its cap which belong to another building next to this tower.

from entering the tower, screens have been employed in some of the newer units. To reduce the amount of dust coming into the building the airflow area is increased immediately under the tower, and dust "pockets" or shelves are provided in this space. Figure 19 shows the change of cross section under the tower and the dust shelves in this space. Tall air towers, while admitting high-velocity winds through the windward openings, induce less dust but are expensive to build and maintain.

(2) Wind towers are for summer use only. If not closed properly in winter, they can increase the infiltration heat losses appreciably.

FIG. 15 A wind tower in Bam with its openings facing north. Almost all of the wind towers in Bam intercept northerly winds and are of low height.

9.4 AIR VENTS

It is a well-known fact that when air is flowing over cylindrical or spherical objects, the velocity increases while pressure decreases on the apex of these objects (Schlichting, 1960).

Cylindrical and dome roofs have been employed in arid regions for a long time. They provide the following advantages over flat roofs:

(1) They are of lighter weight and do not require wooden beams, which are generally scarce in the area.

(2) They have a higher heat transfer coefficient (Kreith, 1965) and greater area as compared with flat roofs of the same base area. The solar energy absorbing area of the flat and curved roofs of the same base area is

FIG. 16 A wind tower in Shiraz. The openings have a height of 2.5 m, and the tower is about 12 m tall.

FIG. 17 The bottom view of the wind tower in Fig. 13. Note the extent of the dividing walls, which adds to the heat transfer area and the energy storage capacity.

FIG. 18 The bottom view of the wind tower in Fig. 13, showing the guide ribs at a corner and walls dividing the airflow into smaller channels.

nearly the same, whereas the convention heat transfer area is higher for the latter. Solar energy gain constitutes a major source of heat gain in the summer.

(3) The air stratification in the room is high, which reduces the heat transfer from the roof to the room.

(4) When a hole is made at the apex of the dome or cylindrical roof, the low pressure created at this point vents the hot air under the roof. Figure 25 shows a cross section of a typical air vent and the air circulation around it.

Figure 26 shows the windows under the dome roof of a house in Bam. Figure 27 shows the openings made in the "cap" placed over the dome

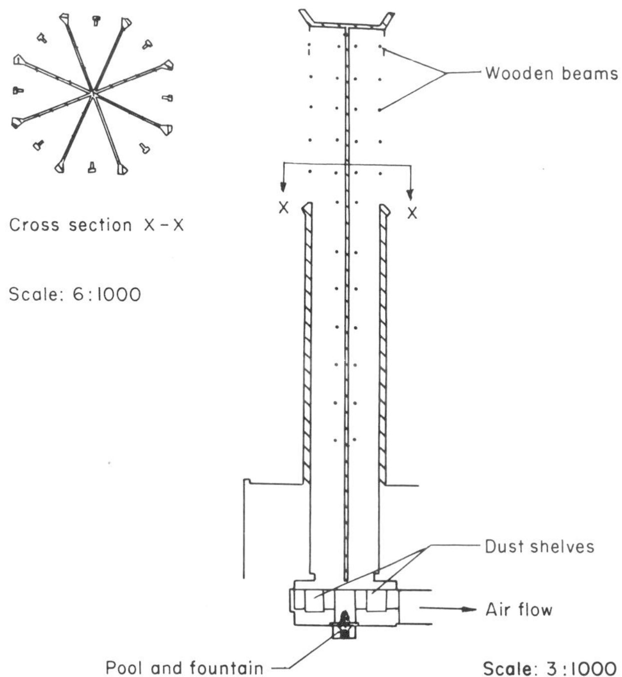

FIG. 19 The cross section of the wind tower in Dowlat-Abad Garden, Yazd. Note the wooden beams, dust shelves, and the pool under the tower.

FIG. 20 The central hallway to which air coming down from the wind tower shown in Fig. 19 flows. Note the eight-sided empty pool and fountain located at the center. This hallway is covered by the dome that is visible in Fig. 9.

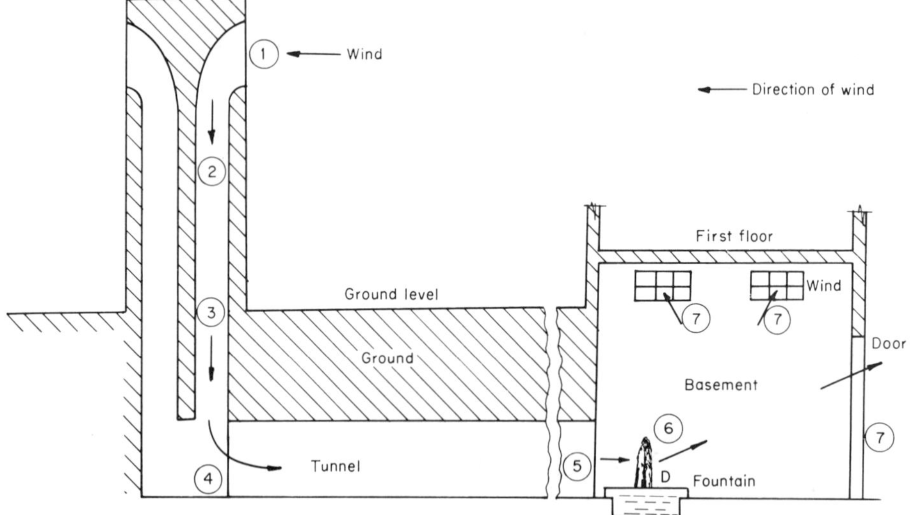

FIG. 21 The cross section of a wind tower connected to the basement by a 50-m-long moist underground tunnel.

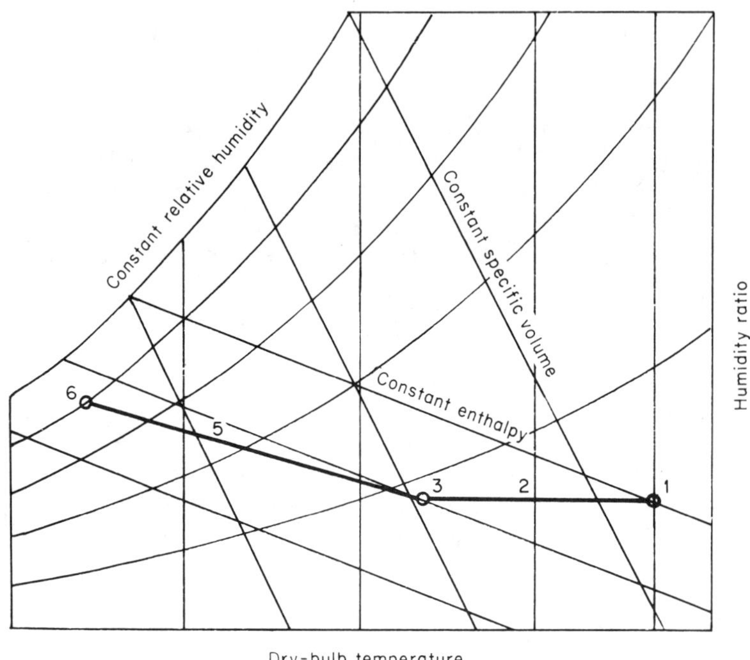

FIG. 22 The psychrometric presentation of the air-conditioning process of the wind tower shown in Fig. 21.

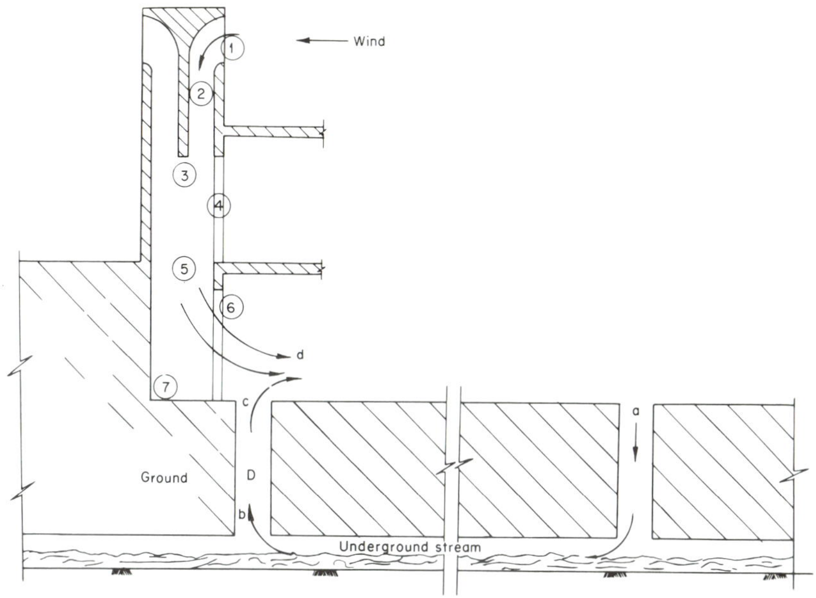

FIG. 23 The cross section of a wind tower used in conjunction with an underground stream.

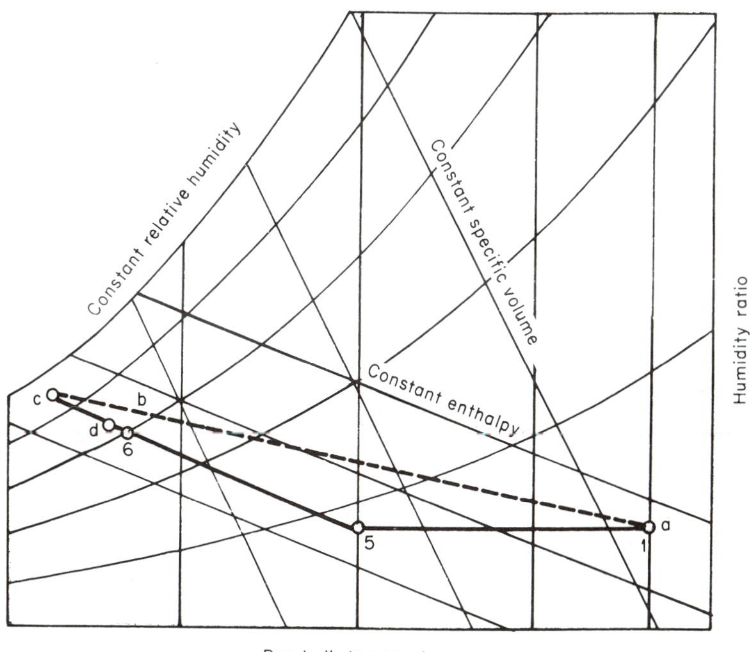

FIG. 24 The psychrometric presentation of the air-conditioning process of the wind tower and underground stream shown in Fig. 23.

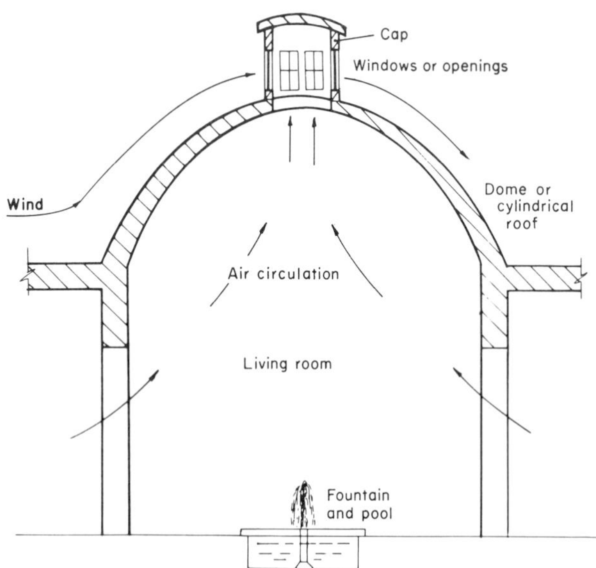

FIG. 25 The air circulation pattern in a room with a curved roof.

FIG. 26 Openings in the cap of the dome roof of a residence in Bam.

FIG. 27 The cap and the openings on the dome of the Dowlat-Abad Garden in Yazd.

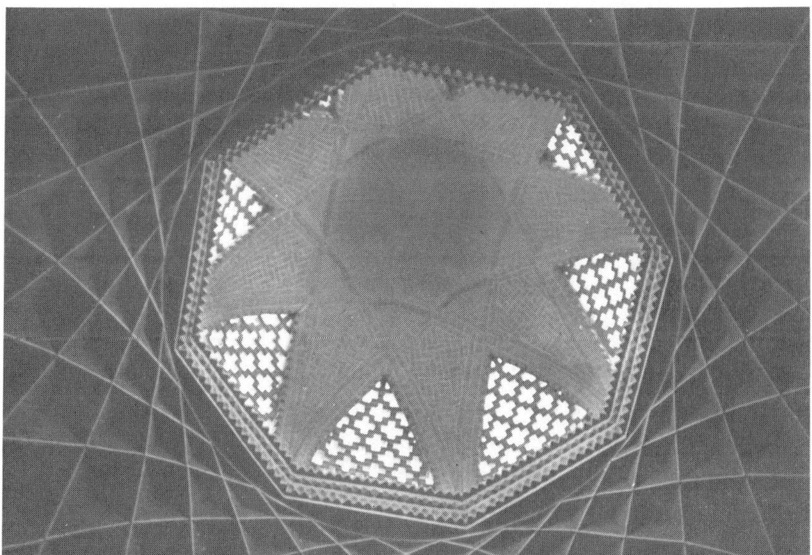

FIG. 28 View of the cap, its openings, and part of the dome roof of the Dowlat-Abad Garden in Yazd, shown in Fig. 27.

FIG. 29 Portion of a building in Bam with its dome roof, the air vent on the dome, and the wind tower.

roof of Dowlat-Abad Garden in Yazd. The cap and the openings can also be seen in Fig. 9. Figure 28 shows a view from beneath the openings in the "cap" and part of the dome roof. Figure 29 shows a portion of a house in Bam equipped with a wind tower, dome roof, and air vent over the roof. The rooms with air vents shown in Figs. 25, 26, and 29 are generally used as living rooms and have a small pool of water situated immediately under the dome.

In areas where wind direction is generally constant, the roofs are cylindrical with the axis normal to this direction, whereas when the wind blows in almost all directions, one finds dome-type roofs.

9.5 CISTERNS

Energy storage from one season to another is an excellent method of passive cooling or heating. As mentioned earlier, the arid zones are characterized, among other things, by their very cold winter nights. The people living in these regions made use of this cold and the fact that deep earth provides good insulation by storing cold water or ice fabricated in winter for summer use.

The making of ice in winter and storing it for summer is discussed in Section 9.6.

Cold water for summer use was stored in cisterns 10–20 m deep. The cisterns were filled during the cold winter nights. To keep this water cold for summer use, a natural evaporation of the water surface was provided by the draft produced by several wind towers. Figure 30 shows the operation of a cistern with wind towers.

Air circulation through the wind tower has been explained earlier. When there was an opening at the apex, air flew down through the tower and out through opening A, entraining with it the saturated air that existed under the dome. When no hole was provided in the apex of the dome (to prevent birds and dirt from falling in), the air passage in the wind towers was short-circuited, with air entering at the windward, entraining some of the air over the water surface, and then leaving at the leeward openings.

The dome roof, in addition to its structural benefits, increases the convective heat losses, as explained in the previous section. Figure 31 shows the dome of a cistern equipped with six wind towers. Figure 32 shows the entrance to the staircase for water pickup from the same cistern. This cistern, located in Yazd, is about 12 m deep with six wind towers about 12 m

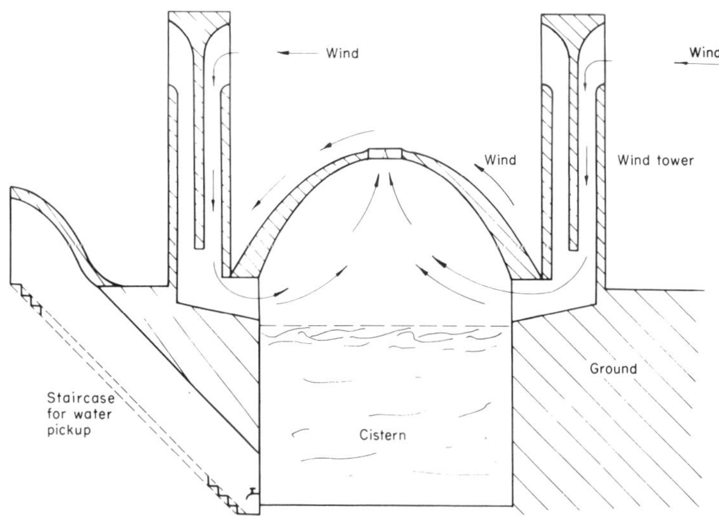

FIG. 30 Evaporative cooling of a cistern by wind towers.

FIG. 31 A six-wind-towered cistern in Yazd with a dome roof.

FIG. 32 The entrance to the water pickup of the six-wind-towered cistern in Yazd.

FIG. 33 The air vent of a cistern in Kerman. The photograph was taken from the floor of the cistern (now being remodeled).

high. Figure 33 shows the air vent of a cistern in Kerman. This cistern (now being converted to a museum) is about 8 m deep and has a water storage capacity of about 1600 m^3. For health reasons, the use of cisterns has now almost been abandoned in Iran.

9.6 NATURAL ICE MAKING

Another means of storing energy from one season to another is by the production of ice during winter nights and its storage in deep underground storage pits for summer use.

A. Construction of an Ice Maker

A natural ice maker consisted of shallow rectangular ponds dug in the ground, as shown in Fig. 34. The ponds might be about 10 to 20 m wide in the north–south direction and several hundred meters long in the east–west direction (Bahadori and Kosari, 1978). The ponds were protected from solar radiation by tall adobe walls that were built to the south of the ponds. The height of these walls was related to the width of the ponds so that the ponds were always in the shade during the ice-making season. To protect the ponds from early morning and late afternoon solar radiation

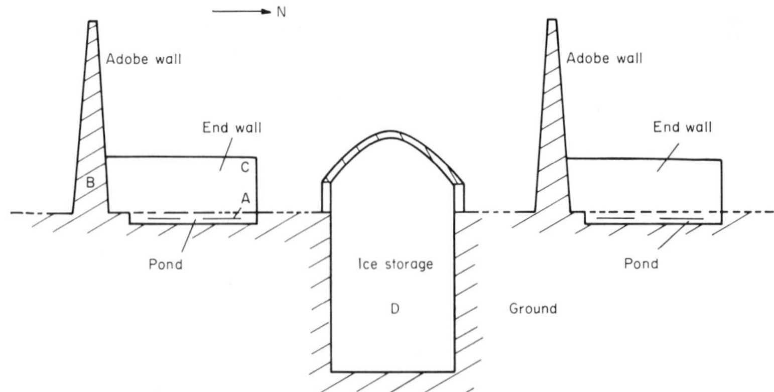

FIG. 34 A natural ice maker consisting of several shallow ponds and a large storage pit.

and from winds in the east–west direction, end walls were built at the east and west ends of the ponds. A natural ice maker might consist of one or several ponds with the protective walls in parallel, serving one large ice storage pit 10–15 m deep and of sufficient volume.

B. Operation of the Ice Maker

The operation of natural ice makers relied heavily on the thermal radiation losses to the sky during winter nights. During winter nights when the sky was clear and cloudless, the ponds were filled with water which was left to freeze. The ambient air was generally a few degrees above the freezing point. The water in the pond received heat by convection from the air and by conduction from the ground, but lost heat to the sky by radiation. The tall walls surrounding the ponds, especially when there were several ponds in the field, reduced the wind effect and hence the heat gain by convection. Depending on the weather conditions, and relying purely on experience, the ponds were filled to a height deemed sufficient for freezing. To speed up the ice making, the ponds might be filled a few centimeters at a time several times during the night. Part of the ice at the bottom of the pond received heat from the ground during the night, thus making the collection of the ice from the pond easier. The ice made in the ponds during the night was cut into pieces the next day and transferred to the storage pit for use the next summer. Figure 35 shows the storage pit and the walls and Fig. 36 the cross section of one of the protective walls of a natural ice maker in Kerman. This wall is about 12 m high and 3 m thick at the base. Figure 37 shows the wall and the storage pit and Fig. 38 a portion of the dome of the storage pit of a ruined ice maker in Bam.

FIG. 35 The storage pit and walls of a natural ice maker in Kerman.

FIG. 36 The walls of a recently renovated natural ice maker in Kerman.

FIG. 37 The walls and storage of a ruined natural ice maker in Bam.

FIG. 38 Portion of the dome of the ice storage pit of the ruined natural ice maker in Bam, shown in Fig. 37.

The operation of natural ice makers has been stopped because of health reasons, and to the writer's knowledge there is no unit now operating in Iran.

9.7 CONCLUSIONS

The passive cooling systems qualitatively described in this chapter have potentially widespread applications in similar climates throughout the world. Natural ice making and deep cisterns, now abandoned in Iran for health reasons, can be employed for uses other than direct human consumption, for example, for industrial, commercial, or household air conditioning.

Modern architects and engineers should acquire a thorough knowledge of passive heating and cooling systems, be familiar with the climate in which a building is to be used, and design a dwelling which, while incorporating today's conveniences, uses a very small amount of energy for heating and cooling.

ACKNOWLEDGMENTS

The author wishes to thank Pahlavi University Research Council for supporting this research and the Cultural Offices of the cities of Yazd, Kerman, and Bam for facilitating his visits to the buildings of interest in those cities. I am also thankful to Dr. S. H. Izadpanah for making similar arrangements for the city of Semnan.

10

Hydronic Solar Heating and Cooling in Georgia

J. RICHARD WILLIAMS

COLLEGE OF ENGINEERING
GEORGIA INSTITUTE OF TECHNOLOGY
ATLANTA, GEORGIA

10.1 INTRODUCTION

The United States has embarked on a massive program for the development of solar energy systems, with the expectation that within 10 to 20 years solar energy will be able to play an important role in meeting the energy needs of modern society. In Georgia the full range of solar energy activities is represented, including low-temperature agricultural applications, solar water heating, heating and cooling for buildings, industrial process heat supply, electric power plants, and total energy systems. Georgia is a very poor state in terms of conventional energy sources. Over 99% of the energy requirements are imported from outside the state. This means that the rapid increase in energy costs does not bring the state any new income but is only a financial drain on the people of Georgia. As a result, the legislature and governor of Georgia have shown a keen interest in developing alternative energy sources, particularly solar. Two of the world's largest solar-heated and air-conditioned buildings and one of the two Solar Total Energy Large-Scale Experiments in the United States are located in Georgia. The Georgia Institute of Technology, Georgia's only engineering institute, was instrumental in carrying out these projects. Georgia Tech currently has a higher level of funding for solar energy projects than any other American university.

ISBN 0-12-620860-3

10.2 SOLAR HEATING AND COOLING TECHNOLOGY

At present there are about 20 solar homes and 4 commercial solar buildings in Georgia. All use liquid-heating, flat plate solar collectors with thermal storage in a tank of water. The sun heats a fluid flowing through the solar collector, and the heat is then stored in a tank. Whenever heating is needed, hot fluid is circulated from the tank through the building heating system. For air conditioning, 3 commercial projects use a water-fired absorption chiller.

A. Flat Plate Solar Collectors

Flat plate solar energy collectors are used for providing low-grade heat. A blackened surface is covered by one or more transparent cover plates of glass or plastic, and the sides and bottoms of the box are insulated. Solar radiation is transmitted through the transparent covers and absorbed by the blackened surface beneath. The cover tends to be opaque to infrared radiation from the plate and also to retard convective heat transfer from the plate; so the black plate is heated by absorption of the solar radiation and, in turn, heats the fluid flowing under, through, or over the absorber plate. Water is most commonly used, since the temperatures involved are usually below the boiling point of water. For some installations silicone oil or mineral oil have been used.

In order to design and construct solar collectors for heating and cooling projects, detailed properties of the materials and characteristics of the various components must be known in order to predict the performance and durability of the collector. Needed property data can generally be classified into three categories: thermophysical properties, physical properties, and environmental properties. Thermophysical properties include thermal conductivity, heat capacity, and radiant heat transfer characteristics. Physical properties include density, tensile strength, melting point, and modulus of elasticity. Environmental properties include resistance to ultraviolet degradation, moisture penetration, and degradability due to pollutants in the atmosphere. All these data are required in order to develop collectors that are reliable, durable, and efficient. Durability is the criterion most often overlooked by the novice in constructing collectors.

The collector absorber plate should have high thermal conductivity, adequate tensile and compressive strength, and corrosion resistance. Copper is generally preferred because of its extremely high conductivity and resistance to corrosion. Collectors are also being constructed of aluminum, steel, and various thermal plastics. Aluminum and steel require a corrosion-inhibited heat transfer fluid. Since most potable waters contain chloride ions and heavy metal ions such as copper and iron, these would cause pitting in aluminum channels. Also, if aluminum is used, one cannot mix copper plumbing and aluminum collectors without taking adequate precautions to ensure that the copper ions from corrosion of the piping and chlorides from the soldering fluxes do not destroy the aluminum. The galvanic effect can be an important factor in multimetal systems; so electrical insulation should always be provided between dissimilar metals. If the flow rate is too high, corrosion can be also produced by the simple erosion process resulting from high-flow rates and turbulence in the fluid passages. Partial blockages of the flow passages can also cause localized high velocities resulting in this type of degradation.

Until recently, absorber plates for flat plate solar collectors were usually constructed with tubes soldered or welded onto a metal plate, which was then blackened. Some of the earlier solar water heaters actually had tubes fastened to the plate without soldering, resulting in poor heat transfer and poor thermal performance. The standard procedure for fabricating an absorber plate was to take a sheet of copper or aluminum and solder tubes to it; sunlight falling on the plate would heat the metal plate, and the heat would be transferred to water flowing through the tubing. One of the more important advances in solar technology has been the development of internal tube collector plates such as the Roll-Bond panel by the Olin Brass Company and the tube-in-strip collector plate by Revere Copper and Brass. The internal tube collectors have superior heat transfer characteristics and also can be mass-produced so that the laborious process of soldering and welding tubes onto a flat plate is eliminated. Also, the tubes cannot come loose from the plate. Because of these and other desirable features of the internal tube absorber plates, such absorbers are being incorporated into many modern collector designs.

A few installations in Georgia have used a trickle collector employing a galvanized iron pan with silicone oil or mineral oil flowing over the top of the blackened metal surface. This type of collector is not as efficient as most conventional flat plate collectors when the fluid temperature is considerably higher than the ambient temperature.

The cover plate or plates through which the solar energy must be transmitted is also extremely important to the function of the collector. The purposes of the cover plates are

(1) to transmit as much solar energy as possible to the absorber plate,

(2) to minimize heat loss from the absorber plate to the ambient,

(3) to shield the absorber plate from direct exposure to weathering,

(4) to receive as much of the solar energy as possible for the longest period of time each day.

The most critical factors for the cover plate materials are strength, durability, nondegradability, and solar energy transmission.

Tempered glass is the most common cover material for collectors because of its proven durability and because it is not affected by the ultraviolet radiation from the sun. Experience has shown that, unless the glass is tempered, the day-to-day thermal cycling of the cover plate tends to cause breakage. Tempered glass, properly mounted onto a flat plate solar collector, is highly resistant to breakage from both thermal cycling and natural events. Glass is also effective in reducing radiated heat loss because it is opaque to the long-wavelength radiation reemitted by the hot absorber plate.

Plastic materials may also be used for cover plates, for example, the acrylic polycarbonate plastics, plastic films of Tedlar and Mylar, and commercial plastics such as Lexan. Plastic materials tend to have limited lifetimes because of the effect of ultraviolet light in reducing the transmissivity of the plastics. Also, they are usually partially transmitting to long-wavelength radiation and are therefore less effective in reducing radiated heat losses from the absorber plate. Some plastics also are unable to withstand the maximum equilibrium temperatures that are encountered in flat plate collectors, especially when the collector is dry. The main advantages of plastic materials are the resistance to breakage, reduction in weight, and in some cases, a reduction in cost.

Most glass and plastic materials of interest have refractive indices of about 1.5; unless special coatings or surface treatments are applied, this results in approximately 8% of the normal incidence solar radiation reflecting from the glass away from the absorber plate from each sheet of glazing, and a greater fraction is reflected at higher incidence angles. This means that the maximum transmittance is 92% for a single, perfectly clear, nonabsorptive sheet of glazing material. In multiglazed panels the reduction in transmission is about 8% more for each additional sheet. In addition, there is a transmission reduction due to the absorption of solar radiation within the material itself. The amount of solar radiation reflected can be reduced considerably by etching or by applying an antireflective coating to the surface. Etching produces a surface coating of a refractive index lower than 1.5, which results in less reflection.

The transmissivity of glass depends on the iron content. A normal sheet of window glass will look green when viewed through the edge because of the iron oxide within the glass. Water-white crystal glass has the lowest iron content and therefore the highest transmission of solar energy. Water-white crystal is available either in annealed or tempered form. Tempered glass has about 5 times the impact and thermal shock resistance of ordinary annealed glass. In selecting the glass for cover plates, the mechanical strength must be adequate to resist breakage from the maximum expected wind load and snow load, and the normally expected impact. The mechanical strength is proportional to the square of the thickness of the glass; so cover plates for solar collectors should normally be at least $\frac{1}{3}$ cm thick.

Thermal shock to the glass cover plates must also be taken into account. It is caused by several different processes. First is the day-by-day heating and cooling from the increase in solar intensity on the collectors during the morning hours and subsequent decrease in the afternoon. In addition, in partly cloudy weather glass temperatures can rise and fall by 50°C or more in a matter of minutes as clouds pass overhead. The central area of the collector is subjected to greater heating than the edges of the glass, since normally the edges are enclosed in flashing and are not exposed at all to direct sunlight. This results in a thermal stress in the glass at the edges that may be estimated at 60,000 kg/m^2 per degree celsius of temperature difference between the heated center and the cooler edge of the glass plate. Thicker glass plates are more subject to thermal shock than thinner plates. Additional stress can occur when a single collector is partially shaded. In this case part of the glass plate is subjected to high temperatures, while the shaded area is not. These various processes can easily result in breakage of nonannealed glass and account for the use of tempered glass in solar collectors.

The rigidity of the cover plate is also important. Rigidity is proportional to the cube of the thickness of the plate. The resistance to fracture under mechanical stress is especially important when the collector is doubled glazed. Some flexure may be desirable to accommodate the expansion of air within the gap when this type of collector is heated.

In Georgia the effect of accumulated dust or dirt on the performance of flat plate solar collectors is minimal. Most solar projects to date have used collectors at about a 45° slope, and in our area the frequent rainfall serves to keep these collectors reasonably well washed. Thus the accumulation of dust over an extended period of time is minimal.

Several solar heating and cooling projects in Georgia use a black chrome selective coating on the absorber plate. Such a selective coating absorbs solar radiation while considerably reducing the emission of infrared heat; this increases the collector efficiency because less heat is lost by reradiation from the absorber plate. At present, black chrome is the selective coating most often used for flat plate solar collectors because of its highly desirable optical characteristics (absorptivity, 0.95; emissivity, 0.07), its durability and resistance to attack by moisture or high temperature, and its moderate cost.

Black chrome has been used as a commercial decorative finish for many years, and the technology for plating black chrome is well in hand. Figure 1 shows the reflectance as a function of wavelength for several types of selective coatings. Black chrome is most commonly made from the Harshaw Chrom-Onyx electroplating bath and is electroplated at about 2000 A/m^2 for about 1 min. Black chrome can be electroplated on steel, copper, or aluminum by commercial electroplators using the same tanks, chemicals, and procedures as for decorative finish black chrome, commonly applied to table legs, doorknobs, metal chair trim, and so forth.

Figure 2 illustrates a conventional flat plate collector design with two transparent sheets of glass mounted over the blackened metal absorber plate. The glass is sealed so as to prevent moisture from entering the collectors or becoming trapped between the two sheets of glazing; however, a small gap at the edges of the glass accommodates thermal expansion. Water, or another heat transfer liquid, circulates through tubes in the plate. A typical trickle-type collector is illustrated in Fig. 3. In this design water is allowed to trickle over the sloping absorber plate surface, thereby becoming heated. These collectors can be manufactured more cheaply than conventional flat plate collectors and are nearly as efficient as long as the difference in temperature between the fluid and the ambient temperature is less than 40°C. However, for larger temperature differences the efficiency of the trickle collector drops off rapidly. Trickle collectors cannot be used for solar air conditioning because of the high fluid temperature required by the air-conditioning equipment.

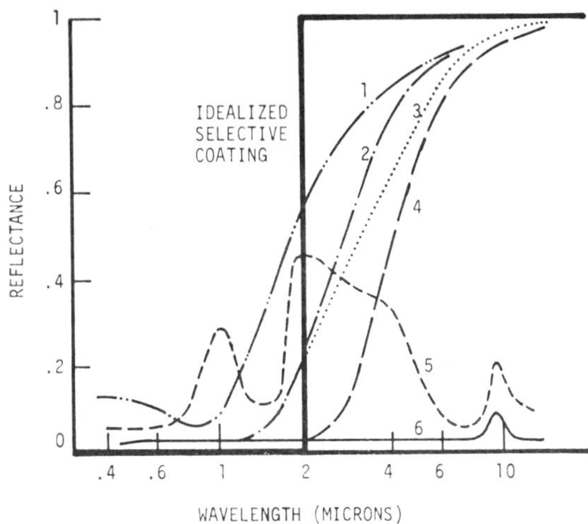

FIG. 1 Reflectance of selective coatings: 1, black nickel; 2, zinc; 3, black chrome; 4, black copper; 5, black ceramic enamel; 6, Nextel black paint.

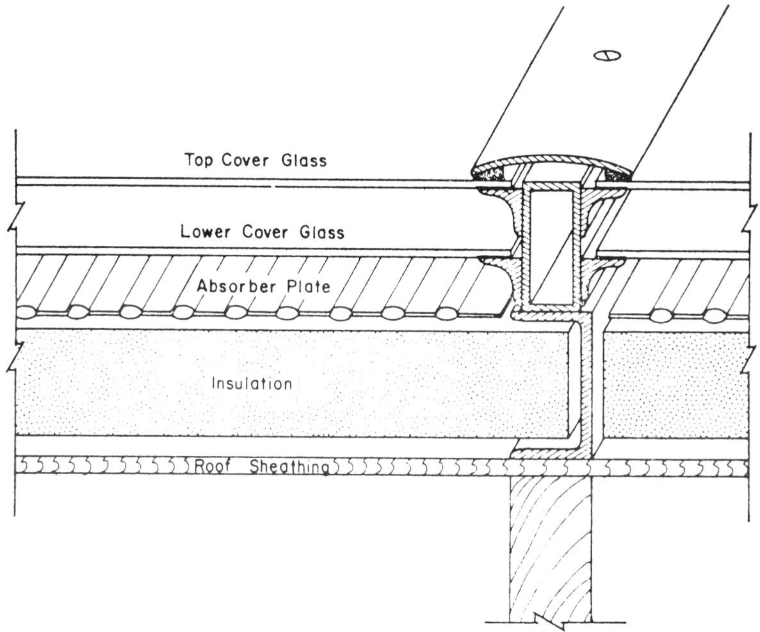

FIG. 2 Cross section of typical conventional flat plate solar collector installation.

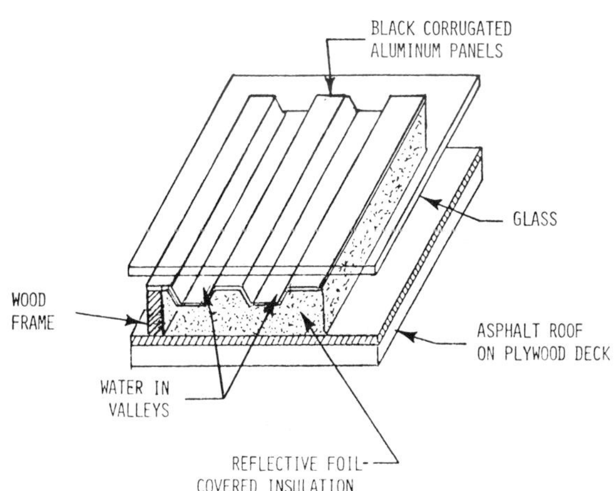

FIG. 3 Typical trickle-type flat plate solar collector installation.

B. Collector Performance Evaluation

Theoretical analyses of the performance and operating characteristics of flat plate collectors have been carried out by numerous researchers since the original Hottel and Woertz paper (1942). It has been shown that the instantaneous performance of a flat plate solar collector operating under quasi-steady state conditions can be described by

$$q_u/A_a = E(\tau\alpha)_e - U_L(\overline{T}_p - T_a) \tag{1}$$

where q_u is the useful heat collected (W), A_a the aperture area (m^2), E the solar radiation intensity on the collector (W/m^2), τ the overall transmisivity of the cover plate, α the hemispherical absorptivity of the absorber plate, $(\tau\alpha)_e$ the effective $\tau\alpha$ product, U_L the heat loss coefficient (W/°C), $\overline{T}_p$ the average collector plate temperature, and T_a the ambient temperature. One may introduce the heat removal factor F_R, where

$$F_R = \frac{\text{actual useful energy collected}}{\text{useful energy collected if the entire collector surface were at the temperature of the fluid entering the collector}} \tag{2}$$

Introducing this factor into Eq. (1) results in a new performance equation,

$$q_u/A_a = F_R\{E(\tau\alpha)_e - U_L(T_{f,i} - T_a)\} \tag{3}$$

where $T_{f,i}$ is the collector fluid inlet temperature. If the instantaneous solar collector efficiency is defined as

$$\eta = \frac{\text{actual useful energy collected}}{\text{solar energy incident on or intercepted by the collector}}$$

then

$$\eta = (q_u/A_a)/E \tag{4}$$

The instantaneous efficiency of the flat plate collector is then found by substituting (3) into (4) to get

$$\eta = F_R(\tau\alpha)_e - F_R U_L[(T_{f,i} - T_a)/E] \tag{5}$$

This indicates that if the efficiency is plotted against $(T_{f,i} - T_a)/E$, a straight line will result where the slope is $F_R U_L$ and the y intercept is $F_R(\tau\alpha)_e$. This is the way actual performance data for solar collectors are presented, with η plotted against $(T_{F,i} - T_a)/E \cdot F_R(\tau\alpha)_e$ is the efficiency the collector would have if the fluid inlet temperature was equal to the ambient temperature. Experimental collector performance data for a variety of flat plate collector types are presented in Fig. 4. Table 1 gives the performance ranking and expected efficiency for these collectors for four different types of applications.

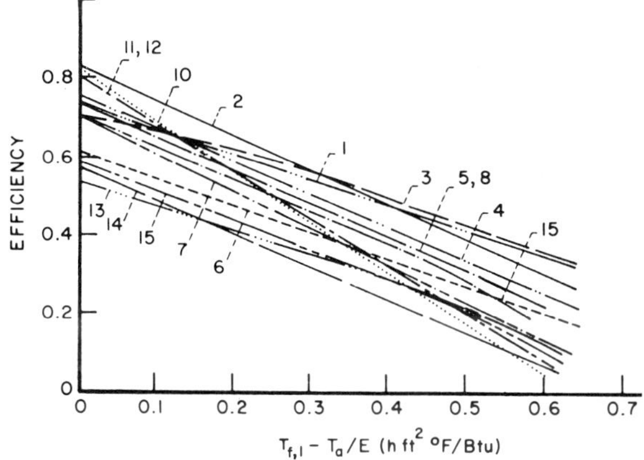

FIG. 4 Results of collector performance testing (see Table 1). (For units of m² °C/W, multiply horizontal scale units by 0.176.)

In general, the energy conversion performance of flat plate collectors can be improved by two methods:

(1) Increasing the transmission of energy through the collector to the working fluid. This may be done by improving

(a) the transmittance (in approximately the 0.4–1.9 μm spectral range) of the transparent (glass or plastic) cover plates,

TABLE 1

Collector Efficiency (Typical Values)

Number	Type	Rankine cycle η (%)	Air conditioning η (%)	Hot water η (%)	Pool heating η (%)
1	Black chrome, 2 glass	34	45	62	83
2	Black chrome, 1 glass	32	43	60	83
3	CuO honeycomb, 1 glass	27	41	58	80
4	CuO, 2 glass	22	35	55	80
5	Black Ni, 1 glass	18	33	53	75
6	Black Ni, 2 glass	17	31	52	73
7	Black Ni, 2 glass	11	28	51	73
8	Black Ni, 1 Tedlar	10	25	51	71
9	Black Ni, 2 Tedlar	7	24	51	70
10	Black paint, 1 glass	5	23	48	61
11	Black paint, 1 glass	4	23	45	59
12	Black paint, 2 glass	3	21	41	57
13	Black paint, 2 glass	0	17	39	54
14	Black paint, 2 glass	—	—	37	—
15	Black paint, 2 glass	—	—	—	—

(b) the absorptivity of the absorber plate to the incident solar radiation (absorptivities approaching 1.0 are obtained by appropriate black or selective coatings),

(c) the heat transfer coefficients from the absorbing surface to the fluid; these depend on the thermal conduction resistance through the absorber plate (thermal conductivity, plate thickness) and on the nature of convection in the flow channels, such as whether the flow is turbulent or laminar, and surface roughness.

(2) Decreasing the thermal losses from the collector to the ambient by reducing:

(a) Conductive losses: Those that occur through the back and sides of the collector can be reduced to a negligible amount by using a sufficiently thick layer of thermal insulation. The main problem is in the front where heat is conducted from the absorber plate through the air layer(s) between the plate and the transparent covers and on out to the ambient air. The increase in the thickness of the air gaps reduces the loss up to a certain limit, where further increases allow significant natural convection. Natural convection transfers heat at higher rates than conduction; this leads to higher, rather than lower, heat losses. Alternatively, several transparent panes could be used to create a number of narrow air gaps. This, however, reduces the transmission of solar energy to the absorber. The single-pane collector is the most efficient when the absorber temperature is not much higher than that of the outer window pane (transmission dominating over heat losses), but becomes rapidly less efficient when this temperature difference increases. Therefore, high-temperature collectors require multiple panes in the window.

(b) Convective losses: These separate into internal convective losses from the absorber plate to the outer window pane, and external losses from the outer window pane to the ambient air. In the absence of wind the external convective losses are due to natural convection. Even low winds, however, dominate convection when they occur. While means could and should be introduced to reduce external convective losses, it would be most useful to reduce the internal losses and thus also to reduce the temperature of the outer window pane. While the available information on natural convection in vertical air gaps is not conclusive, the convection is very small and comparable in its effect to conduction for small values of the Grashof (or Rayleigh) number, and it becomes significantly greater for large values of these numbers. The nature of the convection also depends on the specific boundary conditions and the geometrical aspects of the enclosure.

Besides the maintenance of narrow air gaps to decrease convection, a cellular structure can be placed between the absorber and the cover plate. The major problems associated with the incorporation of cellular structures are that they reflect a part of the solar radiation, thus preventing it from reaching the absorber plate, and that they increase the thermal conductivity of the space between the absorber and the cover plate, as well as add to the cost of the collector. The evacuation of the space between the absorber and the window pane has been tested. Evacuation of the air eliminates the internal convection and conductive losses. This can normally be done only for tubular collectors.

(c) Radiative losses: The radiative losses from the absorber to the ambient can be reduced by a spectrally selective coating on the absorber plate. These coatings have a high absorptivity in the solar spectrum but have a substantially lower emissivity, usually of the order of one-tenth, in the infrared spectrum in which most absorber plates radiate. The selective absorbers thus decrease heat losses and increase collector efficiency.

C. Solar Air Conditioning

Three types of solar air-conditioning units are under development: absorption–desorption cycle units, Rankine cycle units, and dessicant units. All three types have been built and operated. The absorption cycle and the Rankine cycle have in common the evaporation and condensation of a refrigerant liquid, these processes occurring at two pressure levels within the unit. The two cycles differ in that the absorption cycle uses a heat-operated generator to produce the pressure differential, whereas the Rankine cycle uses a compressor; the absorption cycle substitutes physiochemical processes for the compressor unit. Both cycles require energy for operation: Heat is required for the absorption cycle, mechanical energy for the compression cycle. The desiccant systems utilize a desiccant to dry the air and evaporative cooling to rehumidify and cool the air. The desiccant is then regenerated using heat from the solar collector.

The air-conditioning systems installed in Georgia have all used the absorption cycle. In the absorption system the working fluid is a solution of lithium bromide and water. Lithium bromide is a salt that strongly absorbs water, especially at lower temperatures. Water-fired lithium bromide absorption chillers suitable for solar air-conditioning systems are currently available in the United States from York, Trane, and Carrier for applications requiring 350 kW (100 tons) or more of cooling. Units rated at 10 kW (3 tons) and 90 kW (25 tons) are available from Arkla-Servel. The cost of the 10-kW Arkla-Servel Solaire-36 was $3000 in 1977.

10.3 SOLAR HOMES AND BUILDINGS IN GEORGIA

Solar heating and cooling projects in Georgia include about 20 solar-heated homes, 1 solar-heated and 3 solar-heated and cooled commercial buildings, and a sizeable number of solar hot water installations.

A. Private Homes

Several enterprising Georgians have installed solar heating and hot water systems in their homes. One system uses 22 m^2 of solar collectors, with a 16,000-liter steel underground storage tank, to provide about 70% of the heating for a home. Another has a 24-m^2 collector on a home and two 1000-liter aircraft dump tanks for heat storage to provide about 70% of the heating demand. Another has an 80-m^2 collector and a concrete tank for heat storage. Still another uses 50 m^2 of collectors and a 6000-liter storage tank to provide 75% of the heating requirements.

A commercially available, prepackaged solar heating and hot water system was installed in a home in Alpharetta, Georgia. The system utilizes 40 m^2 of copper-laminated plywood to which rectangular copper tubes are clamped and bonded, coated with black paint, and covered with plastic glazing. The system utilizes two 4000-liter concrete storage tanks buried under the garage and provides two separate heating zones. The total design heat load for the house is 16 kW.

B. Government-Funded Solar Homes in Georgia

The solar home in Shenandoah, Georgia (Fig. 5), was one of the first government-funded (HUD) housing projects to be completed in the United States. The 140-m^2 home (plus full basement) utilizes a 27-m^2, double-glazed, copper, flat plate collector array integrated into the south-facing roof of the home. The collectors utilize copper-laminated plywood as the absorbing surface to which copper rectangular tubes are clamped and bonded, and the copper surface is coated with 3M Nextel Black Velvet paint. The sealed insulating glass covers are supported by copper battens attached to copper sealing strips. The perimeter of the collector array is flashed with 16-oz copper sheets. The 4000-liter steel thermal storage tank located in the basement is surrounded by over 25 cm of blown-in fiberglass insulation. The tank sits on a 5-cm slab of foam glass over 5 cm of styrofoam. The collectors drain naturally whenever the pump is off; so there are no problems with freezing in the collector even though no antifreeze is used in the system. The solar energy system utilizes a 2-ton (7-kW) water source heat pump to provide backup heating in either the solar-assist or conventional mode and summer air conditioning.

FIG. 5 Shenandoah solar house.

A conventional home of standard construction was also adapted to solar energy by mounting modular collectors on the roof with the thermal storage tanks located in the basement. The solar collector array consists of twenty-five 1- × 2-m modular copper collectors with seven panels installed on the garage roof and eighteen on the roof of the main house structure. The collector circulating pump is operated by a differential thermostat controller. Two 4000-liter steel thermal storage tanks are located in the basement and insulated with fiberglass. Auxiliary energy for space heating, hot water, and air conditioning is provided by natural gas. A boiler provides hot water for space heating and domestic hot water heating, and an Arkala chiller provides chilled water for air conditioning. Domestic water is heated in a separate heat exchanger.

The home utilizes a number of energy-conserving features including 15-cm stud walls with fiberglass insulation plus 2.5 cm of styrofoam. Clock thermostats provide nightly setback, and a hot water coil in the fireplace chimney heats circulating water to provide additional heat to the solar storage tank. Domestic hot water is heated by circulating water from the thermal storage tanks through a copper coil in the domestic hot water tank. The solar collector array is 5 m high and 8 m wide and integrated into the roof of the home. A unique feature of this particular home is the total reliance of the homeowner on solar renewable fuels for supplying 100% of the heating and domestic hot water needs of the home. The only backup is a wood-burning fireplace.

A solar home supported by a HUD grant has been completed in Dacula, Georgia (Fig. 6), using a commercially available packaged solar hy-

FIG. 6 Lindsey solar house in Dacula, Georgia.

dronic space-heating and hot water system. The system utilizes 23 m^2 of double-glazed, copper, integrated roof solar collectors to provide space heating and domestic hot water for the 100-m^2 home. The collectors are of new construction panels with rectangular copper tubing clamped and bonded to copper-laminated plywood covered by double-glazed, sealed, insulated glass sandwiches supported on copper battens. The collectors are flashed on all sides with 16-oz copper sheets so that both the glazing and absorbing surfaces individually form leak-tight roofs. The absorbing surface is 3M Nextel Black Velvet paint. The 2000-liter thermal storage tank is constructed of steel with all seams double welded, inside and out. It is 2 m in diameter and 1 m high buried in the ground at the end of the house. The tank is insulated with gravel and 8 cm of styrofoam, with the top covered by two layers of polyethylene plastic moisture barrier with 25 cm of dirt on top. The dirt forms a slight mound to ensure that rainwater flows away from the site of the tank.

Domestic hot water for the HUD houses is heated by water circulating from the thermal storage tank through a coil in the domestic hot water tank. Four pumps are used in the system; two circulating pumps (one below ground) provide circulation through the collectors, a $\frac{1}{12}$-hp pump circulates water through the heating coil at the inlet to the furnace, and a $\frac{1}{20}$-hp pump circulates hot water through the coil in the domestic hot water tank.

HUD also supported the installation of independent solar systems in a three-unit apartment building in Swainsboro, Georgia (Fig. 7). Water source heat pumps provide auxillary heating in either the solar-assist or conventional mode. In addition, domestic hot water is heated by water

FIG. 7 Solar apartments.

circulating from the thermal storage tank through a copper coil in the domestic hot water tank. Table 2 lists the collector array size, the energy for heating and hot water provided by the solar systems, the percentage of the total load provided by the solar energy, the cost of the solar system, and the net present value of the savings realized by the homeowner from the solar installation.

The solar collector arrays form the south-facing roof of the apartment building and are constructed of copper-laminated plywood with rectangular copper tubes clamped and bonded to the surface, which is coated with black paint. Both the copper-laminated plywood and the tempered glass glazing form leak-tight roofs for the structure. The glazing is supported by all-copper solar battens mounted on copper receiving strips and flashed

TABLE 2

Features of Independent Solar Systems for Three-Unit Apartment Building in Swainsboro, Georgia

	West unit	Center unit	East unit
% Solar	75.3	73.5	75.1
Heating (MBtu)	30.5	24.8	27.0
Hot water (MBtu)	20.7	20.7	20.7
Collector area (m^2)	31	23	27
Solar system cost (thousand $)	11.2	9.2	10.1
Cost savings (thousand $)	10	8	9

around the outside with 16-oz copper sheets. The 4000-liter thermal storage tanks are 2 m in diameter and 1.7 m high, constructed of steel with all seams double welded, with a 40-cm sealable manhole on the flat top of the tank. The inside of the tanks are sandblasted and coated with coal tar epoxy polyamide, which is rated for a 250°F continuous wet temperature. The collector arrays drain naturally whenever the pump is off; so there is no danger of water freezing in the collectors.

All pumps, valves, and electronic controls for each system are contained within a single solar control unit, 1 × 1 × 1 m. The three solar control units for each system are located in a common solar equipment room at the gable end of the building to provide accessibility while conserving costly interior space. Each of the three systems was installed in less than 1 week.

Three solar homes have been constructed under a HUD grant at the same site north of Atlanta using identical solar energy systems (Fig. 8). The installation of identical solar systems in three homes of different architectural design is providing an opportunity to evaluate the effect of architectural considerations on solar system performance. Each system incorporates 23 m^2 of solar collectors, a 4000-liter coal tar epoxy-coated steel thermal storage tank, and a 320-liter domestic hot water tank. The collectors are single glazed at an angle of 45°. These homes incorporate numerous energy conservation features including flow control devices to reduce hot water consumption, fluorescent lights, extra insulation, and a clock thermostat. The design heating load is 5 kW, about half that of a typical house of the same size.

The solar collectors use copper absorber plates with rectangular

FIG. 8 Solar home north of Atlanta, Georgia.

copper tubes bonded to the surface, painted black, and covered with a single sheet of double-weight tempered glass, which is supported by copper solar battens mounted on sealed copper receiving strips. The 4000-liter storage tanks are constructed of steel with all seams double welded and are 2 m in diameter and 1.1 m high with a 40-cm sealable manhole on the flat top of the tank. One-inch FPT fittings are provided in the sides of the tank near the top and bottom for solar connections. The interiors of the tanks are coated with coal tar epoxy polyamide. The epoxy is applied after an SP 10 white metal sandblast with airless spray to provide a 9-mil coat; a second coat is applied within 12 h. The tanks are tested against leakage before installation by pressurizing the air.

Each solar system utilizes a single hydronic control unit that contains all the pumps, valves, and electronic control devices required to regulate that solar system. Since the complexities of the solar system are contained within this single unit, on-site installation took less than 1 week for each system. The control unit is contained within a rigid metal steel cabinet 1 × 1 × 1 m. All piping for the system, both within and outside the control unit, is copper and is insulated with 2-cm rubberized foam insulation.

C. Solar Heating and Cooling System Demonstration for the School of Mechanical Engineering at Georgia Tech

An array of 48 flat plate solar collectors are in use for a demonstration project providing hot water, space heating, and chilled water cooling for a portion of a classroom building of the School of Mechanical Engineering at Georgia Tech in Atlanta, Georgia (Fig. 9).

The solar collectors comprise a south-facing array of two rows containing 24 modular collectors per row for a total aperture of 78.97 m^2. The collectors are tilted 45° from the horizontal. Space is reserved between the rows for future installation of reflector augmentation for the north row. The collectors, produced by four different manufacturers, are both single and double glazed and utilize a variety of absorber coatings to permit an evaluation of the instantaneous efficiency of each collector type. Twelve aluminum collectors in the array are operated in an isolated loop containing water/ethylene glycol solution separated from the principal water loop by a shell-and-tube heat exchanger. The aluminum collectors are further protected by magnesium anodes.

Heating and cooling is provided through an integral fan coil unit mounted in the conditioned space. During the heating season the collectors are operated at around 65°C. Hot water storage is provided by a well-insulated 5600-liter tank mounted at ground level.

FIG. 9 Flat Plate Collectors in School of Mechanical Engineering at Georgia Institute of Technology.

An Arkla Solaire-36, 10.5-kW (3-ton) LiBr absorption cycle unit produces chilled water for cooling. During the cooling season, the collectors operate at about 90°C. The hot water tank is used for thermal storage, and another 5600-liter tank is used for cold water storage. Heat is rejected through a Marley model 4615 cooling tower of 28-kW (8-ton) nominal capacity. Control of the system is effected through four 4-cm solenoid valves, seven 2.54-cm motorized valves, and five circulating pumps sized from 100 W to 1.12 kW. The system can be operated in several modes for storage of hot or chilled water, heating or generation of chilled water using stored heat, or cooling using stored chilled water.

Distribution is provided through 2.54-cm type M copper pipe. Approximately 450 m of exposed pipe is involved in this system because design restraints required locating the storage tanks at grade and the array on a roof remote from the conditioned space. Rubatex closed cell insulation is used on the piping, and polymer foam insulation is used on the storage tanks.

D. Heating and Cooling of Large Buildings

Georgia Tech was a subcontractor to Westinghouse for the George Towns Elementary School solar retrofit project. This school, located in Atlanta, was chosen for the first large solar heating and cooling project because of the balanced heating and cooling loads in this area. The school is a 1-story, 3200-m² building serving about 500 students. It utilized 576 standard size aluminum PPG collectors in a 1000-m² array with reflector

FIG. 10 George Towns Elementary School solar project.

augmentation. Three 60,000-liter steel tanks provide both hot and chilled water storage. This system has been operating for more than a year (see Fig. 10).

Georgia Tech was the prime contractor to ERDA for the development of the Shenandoah Solar Community Center, currently the world's largest solar-heated and cooled building. This 6000-m^2 multipurpose community center utilizes 1100 m^2 of black-chrome-coated, double-glazed, copper solar collectors augmented by 2100 m^2 of Alzak reflectors. Thermal storage is provided by one 60,000-liter hot water storage tank and two 120,000-liter chilled water storage tanks. One unique feature of this build-

FIG. 11 Shenandoah Solar Community Center.

ing is the large 2.5- × 6.5-m collector modules that were used; only 63 were needed for the project and all were installed in only 1 day by helicopter. The solar energy system is providing almost all of the cooling load, 95% of the heating load, and most of the domestic hot water. In addition, a large swimming pool adjacent to the building is heated from the collectors whenever the heat available from solar energy exceeds the demands of the building. This system has been operating since April 1977 (see Fig. 11).

10.4 SOME SOLAR SYSTEMS USED WITH GEORGIA PROJECTS

Three types of solar energy systems that have been used successfully with Georgia solar energy projects are described. All these projects are hydronic; that is, water is used as the heat transfer fluid. The solar collectors have copper absorber plates with one or two covers of tempered glass. The storage tanks are steel, with a coal tar epoxy coating. The systems are designed so that the collectors drain by gravity whenever the pump is not running; therefore water cannot freeze in the collector.

A. Solar Heating and Hot Water Systems

Figure 12 illustrates a basic solar heating and hot water system, with solar energy providing heat and domestic hot water, and some type of conventional (or backup) system for use when the sun's energy is insufficient to meet the heating demand. Two differential thermostats control the solar collection and hot water heating processes, and a room thermostat activates the pump that circulates hot water to the heating coil in the furnace. This system was installed late in 1976 in the solar home in Dacula, Georgia, and later in the three homes at the same site north of Atlanta, as described previously.

Whenever the solar radiation intensity is sufficient to heat the collectors hotter than the water in the bottom of the thermal storage tank, the differential thermostat turns on the main circulating pump, water is circulated from the bottom of the thermal storage tank through the solar collectors, and solar-heated water returns at the top of the tank. If at the same time there is a demand for heating, the room thermostat activates the heating pump, which circulates water from the upper part of the thermal storage tank through the heating coil in the return duct to the furnace. In this manner, solar-heated water is used to heat air directly without any fuel being used by the furnace.

If the solar radiation intensity is insufficient to heat water in the collectors hotter than the water in the bottom of the thermal storage tank, the

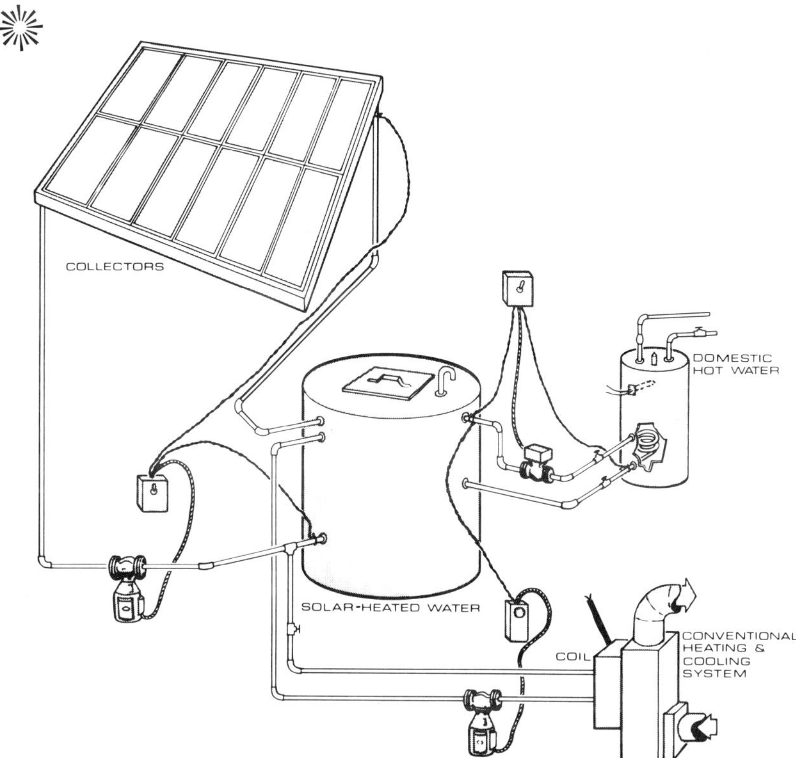

FIG. 12 Solar heating and hot water system.

differential thermostat turns off the main circulating pump. If there is at the same time a heating demand and the temperature of the water in the thermal storage tank is sufficient to meet this demand, then the space heating pump circulates water from the upper part of the thermal storage tank through the heating coil and back into the bottom of the thermal storage tank.

When the solar radiation intensity is great enough to heat the water in the collectors hotter than the water in the bottom of the thermal storage tank, the main circulating pump is turned on to circulate water from the bottom of the thermal storage tank, through the copper tubes in the collector absorber plates, and back into the upper part of the thermal storage tank. If there is no space heating demand, the water in the thermal storage tank is continuously recirculated and heated to higher and higher temperatures by solar energy.

When the temperature in the thermal storage tank is insufficient for direct solar heating, the conventional system operates in addition, so that the solar system is then preheating air entering the furnace, thereby reducing fuel requirements for the furnace. When the water in the thermal storage tank is below 80°F and heating is required, the furnace operates in the conventional manner without solar preheating.

The control system activates the domestic hot water circulating pump whenever the water in the thermal storage tank is hotter than the water in the bottom of the domestic hot water tank, except that the pump is deactivated if the water at the bottom of the domestic hot water tank reaches 140°F, in order to prevent scalding. Heating the domestic hot water from the thermal storage tank enables the solar system to operate with maximum efficiency year round, since the collectors are always operated at the lowest temperatures at which useful heat can be collected for space and hot water heating. The collectors drain by gravity whenever the pump is off, thereby eliminating any possibility of freezing in the collectors. This has several advantages:

(1) Assured freeze protection.

(2) Reduction of collector thermal inertia, which can considerably enhance collector performance. When the differential thermostat turns off the pump, the hot water drains from the collectors into the insulated thermal storage tank instead of remaining behind in the collectors to cool off.

(3) Elimination of the possibility of boiling and pressure buildup in the collectors during a power failure or pump malfunction.

B. Solar Heating System with Heat Pump Backup

This solar energy system (Fig. 13) provides solar space heating and solar domestic hot water, with electrical auxiliary heating, hot water, and air conditioning. The system is designed so that an absolute minimum amount of electrical energy is required for backup. This is achieved by utilizing a solar-assisted electrical heat pump to provide efficient auxiliary heating when the amount of solar energy available is insufficient to provide direct solar heating. This system was installed in the Shenandoah solar home in June 1976 and later in the three-unit apartment building in Swainsboro. Each apartment unit had a completely separate system.

This system operates exactly as before except that when the water in the thermal storage tank is between 50 and 50°F and heating is required, the heat pump operates in the solar-assist mode to extract heat from thermal storage for heating the home. Water is circulated between the heat pump and a coil in the thermal storage tank. Also, the heat pump cools the home in the summer with heat rejection by the outside fan coil unit in the conventional manner.

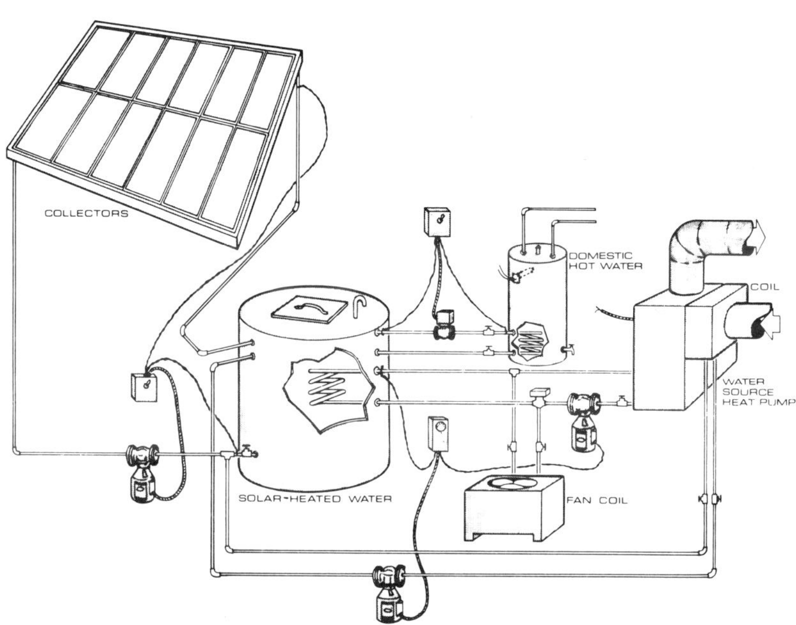

FIG. 13 Solar heating and hot water system with solar-assisted heat pump for backup heating and summer cooling.

C. Solar Air-Conditioning Systems

This system (Fig. 14) provides space heating, air conditioning, and hot water from both solar and conventional sources. The system schematic illustrates a single domestic hot water tank and a single fan coil unit for conditioned air distribution; actually any number of these units can be used with the solar system, extracting heating and cooling from common sources. The system employs a collector array, a thermal storage tank, a boiler, a water-fired absorption chiller, a chilled water tank, any number of domestic hot water tanks and fan coil units for conditioned air distribution, and miscellaneous pumps, valves, thermostats, and other components. This system was installed in the School of Mechanical Engineering at Georgia Tech.

In the solar heating mode, hot water from the collectors is pumped directly through heating coils in the fan coil units that provide conditioned air distribution. The cooler water from the fan coil units returns to the collector array where it is heated. The pumps and valves associated with the fan coil units are activated by their respective room thermostats. The main circulating pump for the collectors is separately controlled by a dif-

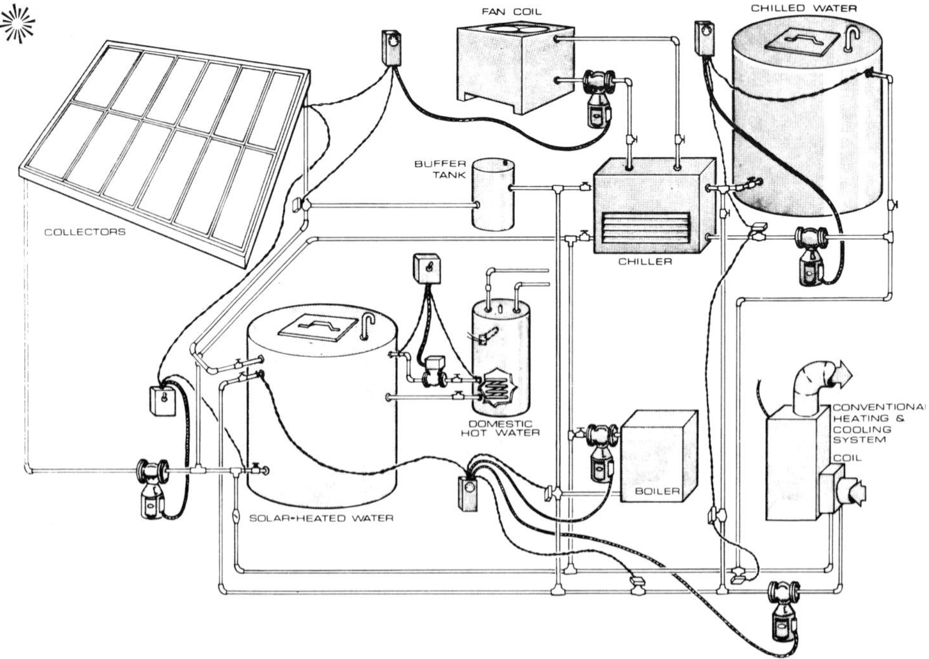

FIG. 14 Solar heating and air-conditioning system.

ferential thermostat. Whenever the solar radiation intensity is insufficient to heat the water flowing through the collectors, the differential thermostat turns off the main circulating pump and the collectors drain by gravity. With the main circulating pump off and the collectors drained, hot water for space heating can be pumped from the thermal storage tank.

The fan coil units and their respective valves and pumps are activated by the respective room thermostats. If the main circulating pump for the solar collectors is off, then solar-heated water flows from the upper part of the thermal storage tank through the coils in the fan coil units, and back to the bottom of the thermal storage tank. In this way space heating is provided from solar energy storage in the evenings and on heavily overcast days.

Heat is stored when there is no demand, and the solar radiation intensity on the collectors is great enough to heat water from the thermal storage tank. The differential thermostat activates the main circulating pump, and water is pumped through the collectors where it is heated. This solar-heated water returns to the upper part of the thermal storage tank while somewhat cooler water continues to be pumped from the bottom of the tank to be heated in the collectors. The differential thermostat has a sensor located on the collectors and another sensor at the bottom of the thermal storage tank; it is the measurement of this temperature that operates the differential thermostat, which has a cut-on/cut-off temperature differential to prevent unnecessary cycling of the main circulating pump. Thus, solar energy is collected whenever possible, regardless of whether or not there is a heating demand, and solar energy can be supplied to the structure regardless of whether or not the sun is shining.

Conventional space heating is provided by hot water from the boiler. The boiler is turned on whenever there is a demand for hot water either for space heating or for operating the chiller and the solar energy from the collectors and storage is insufficient to meet the demand. When there is a heating demand, the boiler diverting valve directs the boiler output to the fan coil units.

When there is a cooling demand and the temperature of water from the solar collectors is too low to operate the chiller, the boiler is turned on, and its diverting valve directs the hot water from the boiler to the generator section of the chiller. The condensing water pump and the chilled water pump also operate so that condensing water is circulated through the condenser portion of the chiller, while cool water is circulated through the evaporator section where it is chilled. Space cooling is derived by circulating this chilled water through the fan coil units.

Solar air conditioning is achieved by operating the chiller directly from solar-heated hot water. When the solar intensity is high enough to operate the chiller, a diverting valve directs the flow of solar-heated water from the collectors directly through the generator section of the chiller and back to the collector array. The buffer tank prevents excessive chiller cycling. Efficient chiller operation is achieved since the solar-heated water is utilized when it is at its hottest, not stored for later use when its temperature is lower. The chiller is operated in the cooling season anytime the solar intensity is great enough to provide efficient operation, and whenever the thermostat calls for cooling, chilled water is circulated through the fan coil unit involved to provide cooled air. If the chiller can produce more chilled water than is required to meet the cooling demand, the excess chilled water flows into the chilled water storage tank to be used later for space cooling.

Chilled water is stored whenever the solar intensity is sufficient to operate the chiller and the system is in the summer cooling mode. Solar-heated water from the collectors is circulated directly through the generator portion of the chiller, and the chiller operates to cool water circulating through the evaporator section. If the demand for chilled water for space cooling is less than the amount of chilled water being produced by the chiller, the excess chilled water flows into the chilled water storage tank. Similarly, if the demand for chilled water at any time exceeds the chilled water available from the chiller, additional chilled water flows from the bottom of the chilled water storage tank to the fan coil units. Thus the chilled water storage tank provides additional cooling capacity when it is needed and storage when the supply from the chiller exceeds the demand for cooling. The chiller is operated from the boiler only when the demand for cooling exceeds the cooling available from solar operation of the chiller and chilled water in the chilled water storage tank.

The building is cooled from storage whenever space cooling is required and the chiller is not operating. Chilled water is pumped from the chilled water storage tank into the fan coil unit and returns to the upper portion of the chilled water storage tank. If the temperature of water at the bottom of the chilled water storage tank is too high to provide space cooling, then the boiler provides the necessary hot water to operate the chiller for space cooling.

11

Some Solar-Heated Buildings in Canada

T. A. LAWAND AND B. SAULNIER

BRACE RESEARCH INSTITUTE
MACDONALD COLLEGE
MCGILL UNIVERSITY
MONTREAL, QUEBEC, CANADA

11.1 INTRODUCTION

The heating of houses in Canada using solar energy is by no means a novel idea. Solar energy, of course, has always contributed to the heating of the buildings through direct solar radiation transmitted through windows on the south- as well as on the east- and west-facing sides of the building. The first efforts in Canada at actually utilizing the sun for the purposes of increasing the available amount of heat were the studies undertaken in the 1940s by Frank Hooper, a young engineer in Toronto (Hooper, 1955). He was particularly concerned with developing methods of reducing the excessive heating bills faced by most Canadian families owing to the large number of heating degree days experienced in this country. Although it took more than 30 years for him to realize these projects in actual field application, it is gratifying to know that Canadians, to a certain extent, pioneered efforts in capturing renewable energy for house heating, which consitutes a heavy demand on the national energy budget (Bolton, 1975).

It has been estimated that approximately 25–30% of the national energy consumption in Canada is expended on house heating. While admittedly energy conservation can, to some extent, reduce this demand, given the extremely cold climate and the general inadequacy and lack of adaptation of Canadian houses to environment, Canadians will be faced with a

ISBN 0-12-620860-3

large expenditure of energy for the provision of space heating for quite some time (Anonymous, 1976).

This chapter examines briefly a number of houses that have been built to date using solar energy as an auxiliary heat source. As a prelude, a brief section has been prepared on solar radiation levels in Canada, providing as well some estimate of the heating requirements.

The populated regions of Canada stretch from the forty-third parallel to even higher than the sixtieth parallel. North of the sixtieth parallel, however, the population is very small indeed. In fact, it can be said that the majority of Canadians live in a narrow band south of the fiftieth parallel. A summary of the average solar radiation intensities on the horizontal in this region is given in Table 1. It must be recalled, however, that almost all solar collector systems in Canada are placed on tilted surfaces, either on roofs or vertical walls. In this respect, it is possible to benefit from higher levels of solar radiation. Furthermore, there are estimates that have indicated that unobstructed south-facing collectors, located in front of a snow-covered field or lawn, will receive up to one-third greater solar radiation owing to the reflected radiation from the ground (Armet and Nardini, 1976).

The heating degree days are given for a number of specific locations. With the increased costs of energy from fossil fuel, nuclear power, and even hydrogenerated electricity, it becomes increasingly evident that any effort to reduce the heating demands, respectively, through

TABLE 1

Estimated Solar Radiation Intensities at Selected Canadian Cities, Bright Sunshine Hours and Heating Degree-Days

City	Average annual solar radiation intensity measured on the horizontal[a] W h/m² day	Bright sunshine[b] (h)	Heating factor[b] (degree-days below 18.4°C)
Dartmouth, Nova Scotia	3470	1876[c]	4200[c]
Montreal, Quebec	3520	1811	4520
Toronto, Ontario	3610	2047	3890
Winnipeg, Manitoba	3850	2136	5910
Churchill, Manitoba	3230	1646	9380
Lethbridge, Alberta	4030	2200[d]	5280[d]
Vancouver, British Columbia	3070	1784	2990

[a] Anonymous (1974a).
[b] Anonymous (1960).
[c] Data for Halifax.
[d] Data for Calgary, Alberta.

(1) energy conservation methods,
(2) environmental design and adaptation of the structure to its milieu, and
(3) the use of solar collection and storage systems

requires considerably increased attention.

These are matters which have not interested Canadians in recent times only. Since 1967 the National Research Council has published data indicating the solar heat gain factors for clear days on various surfaces, and the incident solar radiation on surfaces of buildings with different orientations in Canada. The Federal Department of the Environment has expended a considerable effort in ensuring that solar radiation is properly and extensively measured throughout the country. Hourly solar radiation data are gathered at approximately 50 stations, while total and diffuse solar radiation are measured at 4 stations. In addition, net radiation is measured at a number of locations. These data are published in the series of Monthly Radiation Summaries (Anonymous, 1971). The prairie regions of Western Canada and Southern Ontario receive the most solar radiation, although the differences across the country are not very large (Hay, 1975a, 1976a,b).

Table 2 indicates the estimated optimum tilt for solar collectors in Canada. It can be seen that during the heating season, the angle of tilt of solar collectors is rather significant (Stephenson, 1967).

TABLE 2

Optimum Tilt for South-Facing Solar Collectors for Canada, as a Function of Time and Latitude[a]

	Optimum tilt (deg) of solar collector for:			
Date	Latitude 45° N	Latitude 50° N	Latitude 55° N	Latitude 60° N
21 January	65	70	75	80
21 February	56	61	66	71
21 March	45	50	55	60
21 April	33	38	43	48
21 May	25	30	35	40
21 June	22	27	32	37
21 July	24	29	34	39
21 August	33	38	43	48
21 September	45	50	55	60
21 October	56	61	66	71
21 November	65	70	75	80
21 December	68	73	78	83

[a] Based on optimum tilt = latitude − solar declination.

TABLE 3

Total and Diffuse Solar Radiation on Horizontal Measured in Quebec Region ($W\ h/m^2\ day$)[a,b]

Location	Jan	Feb	Mar	Apr	May	June	Jul	Aug	Sept	Oct	Nov	Dec	Avg.
Fort Chimo—total	677	1622	3349	4761	5135	5602	4038	4014	2007	1190	525	303	2769
Goose Bay—total	980	2089	3186	4458	4948	5111	4336	4026	2906	1657	887	759	2995
Goose Bay—diffuse	572	1015	1751	2626	2754	2684	2509	2147	1587	992	584	467	1636
Goose Bay—diffuse/total	60	55	56	59	57	53	51	53	55	60	67	62	57 (3%)
Inoucjouac—total	700	1727	3664	5497	5893	5788	5041	4680	2509	1237	467	443	3137
Montreal—total	1670	2439	3478	4470	5287	5602	5882	4750	3781	2322	1225	1050	3480
Montreal—diffuse	817	1225	1680	1821	2299	2579	2567	2241	1645	1109	735	630	1612
Montreal—diffuse/total	51	51	48	42	44	47	44	47	44	48	59	60	48 (8%)
Moosonee—total	1216	2381	3606	4680	5322	5570	5088	4341	3011	1599	957	899	3218
Nitchequon—total	1167	2264	3804	5053	5800	5345	4773	4306	2579	1529	910	689	3185
Normandin—total	1412	2567	3991	5193	5205	5858	5450	4610	3291	1695	1015	1074	3503
Ottawa—total	1610	2521	3548	4645	5240	5613	5718	4925	3489	2264	1144	1155	3489
Sept-Iles—total	1330	2451	3501	4481	4808	M[c]	M	M	3524	2369	1062	770	3085

[a] The location of the solar radiation stations for the Quebec region is shown in Fig. 1.
[b] From Villeneuve (1967).
[c] M = Missing.

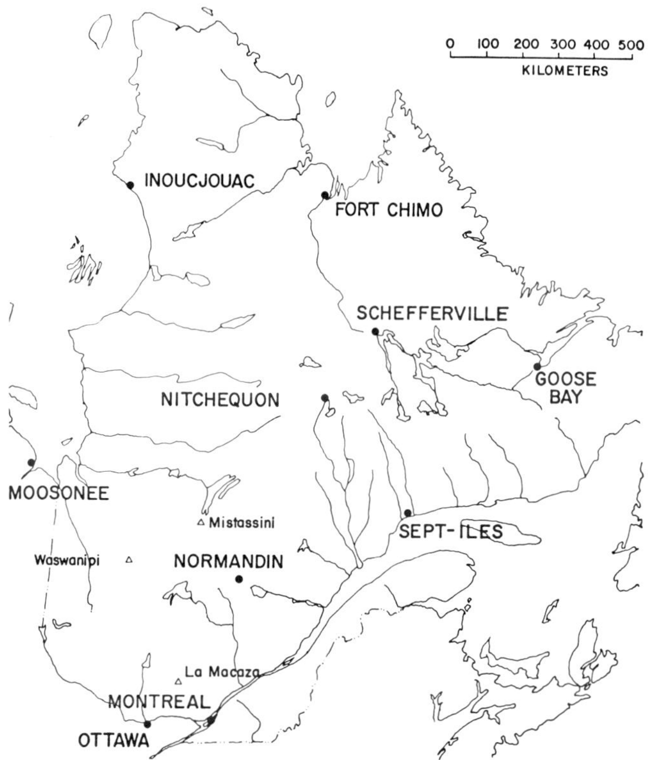

FIG. 1 Solar radiation stations, Province of Quebec. ● = stations; △ = location of houses.

Table 3 indicates the solar radiation intensities for Québec and Labrador. The closer one gets to the ocean, the less the solar radiation intensity. Also, the percentage of diffuse solar radiation increases because of cloudier conditions along the coasts. The annual average solar radiation on a horizontal surface is given in Fig. 2.

Description of Some Solar Houses

A number of solar-heated houses are described in subsequent sections of this chapter.* It is not possible at this time to give detailed performance data because of the relatively recent construction of most of the houses. At this time all of the solar houses described herein are performing satisfactorily. It will be a number of years before their performance can be

* The solar heating systems are designed to provide 40–50% of the heating load of the house (Russell, 1974).

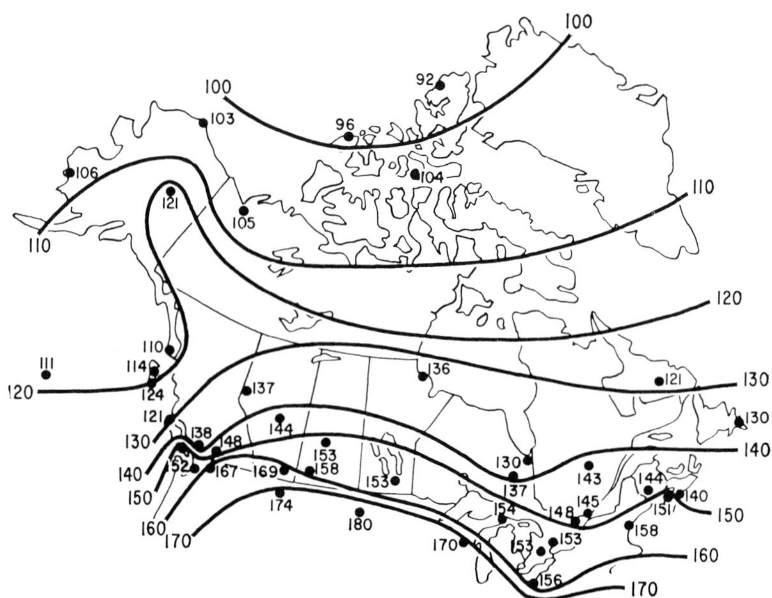

FIG. 2 Annual average (1966–1970) of 24-h solar radiation (W/m^2) on a horizontal surface at various reporting stations in Canada. [From Bolton (1975); data extracted from tables published by the Main Geophysical Observatory, Leningrad, USSR (Anonymous, 1971).

specified owing to the variety of the systems and the variations in the intensity of the winter seasons; thus in a few years, following careful measurement, the exact contribution of the solar heating systems to the total heating demand will be better known (Quirouette, 1975; Sasaki, 1975, 1976; Argue and McCallum, 1976; Allen, 1975).

11.2 DIRECT ENERGY ASSOCIATES HOUSES

The Direct Energy Associates, a group located at Ayer's Cliff in the Eastern Townships, Quebec, has to date designed and built a number of small country houses supplemented with solar air-heating systems. The group's basic premise is "that solar heating can be made available to most people buying a house, at a cost that is low enough to be attractive and economical." Their solar heating systems all incorporate solar air heating combined with rock storage.

All the houses built to date are located in the Eastern Townships, east of Montreal, latitude 45° N. The houses are well insulated and their living area ranges between 40 and 80 m^2. The size of the air-cooled collectors varies from 10 to 40 m^2 in area, according to the house design, and the collectors are either mounted vertically or at 60° inclination to the horizontal. All collectors face due south (Swartman, 1976; Lawand, 1975). (See Figs. 3 and 4.)

A. Solar Collector System

In all the houses the collectors consist of a corregated or flat steel sheet painted with a heat-resistant black paint and covered with tempered double-pane glass. The air passes behind the steel sheet of the collector and, in most houses, is circulated from the collectors to the storage area by means of a fan through simple duct work. For one house with vertical walls, the system is entirely passive; this particular case having its rock storage directly behind the collector absorber plate.

B. Storage System Description

The storage medium consists of crushed rock (4-cm diameter) piled in an insulated container which in three of the houses extends vertically from the basement foundations into the first floor. The storage area holds from 9 to 38 tons of crushed rock, according to the design of each particular house. The stored heat is circulated into most houses using an additional fan, which is separated from the collector fan by dampers. No provision is made for solar-heated domestic water.

FIG. 3 *Solar I*; Direct Energy Associates, Ayer's Cliff, Quebec. Area of house, 37 m^2; one-story house with a small sleeping loft. Collector area is 18 m^2 and is sloped 60° to the horizontal. Storage consists of 9 tons of crushed rock in an insulated area in the basement. The house heating fan and collector fan are kept separate by means of dampers. Auxiliary heating is provided by a 2.2-kW baseboard heater and a high-efficiency wood stove.

FIG. 4 Solar III; Direct Energy Associates, Ayer's Cliff, Quebec. This house, shaped as a cube, with a 7.3-m side, has floors on five different levels. Vertically mounted collector has 35-m^2 area. Storage consists of 36 tons of crushed rock in an insulated container, which extends vertically from the foundations to 1.2 m above ground level. Auxiliary heating is provided by electric warm air furnace. Because of the height of the house, a thermostatically controlled blower moves warm air from the top of the house to the ground floor.

C. Backup System

According to each particular design, the backup heating systems are either baseboard electric heating units, standard electric warm air furnaces, electric elements in ducts, or high-efficiency wood stoves. The electric heating backup systems are energized only when the rock storage temperature is too low.

In one particular case in which a wood stove is used as the backup heating system and the storage compartment protrudes into the house, the chimney pipe passes through the storage compartment. It also has fins welded to it to increase heat transfer to rocks when the stove is operating.

11.3 HOFFMAN HOUSE

Location British Columbia, 49° N (Fig. 5).

Address 5511 128th Street, Surrey (near Vancouver).

Site description Residential suburban area.

Type of building Single family, one story and basement built in 1968.

FIG. 5 Hoffman house. The oldest currently operating solar-heated house in Canada, located in Surrey, British Columbia.

Size of building 130 m^2 plus a 120-m^2 basement that is partially heated.

A. Solar Collector System

Operating mode Active (electrical circulating centrifugal pump).

Type Retrofitted, closed cycle, water-cooled collector built in 1971. 6-mm-i.d. copper pipes soldered 15 cm apart on thin coppr sheets (0.125 mm) and coated with a nonselective flat black paint. Double glass cover separated by a 2-cm sealed air space. The collector is backed with 15 cm of fiberglass insulation. The collector is drained when not in use. The stagnation temperature is 125°C.

Size The total collecting area is 43 m^2; the effective collecting area, 38.5 m^2.

Orientation South facing with a 58° slope from the horizontal.

Location The collector is placed on the A-frame part of the house roof.

B. Storage System Description for Heating and Domestic Hot Water Purposes

Three thousand liters of water are stored in two tanks (1300 and 1700 liters) that are located in an insulated room in the basement of the house.

During the heat collection period, water is circulated using a 0.2-kW electric pump between the tanks and the solar collectors. The tank walls serve also as water–air heat exchangers, and the access door to the room where the water tanks are located, is opened to allow this warmed air to heat the house when needed (Hoffmann, 1975).

Domestic hot water and swimming pool heating are obtained by means of two heat exchangers located in the tanks. The water tank heat exchanger coil for the domestic hot water operates as a preheater. After passing through the heat exchanger, the domestic hot water enters a standard electric hot water tank where its temperature is raised to the required level when solar heating is insufficient.

C. Controls

Heat collection system An electronic differential thermostat allows the electric pump to work whenever the collector temperature is 15°C warmer than storage, and water circulation is cut when this difference drops to 1°C; the collector is then drained.

Heat distribution from storage Manually operated (access door to the storage room).

Backup system Electric baseboard heaters fully rated for maximum winter load.

The system was designed, engineered, and built by the owner, E. Hoffman.*

Cost $2000 (not including labor).

11.4 LORRIMAN HOUSE

Location Ontario, 44° N (Fig. 6).

Address 2940 Quetta Mews, Meadowsville, Ontario (west of Toronto).

Site description Residential suburban area; no physical obstructions shading the solar collector.

Type of building Single-family, two-story house plus basement.

Size of building 125 m^2 of living area plus approximately 25 m^2 of

* After 7 years of operation, Mr. Hoffman estimates that approximately 50% of the house heating requirements are supplied by this system.

(a)

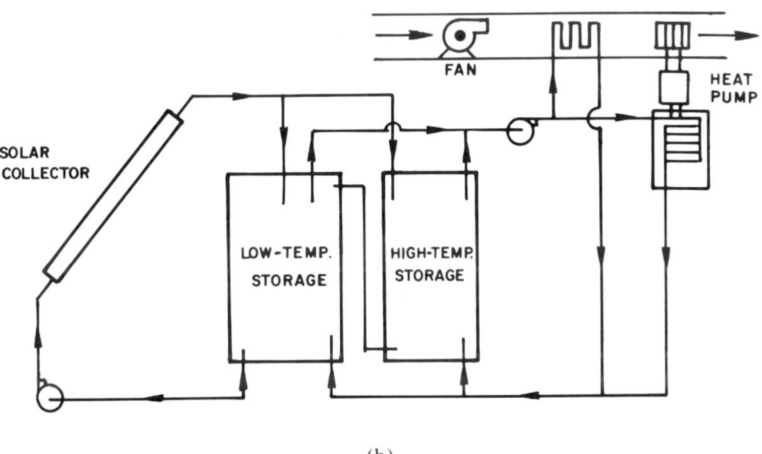

(b)

FIG. 6 Lorriman house; Mississauga, Ontario. (a) Solar collectors and liberal use of south-facing windows for passive solar collection. (b) Solar system schematic diagram.

double-glazed window area, 70% of which is facing south. The house is insulated above normal standards; built in 1976.

A. Solar Collector System

Operating mode Active (electrical circulating pump); passive (large south-facing window area on first floor).

Type Water-cooled collector: 33 Sunworks panels (each 2.2 × 0.9 m). Copper pipes on copper sheet treated with a selective black coating.

The collector has only single glazing, and the system is drained when not in use so that freezing does not occur in the pipes.

Size The total solar collector area is 64 m^2.

Orientation South facing with a 60° tilt to the horizontal.

Location On the sloping roof of the house facing the backyard.

B. Storage System Description

There are two tanks, each with an 11,300-liters capacity. They are made of steel-reinforced concrete and insulated with 7.5 cm of rigid insulation. One tank is for high-temperature storage for house heating; and the other, lower-temperature storage, is used for feeding water to the solar collectors. The collector output water is pumped to either of the tanks or to both according to the output temperature. For house heating, the water is circulated between the tanks to a heat exchanger located in the duct work of the house. The heat is extracted from the heat exchanger and distributed to the rooms by means of a blower (forced-air heat distribution). When the temperature in the tanks is lower than 50°C, a water-to-air heat pump extracts energy from the tanks to deliver it to the duct work where the condenser coil of the heat pump is located (Lorriman, 1975).

Domestic hot water is provided by solar heating at times when the system has excess capacity.

C. Controls

Heat collection system Electronic differential thermostat starts the circulating pump.

Heat extraction and distribution from storage Operated by thermostatic switch with heat extraction priority given to the low-temperature storage (up to 50°C) and then to the high-temperature tank, to ensure the lowest possible inlet temperatures to the solar collectors at all times.

Backup system Electric resistance heaters in ducts leading to rooms.

Cooling in summer The heat pump operation is reversed and the heated water delivered to tanks or to the city water pipes.

The system was designed by Lorriman and Ferguson.

Cost Estimated at $130,000 ($60,000 provided by the federal government).

11.5 LA MACAZA HOUSE

Location Quebec, 46.5° N (Fig. 7).

Address Manitou Community College, La Macaza, Quebec (170 km northwest of Montreal).

Site description College campus, located in isolated area, free from major obstructions.

Type of building Single-family, one-story house, no attic, built in 1976. Squared log construction well insulated. The foundations of the house are externally insulated and bermed with earth on all but the south side, completed in 1976.

Size of building 94 m^2.

A. Solar Collector System

Operating mode Generally active; however, the system operates in the passive mode in cases when there is an electricity failure.

Type Air-cooled collector. The collector is made of a screen of expanded metal sheet located in front of a flat metal absorbing sheet, both painted with a nonselective flat black coating. The purpose of the expanded metal sheet is to provide increased heat transfer area. The collector transparent cover consists of double-layered sealed tempered glass units set in wood frames.

The air is circulated by a 1 hp axial flow fan located in a manifold at the base of the collector. Air is blown upward through ducts at the back of the collector and then downward through the expanded metal sheet air heater. At the bottom of the collector, air enters the manifold leading to the blower, which forces it to the storage area. The air returns from storage to the collector by means of insulated ducts arranged to fit between the floor joists (running north–south). The same blower is used for heating the house from storage. This cycle is operated by manually controlling crank and spring-loaded baffles located in the collector supply and return ducts. When there is an electrical power failure and the sun is shining, the collector glass cover is protected from high-temperature breakage by opening rectangular apertures located at the top of the south wall inside the house, and opening the air supply and return ducts at the base of the collector. The collector then operates similarily to a modified Trombe wall.

Size The collector area is 37 m^2.

FIG. 7 La Macaza house; 94-m^2, one-story house; air heating collector of 37-m^2 area. The rock storage (32 m^3 of fist-size rocks) is located under the bedrooms in the insulated northern half of the house basement. Heat distribution from storage to house is manually operated (forced air or natural convention). Electric baseboard backup heating is used to maintain house temperatures. (a) West view, (b) north view.

Orientation and location The solar collector is mounted vertically and occupies the entire south wall of the house.

B. Storage System Description

The storage area occupies the northern half of the house basement, located under the bedrooms. Rocks (5- to 10-cm diameter) form the storage medium. The rocks are piled on a layer of hollow concrete blocks. These are placed so that their openings line up to act like horizontal distribution ducts. Small spaces left between blocks allow the air to escape and diffuse through the rock pile. The air reaches the storage area just after it has passed the fan and enters the main distributing plenum chamber, at the bottom of which starts the concrete block network described previously.

For house heating the air can be circulated by the fan up through the storage and into any room of the house. In the bedroom floor registers can be opened to allow adequate convective heating much of the time, with or without the need of the fan. No provision is made for domestic hot water heating. The storage holds 32 m^3 of rock weighing 16 metric tons.

C. Controls

Heat collection system Automatic: A differential thermostatic switch starts the blower whenever the collector plate is 30°C above storage temperature; supply and outlet ducts must be opened.

Heat distribution system All-manual: Both supply and outlet ducts to the collector are closed by means of a crank lined to spring-loaded baffles. The heat can then be used passively through bedroom registers or actively by starting the blower (switched manually).

Backup system Electric baseboard heaters totaling 9 kW.

The system was built by the Shelter Systems Group, School of Architecture, McGill University, Montreal, with solar design by the Brace Research Institute, McGill University, Montreal (Anonymous, 1974b; Stein and Wexler, 1974).

Cost The total cost of the solar heating system amounted to $5000.

The solar heating system provides nearly 50% of the house heating requirements.

11.6 PEPPER HOUSE

Location Ontario, 43° N.

Address RR #1, Granton, Ontario (20 km north of London).

Site description Flat ground, rural area; no obstructions from trees or buildings.

Type of building Single-family two-story house built in 1969.

Size of building A total of 200 m^2 of floor area.

A. Solar Collector System

Operating mode Active (electrical circulating pump).

Type Deteached, trickle water system, having a single glass cover; 9-cm fiberglass insulation backing metal absorber plate, painted in flat green color (built in 1975).

Size 6.1 × 12.2 m wide; total 74.5 m^2.

Orientation South facing with a 58° slope to the horizontal in order to achieve maximum solar heat collection during the period November–February.

Location The A-frame collector is separated from the house, and connected to the house by 33 m of insulated piping.

B. Storage System Description

There are 13,700 liters of water stored in four tanks of differing capacities in the house basement. For the heating circuit, a facility is provided for circulating the water from any combination of the four tanks to and from the collectors, depending on the weather conditions and the house requirements. The heat is extracted from the tanks through a heat exchanger installed in the 3500-liter tank for hot water for domestic use and swimming pool heating. The tanks are installed in an insulated basement room that is linked to the forced-air standard duct work of the house. When needed, dampers will operate this system via the water storage and distribute the warm air to the rooms (Pepper, 1976).

C. Controls

All controls are automatic.

Heat collection system The pump will operate if the collector is 5°C hotter than the storage tank temperature.

Storage heat distribution system Electric damper thermostatically activated and controlled by a low-temperature limit switch.

Backup system Forced-air oil furnace and thermostat.

Cooling in summer By interchanging the set of leads on the automatic controller, the pump is operated when the collector is 5°C lower than storage tank water temperature (mostly nighttime operation).

The system was designed, engineered, and built by the owner, C. Pepper.

Cost $1300 using recycled and second-hand materials (glass, steel storage tanks).

Note: The system has provided approximately 50% of the heating requirements of the house.

11.7 PROVIDENT HOUSE

Location Ontario (Fig. 8).

Address Near Aurora Road and Highway 400, located in Carrying Place Estates, Ontario (near Toronto).

Site description Suburban area.

Type of building Single-family $2\frac{1}{2}$-story wood frame house and basement (water storage area) built in 1976. Insulation in walls and roof is 15 cm of fiberglass; earth berm on north side of house.

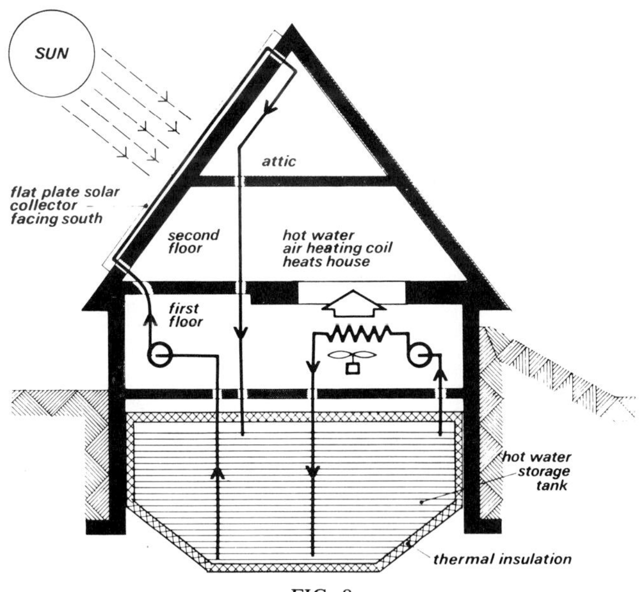

FIG. 8

Size of building 260 m^2 of floor area.

A. Solar Collector System

Operating mode Active.

Type Closed cycle water collector. The collector itself is made of steel pipes and plates, coated with a selective black coating. The collector has a single layer of glass. To prevent scale buildup in the collector, the circulating water is softened and treated. The water is drained when the collector is not in use to prevent freezing problems.

Size There are 67 m^2 of area for the house heating system and 7.5 m^2 for domestic hot water heating.

Orientation Collectors face due south with a tilt angle of 52.5° to the horizontal.

Location On part of the sloping roof of the house.

B. Storage System Description

The entire basement of the house is the water storage tank. A rectangular tank of poured concrete measuring 8.5 × 8.5 × 4 m is insulated on the inside with 15 cm of foamed plastic and lined with a waterproof plastic. This storage tank contains 276,000 liters of water. The tank top consists of plywood, waterproofed on the inside and insulated on the upper side. This system capitalizes on the long hot summer days for collecting the heat that is used throughout the winter. It is an annual cycle storage of the solar heat. The storage contains no partition, but natural thermal stratification occurs and is exploited. By the beginning of winter, the storage temperature is 75°C. As the heating season advances, the heat requirement is in excess of the heat collected so that by the end of March, the storage temperature drops to 40°C. Thereafter, the collection rate exceeds the heating needs and the storage temperature starts rising again. The rooms are heated by a forced-air distribution system drawing the heat by means of a heat exchanger located in the storage tank.

C. Controls

Heat collection system Automatic.

Heat distribution system Thermostat.

Backup system None used.

The house was designed by F. C. Hooper and J. Hix.

Cost The funding for the house came from the Canadian Ministry of Urban Affairs ($90,000) and the Ontario Ministry of Energy ($50,000).

Note: The building is designed to provide 100% of the heating needs of the house.

11.8 SICOTTE HOUSE

Location Quebec, 45°30′ N (Fig. 9).

Address Chemin de Saverne, Lorraine, Quebec (north of Montreal).

Type of building Single-family 1½-story house plus basement; no attic space is provided. The house is insulated above standards in the walls and ceilings. All window areas are double-glass sealed units.

Size of building 150 m² of living area.

A. Solar Collector System

Operating mode Active.

Type Air-cooled collector. The absorber plate is treated to have selective surface properties. The air, blown by a circulating fan, passes at the back of the absorber plate. The collector is covered with two sealed tempered glass panels. The system provides three modes of operation:

(a) Direct heating cycle: The heated air goes directly from collector to house.

(b) Storage buildup cycle: The heated air is circulated from collector to storage when there is no demand for heat from the house.

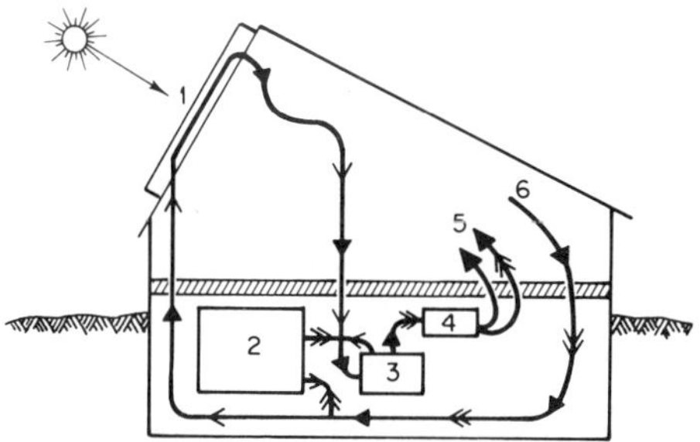

FIG. 9 Sicotte house. The air circulation for the three operating modes of the house are indicated by the arrows: A, collector-to-house cycle (—▶); B, collector-to-storage cycle (→); C, storage-to-house cycle (↠); 1, air heater collector; 2, storage area; 3, air filtration and blower unit; 4, auxiliary electric heating elements; 5, warm air distribution; 6, air return. All cycles operate under automatic control devices.

(c) Indirect heating: The house is heated from the heat accumulated in the storage.

Size There are 28 m^2 of area. Also, on top of the flat roof south-facing veranda, an aluminum reflector measuring 2.3 × 9.5 m is installed and provides increased radiation on the collector through reflection of direct solar radiation.

Orientation Facing due south with a 58° tilt from the horizontal.

Location On second floor roof.

B. Storage System Description

The storage medium is rock interspaced with layers of paraffin wax. The storage area is a reinforced concrete form cast with the basement foundations, apart from its bottom slab. The storage area measures approximately 1.8 × 1.8 × 2.4 m in height and is located in the basement of the house. It contains 15 tons of rock and 400 kg of paraffin wax (melting point 48°C). It is insulated on the inside with 5 cm of polystyrene. The thermal capacity of the storage is 175,000 kcal at 65°C.

The heating system includes an air treatment unit and an auxiliary electric heating element through which the air directed to the rooms passes and is heated to the required temperature level whenever the collector output or the storage temperatures are too low. The heated air is distributed to the rooms by means of a standard forced air blower system and associated duct work.

C. Controls

For operating modes A and B (collector to house and collector to storage), the blower and baffles systems are controlled by electronic differential thermostats and a house thermostat. For operating mode C (storage to house), they are controlled from the house thermostat alone.

Backup system Electric heating elements located in ducting system to the rooms; also a Norwegian wood stove whose chimney passes close to the storage wall.

Cost The cost of the solar heating system was $6500.

The house was designed and built by Thermosolar Inc., Montreal, and has provided two-thirds of the house heating load according to the estimates of Mr. Sicotte.

11.9 THE ARK—PRINCE EDWARD ISLAND

The Ark for Prince Edward Island was initiated, designed, and built by the New Alchemy Institute, Massachusetts, and is located at Little Pond, Prince Edward Island, 46.5° N (Fig. 10). Built in 1976, the Ark is an experimental integrated energy production structure whose construction has been funded by the Canadian Ministry of State for Urban Affairs, the Province of Prince Edward Island, and the New Alchemy Institute. The Ark houses a research laboratory, living unit, family garden greenhouse, and a small commercial greenhouse and fish farm.

A. Greenhouse Heating and Storage

Solar collection 185 m^2 of acrylic double glazing facing due south.

Storage area 90 m^3 of rock located in the basement at the back and separate from the greenhouse. Also 72,000 liters of warm water fish culture. These fish ponds are made of transparent fiberglass cylinders and are located in the northern part of the greenhouse.

Backup heating Resistance coils in air ducts between the greenhouse and the rock storage.

FIG. 10 The Ark of Prince Edward Island. Southeastern view of the building under construction in summer 1976. The production greenhouse occupies two-thirds of the southern half of the building starting from the right foreground on the photograph. The solar-heated living quarters are on the left.

B. Residential Heating and Storage

Solar collection 80 m^2 water-type solar collector consisting of a flat metal plate with selective black surface.

Storage area 80,000 liters of water located in a tank under the living room (northwest room of house). A water–air heat exchanger and blower located in the air distribution duct to the house provides the necessary heat.

Backup heating A wood-burning boiler for the water storage and wood stoves are included as auxiliary heating systems.

11.10 RECENT SOLAR ENERGY HEATING DEMONSTRATION PROGRAMS

In 1976, the National Research Council of Canada undertook a solar energy heating program that consisted of funding the design and construction of 14 solar-heated demonstration homes across Canada. Unlike the earlier efforts, which had been undertaken in Canada by federal authorities, these programs sponsored the design and construction of only the solar space-heating components for new or recently constructed single-family dwellings, and not the houses. The designs themselves were selected to ensure that a significant portion, that is, greater than 30% of the heating load of these houses should come from the solar system. The size of the houses varies from 93 to 186 m^2 of floor area. At the least these houses are required to conform to Residential Standards Canada 1975, which allow for increased insulation. A résumé of the contracts awarded under these programs is listed in Table 4. Most of these solar heating systems were put into operation by the spring of 1977. A description of several of these systems is presented in the following sections.

11.11 IVES HOUSE

This five-bedroom two-story house is located in Hudson, some 25 miles west of Montreal, Quebec (Fig. 11). The house is very well insulated and was part of the program of solar heating of buildings, initiated by the Division of Building Research of the National Research Council of Canada. The building and solar heating system was completed in March 1977 and reduces the auxiliary space heating by at least 30%.

A. Solar Collection System

Solar energy collection is of the water-circulating type; 12 solar panels with a total collector area of 37 m^2 are mounted on the south-facing roof at

TABLE 4

Program on Solar Heating of Buildings Sponsored by the National Research Council of Canada in 1976[a]

Contractor	Location of house	Working fluid in solar collectors
Canadian British Consultants Ltd., Halifax, Nova Scotia	Charlottetown, Prince Edward Island	Water
A. Penny, Halifax, Nova Scotia	Halifax, Nova Scotia	Air
ADI Limited, Fredericton, New Brunswick	Fredericton, New Brunswick	Air
G. G. Murray, Saint John, New Brunswick	Saint John, New Brunswick	Water
Fenco Consultants Ltd., Toronto, Ontario	Montreal, Quebec	Air
The Proctor and Redfern Group, Toronto, Ontario	Thunder Bay, Ontario	Water
Direct Energy Associates, Quebec, Quebec	Ayers Cliff, Quebec	Air
C. J. M. Ives Hudson, Quebec	Hudson, Quebec	Water
Nu-West Development Corp. Ltd., Calgary, Alberta	Calgary, Alberta	Water
R. M. Hardy and Associates, Calgary, Alberta	Calgary, Alberta	Air
J. R. Borden, Regina, Saskatchewan	White City, Saskatchewan	Air
W. L. Wardrop and Associates, Winnipeg, Manitoba	Winnipeg, Manitoba	Air
Delmarco Management Ltd., Langley, British Columbia	Aldergrove, British Columbia	Air
D. B. Thompson, Brentwood Bay, British Columbia	Victoria, British Columbia	Water

[a] Only the solar collectors and storage system components of these houses have been funded under this program (Thompson, 1976; Thompson and Swartmen, 1976; Hamilton, 1976; Hamilton and McConnell, 1976; McConnell *et al.*, 1976; Higgin, 1976; Shurcliff, 1976).

an angle 50° above the horizontal. They are lightweight double-glazed panels proven to withstand freezing conditions; the water circulates in silicon rubber tubes.

B. Storage System Description

The water storage is located in the basement of the house and is comprised of two fiberglass tanks holding 7600 liters of water. The heat is

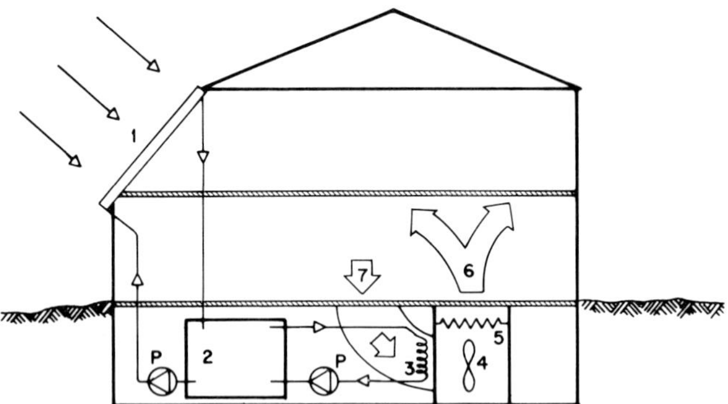

FIG. 11 Ives house. Schematic diagram of the solar heating storage and air distribution systems. 1, solar collector—water type; 2, water heat storage; 3, water–air heat exchanger; 4, blower; 5, electric furnace; 6, distribution of heated air; 7, air return duct; P, pump.

transferred from storage to house through a conventional forced air system by means of a water–air heat exchanger located in the air return duct of the air distribution network.

C. Backup System

The auxiliary heat is supplied by two electric furnaces and two high-efficiency wood-burning fireplaces.

11.12 FENCO CONSULTANTS LIMITED HOUSE

The house is located in Champfleury development, Ste. Rose, Laval, Quebec. The project is a retrofit on a model house in a Campeau Corporation housing development. The house is a 1300-ft^2 bungalow. The solar heating system uses air-heating collectors and a rock heat storage. The collector area is 465 ft^2 and the storage volume is 300 ft^3.

The collectors were manufactured by Solcan Limited of London, Ontario, Canada. Their design incorporates a selective surface on a steel panel that has minimal thermal capacity, urethane foam insulation, and fiberglass covers. This system provides more than 40% of the heating load.

11.13 THE PROCTOR AND REDFERN GROUP HOUSE

This house has been built by Mayotte Limited of Thunder Bay, Ontario, and is located in Thunder Bay. The house was designed by The Proctor and Redfern Group, consulting engineers from Toronto. The house is an 1800-ft^2 two-story building. The solar heating system uses liquid-heating collectors with water as the heat transfer medium. The system is supplemented by a water-to-air heat pump. Two water storage tanks are used, one for the heating system and one for domestic hot water.

The collectors have been manufactured by Solcan Limited of London, Canada (Anonymous, 1976b). A significant portion of the heating loads of the house have been met by this solar-assisted heat-pump system.

12

Solar House Heating with Heat Pipe Collectors

H. GEHRKE

DORNIER SYSTEM GMBH
FRIEDRICHSHAFEN, FEDERAL REPUBLIC OF GERMANY

12.1 INTRODUCTION

The policy of the Federal Republic of Germany aims at reducing heating requirements and developing alternative sources of energy in order to conserve the valuable and expensive fossil fuels, especially petroleum, and to use them in a more advantageous manner. In 1973 about 44% of the end use energy consumption in the Federal Republic of Germany was used by households and small consumers, and of this, 80–90% represents energy for heating purposes.

12.2 CONCEPT

A partial program of developmental activities with a view toward using solar energy comprises the project "Solar House Essen," which was a joint effort of Dornier System GmbH, as system leader, and the RWE subsidiary Energietechnik GmbH, and was sponsored financially by the Federal Ministry for Research and Technology. Dornier developed and installed the collectors on the roof of the trial house, and Energietechnik was responsible for incorporating the use of solar energy into the heating system.

The trial house, a solidly constructed conventional building occupied by three families, is situated on a hillside with a southern exposure on the periphery of Essen. The specific heat requirement according to DIN 4701

ISBN 0-12-620860-3

amounts to 75 W/m^2. The solar installation heats a total floor space of about 190 m^2 via underfloor heating during interseasonal periods. Additional energy is available for the hot water supply. By installing some form of heat storage it is possible to bridge periods of little solar radiation for several days. During times of low radiation a water/water heat pump is utilized to increase the efficiency of the entire installation. Surplus heat is conducted to a swimming pool in the cellar of the house.

12.3 SOLAR COLLECTORS

The roof of the house is inclined 48° and is ideal for utilizing solar energy all year round. Collectors, working according to the heat pipe principle, with an active surface of about 65 m^2, were installed (Fig. 1). The main feature of these collectors is the heat transport mechanism achieved by vapor convection and condensation. Figure 2 shows the schematic structure of the absorber. The individual absorbers consist of seamless extruded aluminum profiles forming a tube about 5.3 m in length, which is closed at both ends. An absorber plate (fin) 0.19 m wide is attached to the pipe, which is evacuated and, compared to its volume, contains a relatively small quantity of freon as the heat-conveying medium that wets the pipe wall evenly via a grooved structure. The solar energy absorbed by the black anodized absorber surface is transmitted to the freon which evaporates more or less depending on the intensity of the sun.

The collector is built so that the upper ends of the heat pipes project into a heat exchanger, through which flows the water that is destined to be heated. The temperature level of the water is slightly below that of the

FIG. 1 Dornier/RWE solar house in Essen with 65 m^2 of heat collectors.

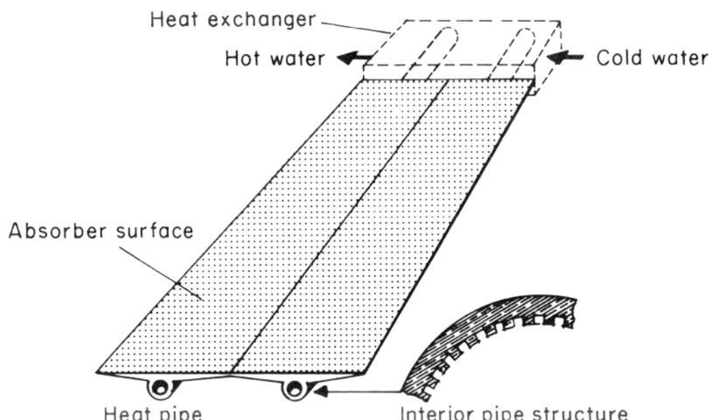

FIG. 2 Schematic of the heat-pipe absorber.

freon vapor which is thereby induced to condense. The heat of condensation is transmitted to the water through the pipe wall. The heat transport within the heat pipe is effected almost without loss by vapor convection and condensation. The temperature difference between the heat-absorbing and the heat-releasing part of the absorber is very low and amounts to about 2°K (by gravitational force). The condensed working medium is transported through the grooves of the pipe wall to the warm side of the absorber, where the cycle starts anew.

The following properties characterize the heat pipe absorber:

(1) low heat capacity and rapid achievement of working temperatures;

(2) thermal diode effect—the heat can be conducted into the water circuit only in an upward direction; it is not possible to cool the circuit medium via the absorber;

(3) increased frost protection via the thermal diode effect; the heat exchanger through which the cycle medium flows can be insulated in an effective way;

(4) reduced risk of corrosion; the surfaces exposed to the cycle medium can be protected easily; no corrosion occurs in the interior of the absorber.

The heat collector is built like a conventional collector, i.e., with a transparent cover on the absorber (in this case, double plexiglass) and polyurethane foam insulation. The collector is mounted in panels about 1.2 × 5 m. Twelve of these large panels cover the roof and are harmoniously integrated into the architectural design.

12.4 ENERGY SYSTEM

Figure 3 is a diagram of the distribution and storage system. Energy in the form of hot water can be conducted to the two 600-liter hot water

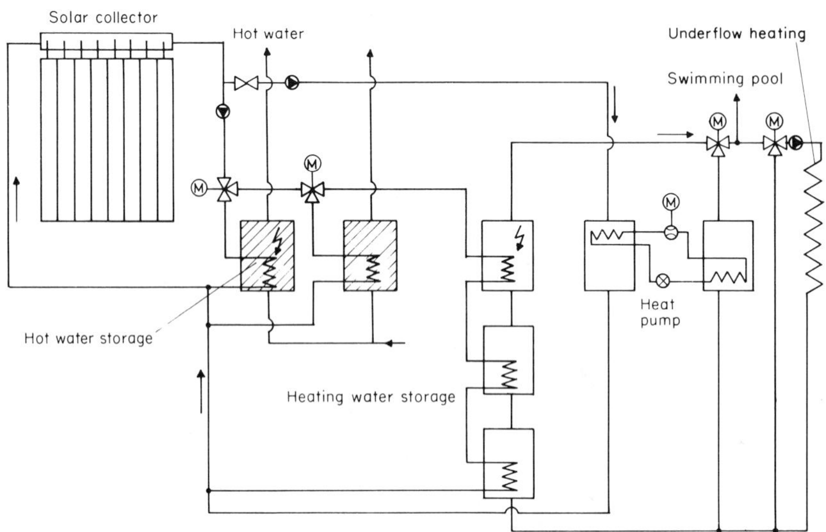

FIG. 3 Simplified scheme of the installation in the Dornier/RWE solar house.

storage tanks or to the three 1200-liter capacity heat sinks. When the irradiation energy of the sun is low, it is possible to switch over to the heat pump storage, which can be cooled down to 5°C. Since the water flows back to the collector at these low temperatures, increased efficiency (energy generation) is realized. On the collector side of the heat pump, there is a 1200-liter storage tank that can be heated up to 65°C. Depending on the temperature difference, a performance value of 2.5 to 3.8 is achieved.

The necessary additional heating of the house is provided by means of electricity. The first hot water storage tank (for the "test apartment") reheats up to 45°C in its upper third with electrical auxiliary heating if necessary, and the water in the second storage tank for the two remaining apartments is reheated with a thermally controlled flow heater.

The reheating of the heat storage is effected by means of a storage heater; a constant temperature is maintained for a certain volume of water that is stored for heating purposes for the next day.

12.5 OPERATING EXPERIENCE

The installation has been in operation since August 1975, and it has been working perfectly since then. During the initial installation an additional cover of infrared reflecting glass was provided, but the loss of efficiency through absorption of solar radiation was greater than the expected benefit from a decrease of the radiation loss from the absorber

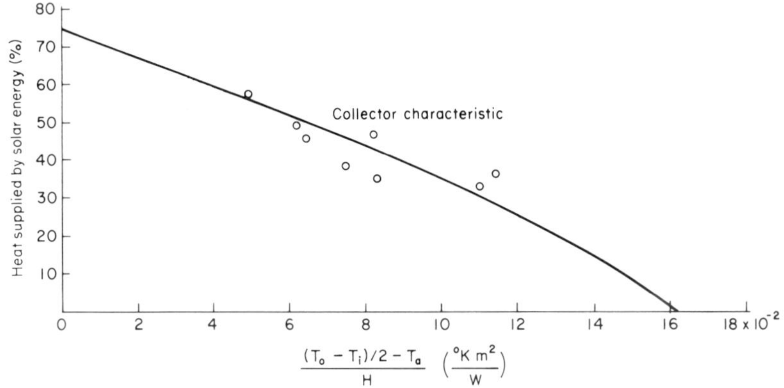

FIG. 4 Standardized efficiency curve for the heat pipe collector (static). (T_0 and T_i, collector outlet and inlet temperature respectively; T_a, ambient temperature; H, global radiation.)

since the supplier was not able to meet the specification requirements. This additional cover has been removed.

When the installation was put into operation, the water circuit was provided with an inhibitor, trademark OC 2002 made by Bayer Leverkusen. After a year there were no signs of corrosion in the circuit even though in the installation a mixture of aluminum and copper, among other metals, had been used for test purposes.

With the incomplete measurements now available, it is possible to draw the standard efficiency curve illustrated in Fig. 4 (influencing factors such as wind and humidity, as well as the dynamic properties of the heat

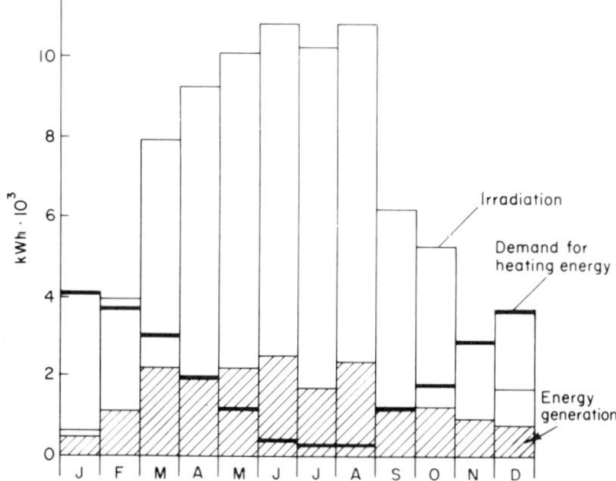

FIG. 5 Irradiation, demand for heating energy, and energy generation by collectors of 65 m² during the course of a year.

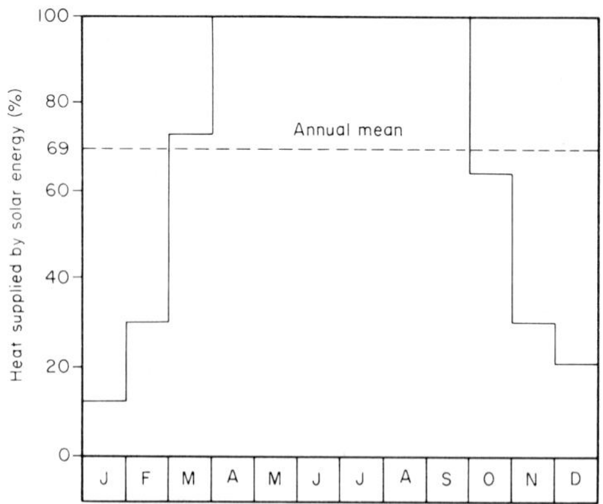

FIG. 6 Proportion of heat supplied by solar energy.

pipe absorber, are not taken into account; the curve relates to the collector having a 16-mm-thick plexiglass web double-plate cover).

With the expected 1350 to 1400 h of sunshine in Essen, the following provisional results were obtained: During one year (November 1975–October 1976) approximately 18,600 kWh (equivalent to 286 kWh/m^2) were produced by means of the collectors; this is 24% of the solar energy that was vertically irradiated on the collectors during this period. The consumption of heating energy during the same period amounted to approximately 24,400 kWh plus the heat requirement of about 6000 kWh for the hot water supply. Figure 5 shows the irradiation, the heat require-

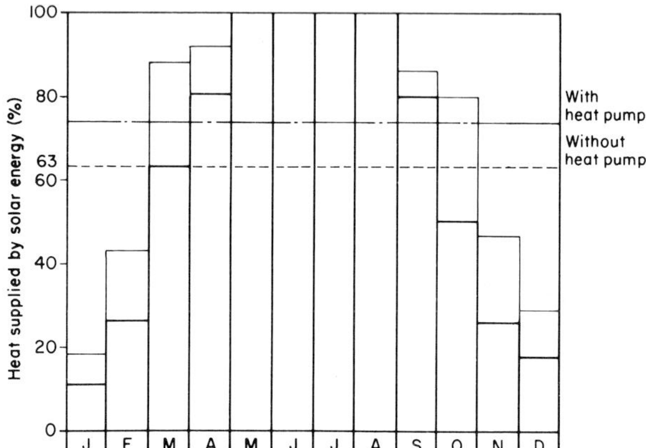

FIG. 7 Overall proportion of heat supplied by solar energy. Thick bars, without heat pump energy demand; thin bars, with heat pump energy demand.

ment (thick bars), and the solar heat generation for each month. The energy absorption of the heat pump is not taken into account.

Figure 6 shows the proportion of heat supplied by solar energy generation during the year. The annual mean is 69%. If the hot water supply is included, 63% of the required heat is supplied, as shown by the thick bars in Fig. 7. When the current consumption of the heat pump is included (this energy is fully used by the heat circuit), the picture changes (thin bars). The supply of heat per year by means of solar energy and heat pump increases to 74%.

12.6 FURTHER ACTIVITIES

Intensive and detailed continued research using a CAMAC data acquisition system at Dornier is planned for the next two years. By means of correlation analyses and similar evaluation methods, the influence of individual weather and installation conditions on the overall efficiency of the installation will be evaluated, and this evaluation will consequently result in the development of an optimal solar installation for Central European needs with respect to alternative possibilities for the provision of hot water heating and hot water.

13

Solar Houses in Japan

KEN-ICHI KIMURA

DEPARTMENT OF ARCHITECTURE
WASEDA UNIVERSITY
TOKYO, JAPAN

13.1 INTRODUCTION

Japan is a country that has a variety of climatic conditions ranging from the subtropical south to the snowy and icy north. The southern pacific coast is blessed with lots of sunshine throughout the year; in particular, the southern islands are so warm that no heating is required. Hokkaido island and the northwestern part of Honshu Island, however, are so cold that nobody there is interested in installing cooling systems. The majority of the Japanese islands have a mild climate, so that people would like to have both heating and cooling if they could afford to do so.

Seven solar houses were built before 1973, five of which were designed by Mr. Yanagimachi. Three solar houses were built in 1974, ten in 1975, and more in 1976. Obviously Mr. Yanagimachi had profound foresight concerning the future energy crisis, and as an experiment built his own solar house in 1958. Table 1 gives a list of the solar houses in Japan.

It is a fact that no technical progress was made after the construction of Yanagimachi Solar House II (Yanagimachi, 1961) (Fig. 1) until 1973. Kimura SH (Kimura, 1975) (Fig. 2) was built in 1972, 16 years after Yanagimachi SH, but in a number of ways it is inferior to Yanagimachi SH. For example, Yanagimachi SH uses no auxiliary heating other than heat pumps with two storage tanks of 40 and 10 m^3, whereas an electric resistance heater is the auxiliary heat source in Kimura SH. The Yanagimachi

ISBN 0-12-620860-3

TABLE 1

Solar Houses in Japan

No.	Name	Period (year, month)	Location	System designer
1	Yanagimachi SH I[b]	1956–1957	Tokyo	M. Yanagimachi
2	Yanagimachi SH II[b]	1958–	Tokyo	M. Yanagimachi
3	Tajiri SH	1960–1970	Funabashi	M. Yanagimachi
4	Kokenomo Villa	1959–1965	Karuizawa	M. Yanagimachi
5	Meikoshi SH	1959–1963	Nagoya	M. Iida
6	Tokai University SH	1969–1970	Kanagawa	S. Tanaka
7	Kimura SH[c]	1972,3-	Tokorozawa	K. Kimura
8	Yazaki SH[d]	1974,7-	Kosai	Yazaki Co.
9	Ohbayashi Laboratory SH[e]	1974,8-	Kiyose	Ohbayashi Co.
10	Soka SH[f]	1975,5-1978	Soka	S. Tanaka; JES
11	Asahi Kuzuha SH[f]	1975,6-	Hirakata	Yazaki Co.
12	PS Factory	1975,3-	Hokkaido	K. Kimura
13	Sekisui-MHI SH	1975,6-	Nara	Mitsubishi Heavy Industry Co.
14	Uchiyama SH	1975,9-	Nagoya	T. Uchiyama

TABLE 1 (*Continued*)

Solar Houses in Japan

No.	Name	Period (year, month)	Location	System designer
15	Kajima Laboratory SH[g]	1975,10-	Chofu	T. Hino
16	Nagano Takemura SH[h]	1975,11-	Nagano	M. Takemura
17	Sanshin SH	1975,11-	Oita	Sanshin Co.
18	KEP/JHC SH	1975,12-	Hachioji	K. Kimura; M. Udagawa
19	MHI SH	1975,12-	Tokyo	Mitsubishi Heavy Industry Co.
20	Toshiba SH	1975,11-	Kawasaki	Toshiba Co.
21	Shiozawa SH	1976,1-	Tokyo	M. Takemura
22	Sharp Eidai SH	1976,3-	Sakai	Sharp Co.
23	Numazu Kanaoka SH[i]	1976,3-	Numazu	Iketani Setsubi Co.
24	Yagihara SH[j]	1976,3-	Hamamatsu	Yazaki Co.
25	Ishibashi SH[j]	1976,5-	Kosai	T. Ishibashi
26	Konishi SH	1976,7-	Kokubunji	Y. Nakajima
27	Yorii SH[k]	1976,8-	Yorii	Nippon Telegraph and Telephone Public Corp.
28	Okuzawa SH[l]	1972,10-	Tatebayashi	K. Okuzawa

TABLE 1 (*Continued*)

No.	Architect	Sponsor	Use	H & C (m²) / Total (m²)
1	Ishikawa; M. Yanagimachi	Owner	Residence	55/80
2	Takenaka Co.; M. Yanagimachi	Owner	Residence	200/224
3	Nitta Co.; M. Yanagimachi	Owner	Residence	74/140
4	S. Hanzawa	Owner	Villa	70/530
5	M. Iida	Nagoya Industrial Research Institute	Experiment	82
6	S. Tanaka	Tokai University	Experiment	6/6
7	K. Kimura	Owner	Residence	150/150
8	A. Oshima	Yazaki Co.	Experimental residence	114/142
9	Ohbayashi Co.	Ohbayashi Co.	Experimental office	80/80
10	K. Kimura; Nunokawa	Science and Technology Agency	Residence	70
11	Japan Architectural Association	Asahi Newspaper Co.	Exhibition hall	432/432
12	(Retrofit)	PS Co.	Factory	
13	Sekisui Heim Co.	Sekisui Co.; MHI Co.	Experiment	106/106
14	T. Mawatari	Owner	Residence	80/203.5

TABLE 1 (*Continued*)

No.	Architect	Sponsor	Use	H & C (m²) / Total (m²)
15	Yajima Co.	Kajima Co.	Laboratory	45
16	N. Takemura	Owner	Residence	234(100) / 235
17	Sanshin Co.	Sanshin Co.	Experiment	30 / 60
18	Japan Housing Corp.	Japan Housing Corp.	Experiment	90 / 90
19	Sekisui Heim Co.	MHI Co.	Residence	50 / 90
20	Toshiba Housing	Toshiba Co.; MITI	Experiment	103 / 103
21	Keikaku Kobo	Owner	Residence	160(100) / 230
22	Eidai Co.	Sharp, Eidai; MITI	Experiment	73 / 119
23	Numazu City	Numazu City	Assembly hall	553.5 / 716
24	Juko Heim Co.	Owner	Residence	97 / 187
25	T. Ishibashi	Owner	Residence	85 / 138
26	S. Konishi	Owner	Residence	60 / 85.7
27	NTTPC	NTTPC	Telegraph and Telephone station	445 / 746
28	K. Okuzawa	Owner	Residence	193 / 224

TABLE 1 (*Continued*)

No.	Solar system	Collector (m^2)	Absorber	Surface	Glazing (no.)
1	H; (C)	20	Al RB	BP	None
2	H; (C); HW	H, C, 98; HW, 33	Al RB	BP	None
3	H; (C); HW	155	Al RB	BP	None
4	H; HW	32	Al RB	BP	None
5	H; (C)	24	Al RB	BP	Glass (1)
6	H	1.6	Al RB	BP	Acrylic (1)
7	H; HW	H, 24; HW, 8.5	Cu tube; Cu plate	BP	Window glass (1)
8	H; C; HW	94	SUS seam weld	SS	½ Temper glass (1)
9	H; C	32.2	Cu Tube; Cu plate	SS	Temper glass (2)
10	H; C; HW	42	SUS seam weld	SS	½ Temper glass (2)
11	H; C	408	SUS seam weld	SS	½ Temper + regular (2)
12	H	20	Steel	BP	Glass (1)
13	H; (C); HW	28	Cu Tube; Cu plate	BP	Plastic (1)
14	H; (C); HW	H, 33.15; HW, 5.76	H, Al RB; HW, Cu	BP	Glass (1)

TABLE 1 (*Continued*)

No.	Solar system	Collector (m^2)	Absorber	Surface	Glazing (no.)
15	H; (C)	15	Al RB	BP	None
16	H; (C); HW	96	Al plate; Cu tube	BP	Cu shingle
17	H; HW	29	Al RB	BP	Glass (1)
18	H; HW	H, 10; HW, 5.5	Steel, Cu, Al plate	BP	FRP (1)
19	H; (C); HW	85	SUS	BP	Glass (1)
20	H; (C); HW	H, 48; HW, 12	Al plate; Cu tube	BP	Temper glass (1)
21	H; (C); HW	60	Al plate; Cu tube	BP	Cu shingle
22	H; HW	21.1	Al RB	BP	Glass (1)
23	H; C	412	SUS seam weld	SS	½ Temper glass (1)
24	H; C; HW	53.4	SUS seam weld	SS	½ Temper glass (1)
25	H; C; HW	55.2	SUS seam weld	SS	½ Temper glass (1)
26	H; HW	9.22	SUS seam weld	SS	½ Temper glass (1)
27	H; C	294.4	SUS seam weld	SS	½ Temper glass (1)
28	H; (C); HW	H, 90; HW, 6	Pool water; SUS	Water; SS	Vinyl film; Water; glass (1)

TABLE 1 (*Continued*)

No.	Heat medium	Tilt (°)	Place	Storage medium	Storage capacity (m^3)
1	Water	90	Roof	Water	5
2	Water	15	Roof	Water	40 + 10
3	Water	15	Roof	Water	35 + 10
4	Water	15, 60, 90	Roof; Fence	Water	90 + 24
5	Water	35	Roof	Water	15 + 15
6	Water	90	Roof	Sand	1.5
7	Water	H, 90; HW, 45	H, South Wall; HW, Roof	Water	H, 1; HW, 0.46
8	Water	25	Roof	Water	H, 4.51; HW, 0.3
9	Water	Changeable	Roof	Water	2.5
10	Water	17	Roof	Water	0.5 + 5
11	Water	25	Roof	Water	10
12	Air	90	South wall	None	0
13	Water	30	Roof	Water	H, 3; HW, 0.25
14	Water	90	Southwest wall	Water	3; 3

TABLE 1 (*Continued*)

No.	Heat medium	Tilt (°)	Place	Storage medium	Storage capacity (m^3)
15	R-12	90	Roof	Water	7
16	EG 80%	50	Roof	Sand; gravel	110
17	Water	25	Roof	Water	3
18	EG 30%	75, 60	Balcony; roof	EG 30%	1
19	Water	55	Balcony	Water	0.46
20	H, Air; HW, Water	40	Roof	H, rock; HW, water	12, 0.4
21	EG 15%	20	Roof	EG 15%	6
22	Water	35	Roof	Water	1
23	Water	25	Roof	Water	20
24	Water	22	Roof	Water	3.5
25	Water	22	Roof	Water	3.5
26	Water	48	Balcony	Water; Gravel; soil	1.37; 1.32; 40
27	Water	25	Roof	Water	15
28	Water	0, 60	Ground; roof	Water; water	68; 0.5

TABLE 1 (*Continued*)

No.	Hot water system	Heating system	Cooling system	Auxiliary heat	Auxiliary capacity
1	Hot water boiler	HP, 2.2 kW	HP	None	
2	HP	HP; radiant	HP	None	
3	HP	HP, 2.2 kW	HP	None	
4	HP, 0.4 kW	HP, 5.5 kW	None	Oil burner	
5	None	Hot air	HP	HP	
6	None	Radiant	None	None	
7	Electric tank	Radiant	HP	Electric heater	H, 2.5 kW; HW, 3.6 kW
8	Preheat	Radiant; FCU	ABS. REF.	LPG boiler	30,000 kcal/h
9	None	FCU	ABS. REF.	LPG burner	30,000 kcal/h
10	Preheat	Hot air	ABS. REF.	HP	1.5 kW
11	None	FCU	ABS. REF.	LPG	150,000 kcal/h
12	None	Preheat of ventilation	None	Existing	
13	Preheat	FCU	HP; FCU	HP	2.2 kW
14	Direct	Radiant; FCU	HP	Gas boiler	

[a] Key: ABS. REF., absorption refrigeration machine; Al RB, aluminum roll-bond; BP, black paint; C, solar cooling; (C), conventional cooling; EG, ethylene glycol; FCU, fan coil unit; H, solar heating; HP, heat pump; HW, solar hot water; SS, selective surface; SUS, stainless steel.

TABLE 1 (*Continued*)

No.	Hot water system	Heating system	Cooling system	Auxiliary heat	Auxiliary capacity
15	None	HP	HP	None	
16	Preheat	Radiant, floor	Radiant cooling	HP	3 HP
17	Direct	Ceiling, FCU	None		
18	2-tank system	Radiant, floor	None	Electric heater	2 kW
19	Direct; electric heater	HP; FCU	HP; FCU	HP	
20	Direct	HP; FCU	HP; FCU	HP	
21	Preheat; kerosine	Radiant, floor	HP; FCU	HP	
22	Preheat; electric	HP; FCU	HP; FCU	HP	
23	Preheat	FCU	ABS. REF.	Kerosine boiler	160,000 kcal/h
24	Preheat	FCU	ABS. REF.	Kerosine boiler	17,000 kcal/h
25	Preheat	FCU	ABS. REF.	Kerosine boiler	17,000 kcal/h
26	Except winter	Natural convection	None	Kerosine boiler	6,000 kcal/h
27	None	FCU	ABS. REF.	LPG boiler	81,000 kcal/h
28	Direct	HP	HP	Electric water heater	

[b] Nocturnal cooling by collector. [c] Movable collector. [d] ABS. REF.—Libr/water, 6000 kcal/h. [e] ABS. REF. 6000 kcal/h. [f] ABS. REF. 4500 kcal/h. [g] Nocturnal radiation. [h] Snow melting. [i] ABS. REF. 67,500 kcal/h. [j] ABS. REF. 4000 kcal/h. [k] ABS. REF. 31,000 kcal/h. [l] Pool water as heat source for heating and cooling.

FIG. 1 Yanagimachi Solar House II.

SH collector is made of blackened aluminum roll-bond without glazing and is installed as an integral part of the roof; the collector panels act as a heat dissipator instead of as a cooling tower during the summer nights. It was determined that collection at very low temperatures does not require a glass covering, since the daytime outdoor air temperature in the Tokyo area is 5–10°C during the winter.

FIG. 2 Kimura Solar House.

13.2 THE HEAT PUMP IN SOLAR HEATING

The heat pump is a very useful tool for solar heating. It allows for low temperature collection with high collector efficiency as seen in Yanagimachi SH. Meikoshi SH (Fukuo *et al.*, 1961) uses the same heat pump system as Yanagimachi SH with two large storage tanks. In Kimura SH solar heat is collected by the movable collector panels installed in the south-facing glass windows and stored in a l-m^3 water storage tank. A 1.5-kW water-to-water heat pump takes the heat out of the storage tank, and the warm water is circulated through pipes embedded in the concrete floor slabs from which radiant and convective heat is discharged to the occupied spaces above and below. These slabs can be considered the secondary heat storage because of their large thermal capacity.

It is generally accepted as one of the merits of the heat pump that it can serve both heating and cooling requirements. This is true, but in most cases the various subsystems for heating and cooling are determined by the conditions for cooling in the overall air-conditioning design, and effective heating cannot be realized. Consumption of electricity during the daytime in summer is not desirable, and for heating only the heat pump

FIG. 3 Mitsubishi Sekisui Solar House.

FIG. 4 Toshiba Solar House.

must be essentially an effective machine for the utilization of low-temperature energy.

The heat pump is widely used in many solar houses built recently in Japan. Mitsubishi Sekisui SH (Sakai *et al.* 1975) (Fig. 3), Toshiba SH (Koizumi *et al.*, 1976) (Fig. 4), Uchiyama SH (Uchiyama, 1976) (Fig. 5), Nagano Takemura SH (Anonymous, 1976) (Fig. 6), and Sharp Eidai SH (Nishijima, 1977) (Fig. 7) are examples that use a heat pump system. The

FIG. 5 Uchiyama Solar House.

FIG. 6 Nagano Takemura Solar House.

FIG. 7 Sharp Eidai Solar House.

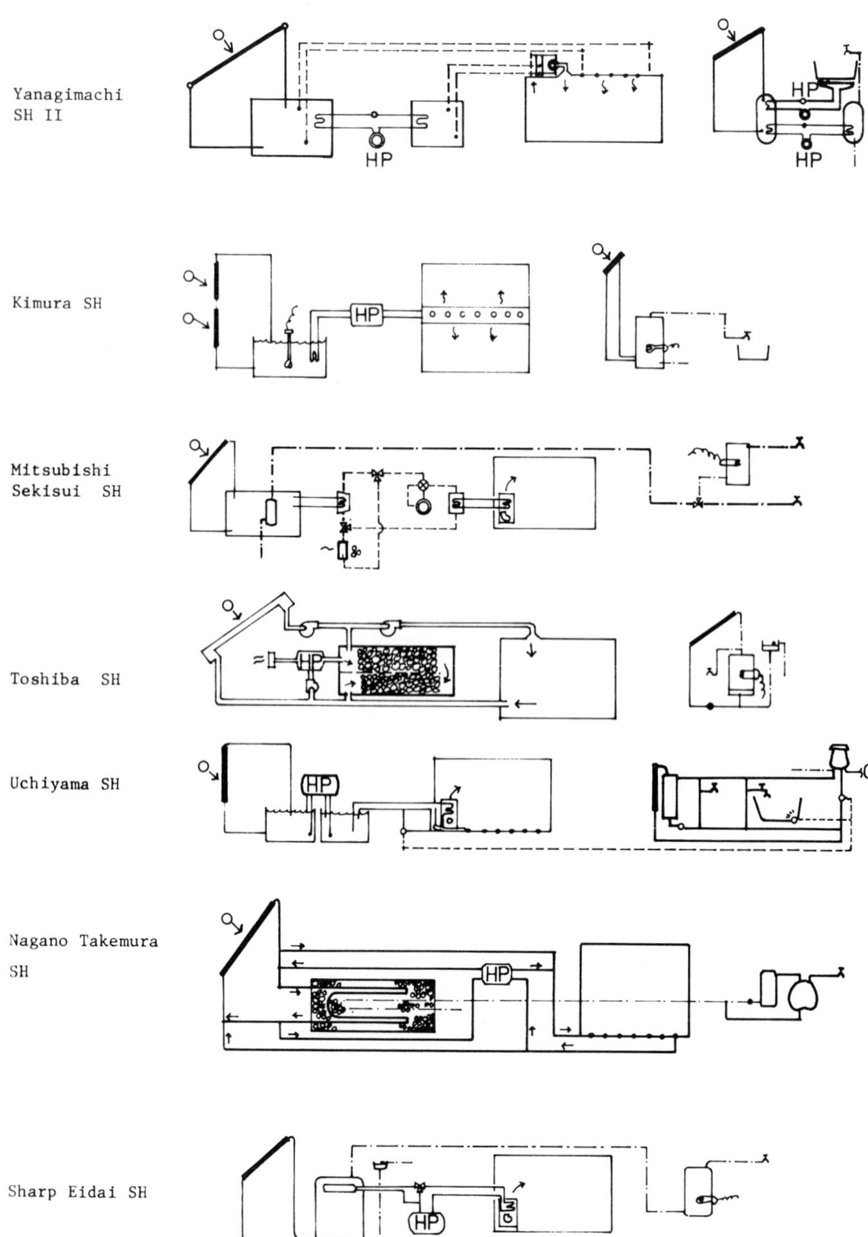

FIG. 8 Solar heating systems employing a heat pump.

heat pump itself is an auxiliary heat source and carries heat from the lower-temperature heat source to the higher-temperature fluid. Basically there are two types of heat pumps in terms of the heat source: water source and air source. The coefficient of performance (COP), defined as the ratio of useful energy to input energy, is higher in water-source-type heat pumps than in air source type. The energy in the outside air can be recovered by the air-source-type heat pump but not by the water source type. Two new types of solar-assisted heat pump systems have now been developed. One is a special type of heat pump that can be used with either a water source or an air source. This heat pump was developed by Mitsubishi Heavy Industries Company and is commercially available. It can take heat from the outside air when there is not sufficient heat left in the storage tank (Mitsubishi Sekisui SH). The other heat pump system is used in Toshiba SH; the air source heat pump works as an auxiliary heating subsystem and the solar air heater is used to store solar heat in the rock bed storage. The idea is that the heat transfer media are all of the air type and common to both heating and cooling operations.

Uchiyama SH uses a water-to-water-type heat pump coupled with a water heat storage tank and water-type collector. Nagano Takemura SH also uses a water-to-water-type heat pump coupled with a rock-bed-type storage tank in which serpentine water pipes are embedded as a heat exchanger. Diagrams of the solar heating systems and heat pumps are shown in Fig. 8.

Okuzawa SH employs a swimming pool as a collector and heat reservoir supplying heat to the heat pump unit in both winter and summer. A plastic sheet covers the water surface during the winter.

13.3 SOLAR COOLING WITH AN ABSORPTION REFRIGERATION MACHINE

Yazaki SH (Ishibashi, 1975) (Fig. 9) is the first solar house in Japan to incorporate solar cooling, and it commenced operation on July 21, 1974. This is probably the first solar cooling operation in a solar house in the world, excluding solar cooling experiments. The operation has been successfully continued with slight modifications and without long-term interruption. The first absorption refrigeration machine used was the one in which the generator was modified from the gas-fired type in the absorption machine which by that time had been produced by the Yazaki Company. This modified model was used in the Ohbayashi Laboratory SH (Nakahara *et al.* 1975), Soka SH (Kimura and Tanaka, 1975) (Fig. 10) sponsored by the Science and Technology Agency of the Japanese Gov-

FIG. 9 Yazaki experimental Solar House I.

ernment, and Asahi Kuzuha Exhibition Center House (Ishibashi, 1976b) (Fig. 11).

The solar cooling system employed in Yazaki SH and other solar houses with solar cooling (except Soka SH) is designed so that auxiliary heat enters the solar tank whenever the tank temperature is not high enough to feed the generator and the absorption chiller operates when the house requires cooling. Though the tank is well insulated, a large storage

FIG. 10 Soka Solar House.

FIG. 11 Asahi Kuzuha Solar House.

unit at a higher temperature of around 90°C must dissipate a considerable amount of heat to the environment throughout the entire period. The solar cooling system employed in Soka SH, however, is designed so that the absorption machine operates when the temperature in the first tank (500-liters) is higher than 90°C, in order to store the chilled water in the second tank (4000 liters). Additionally, a 1.5-kw air source heat pump works as an auxiliary to cool the water in the second tank during the summer as well as to warm it in winter. The system in Soka SH is quite complicated, and the electricity required for the operation of pumps, fans, and control equipment was found to be greater than that needed for conventional cooling by a compressor-type refrigeration machine. It is esti-

FIG. 12 Numazu Kanaoka Solar House.

FIG. 13 Yagihara Solar House.

mated, however, that the electrical consumption in future town houses employing this complicated system could be justified by the larger quantity of homes being supplied. The Yazaki system is rather simple, but it is likely that the total amount of energy expended by solar cooling could exceed that used in conventional cooling from the global ecological point of view. This is true if auxiliary heat were more than 20% of the total used to boost the tank temperature.

The new type of absorption chiller developed by the Yazaki Buhin Company is operable with the heat media at 75°C as the minimum inlet temperature to the generator, and the concentration of LiBr is varied with the temperature of the heat media, which in turn contributes to higher col-

FIG. 14 Ishibashi Solar House.

FIG. 15 Yorii Telegraph and Telephone Solar House.

lector efficiency. This machine has been installed in Kanaoka Assembly Hall in Numazu City (Ishibashi, 1976a) (Fig. 12), Yagihara SH (Fig. 13), Ishibashi SH (Ishibashi, 1978) (Fig. 14), and the Yorii Telephone Station of the Nippon Telegraph and Telephone Public Corporation (Kudo *et al.*, 1976) (Fig. 15). These installations are performing well, and it is said that some machines are also being used in other countries. Figure 16 shows the solar cooling systems with the absorption refrigeration machine used in these solar houses. The dashed lines indicate the heating route when the absorption chiller is not used.

13.4 SPACE-HEATING SYSTEM

An air system is usually applied to interior air conditioning when both heating and cooling are required. When only heating is desired, as is common in most residential houses in Japan, low-temperature radiant heating is regarded as the most suitable for solar heating in terms of the efficient use of both solar and auxiliary energy. In this respect, radiant heating is widely used in solar houses: Yanagimachi SH, Kimura SH, Uchiyama SH, Nagano Takemura SH, a part of KEP SH of the Japan Housing Corporation (Udagawa and Kimura, 1976) (Fig. 17), Yazaki SH, Ishibashi SH, and Konishi SH (Nakajima and Ohashi, 1977) (Fig. 18).

A direct solar heating system without the aid of a heat pump has also been attempted in the Tokai University Experimental SH (Tanaka and Suzuki, 1970) and the Japan Housing Corporation SH. From the economic standpoint, with regard to the initial cost of the solar system,

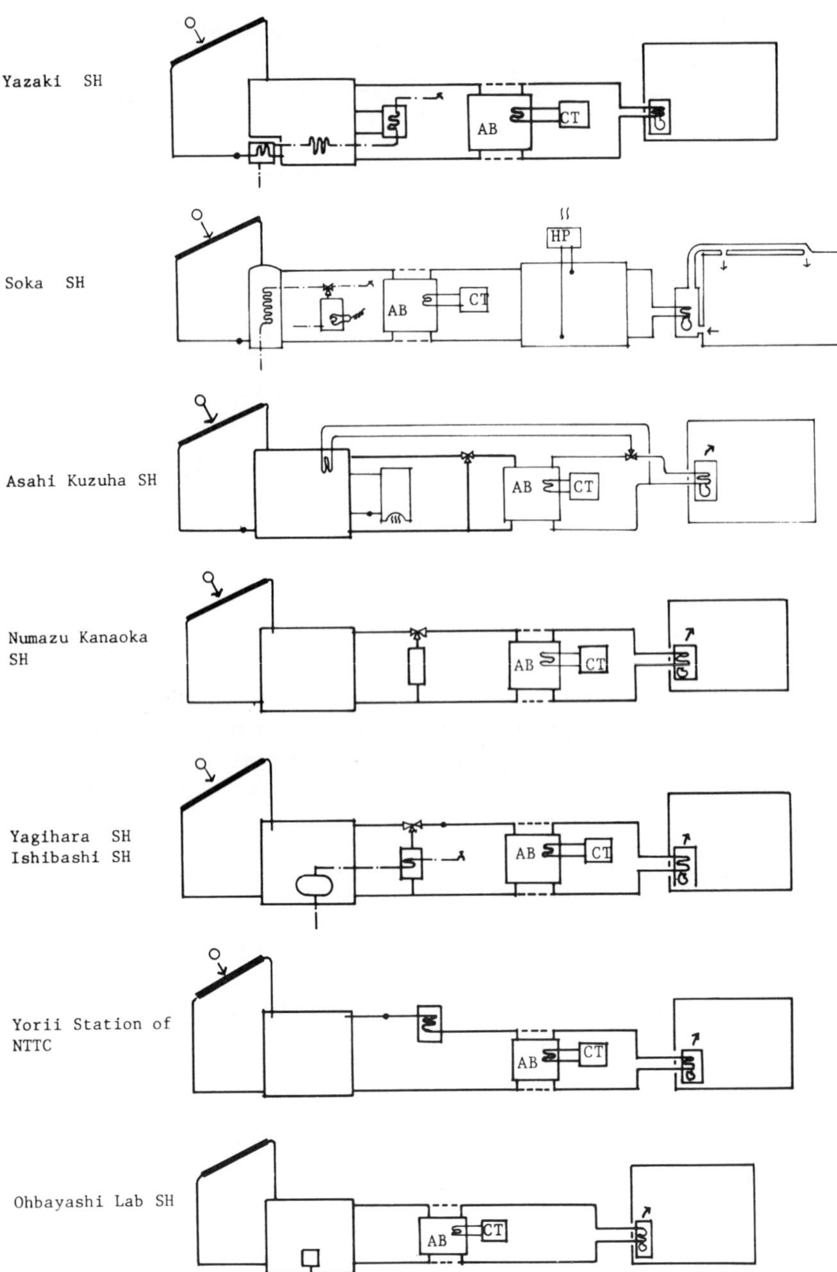

FIG. 16 Solar cooling systems using the absorption refrigeration machine.

FIG. 17 KEP Japan Housing Corporation Solar House.

FIG. 18 Konishi Solar House.

FIG. 19 PS Factory.

simple and efficient direct heating systems must be developed if solar energy utilization is to be increased. It is regrettable, however, that there are always problems with the direct system regarding the diurnal variation of the storage tank temperature level. The larger the capacity of the storage tank, the lower the temperature and thus the greater the amount of auxiliary heat needed to raise the temperature. The smaller the storage tank, the higher the tank temperature and thus the smaller the collection. A sophisticated control in the combination of multilayer storage units may solve this problem but may cause the initial cost to rise. It is preferable to locate the auxiliary heat source at the point nearest the heat discharge to the room; however, in order to reduce the capacity of the auxiliary heat storage, a part of the storage might have to be reheated. This approach is attempted in the first-floor unit of the experimental apartment house (KEP SH) of the Japan Housing Corporation.

One of the most effective uses of solar energy would be the preheating of the intake of outside air used for ventilation. A simple air space made by a blackened steel plate with a glass covering was used in the solar preheating system at the PS factory in Hokkaido (Kimura and Miyazaki, 1976) (Fig. 19). The overall collector efficiency amounted to 70% on a daily total basis in winter owing to a small temperature difference between the outside air and the air in the collector.

Figure 20 shows the direct solar heating systems used in the abovementioned solar houses.

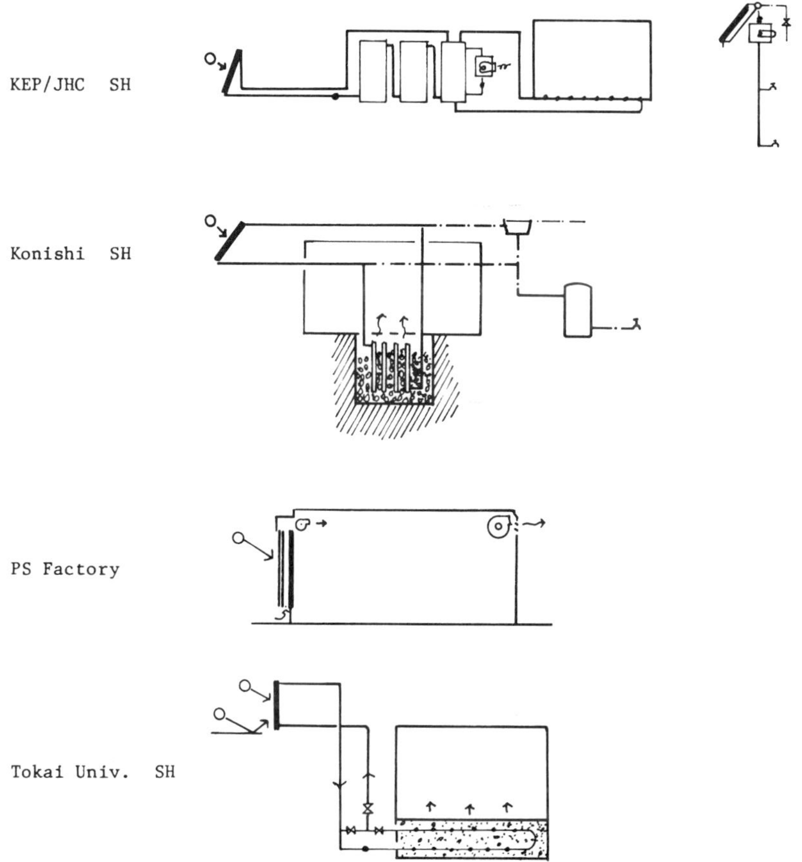

FIG. 20 Direct solar heating systems.

13.5 COLLECTORS USED IN SOLAR HOUSES

The collector is no doubt the essential component of the solar house. An interesting question is where and how the collectors can be installed. The optimum shape of a house integrated with collectors would vary according to different requirements and conditions: whether the system will be used for hot water only, for hot water and heating, or for heating, cooling, and hot water; where the house is situated; whether the house is newly designed or already existing and how much energy must be saved by solar energy utilization. Table 2 reviews types of collectors used on solar houses in Japan.

TABLE 2

Various Types of Collectors in Solar House Design

Type	Basic form	Examples				
Roof mounted	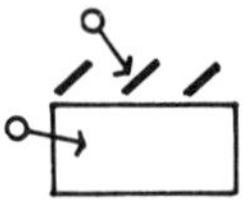	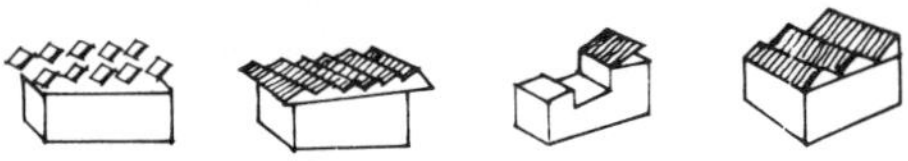 Ohbayashi	Munazu Kanaoka	Meikoshi	Torii	
Gable roof	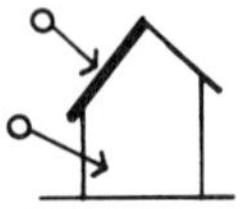	 Mangano Takemura	Toshiba	Sharp Eidai	Ishibashi	Uesaki
South vertical	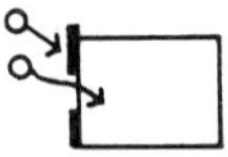	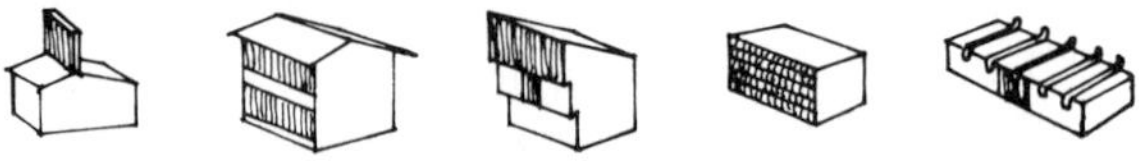 Tokai Univ.	Kimura	Uchiyama	Kajima	PS
One-sided shed roof	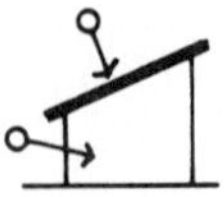	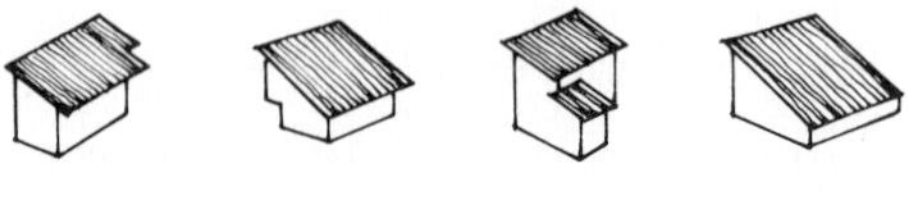 Yanagimachi	Yazaki	Soka	Asahi Kuzuha	
Balcony		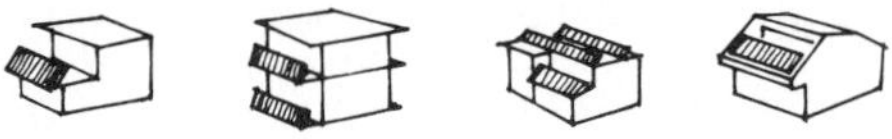 Mitsibshi Sekisui	KEP JHC	Yagihara	Konishi	

The roof-mount-type collector is very common and easily constructed. Both the collector and the roof should be waterproofed and have adequate drainage, as in Kanaoka Assembly Hall and Ohbayashi SH.

The gable roof type, as employed in Toshiba SH, Sharp Eidai SH, and Nagano Takemura SH, provides a large collector area on the south side of the gable roof. The collector tends to obstruct the sunlight into the room on the upper level of the house; so small windows are provided, sacrificing part of the collector area. This roof collector arrangement is suitable for both heating and hot water supply in terms of the tilt angle.

The vertical-type collector is often used for heating purposes only, primarily because it facilitates allocations of interior space, though the total amount of solar radiation on the vertical surface during the winter is about 10% less than that on collectors at the optimum tilt angle. Tokai University SH, Kimura SH, Uchiyama SH, and Kajima Laboratory SH are examples of the use of the vertical collector.

The one-sided shed-roof-type collector is adopted in solar houses with solar cooling, since a large collector area is required to cool the house at the rate of approximately 30 m^2 for 1 RT* of cooling capacity. Yanagimachi SH, Yazaki SH, Soka SH, and Asahi Kuzuha SH use the entire roof area as the collector surface.

Balcony-type collectors are mounted on the balcony in place of railings. The breeze or the view from the window may be obstructed if the collectors are mounted along the entire balcony frontage. Like roof-mounted-type collectors, the balcony type lose heat to the ambient air from the rear side of the collector, though this loss may be small if insulation is sufficient. Other types of collectors integrated into the building structure cause the rear heat loss to contribute to the heat gain inside the house via the insulation in both winter and summer.

13.6 SELECTIVE SURFACES

In 1974 selective surface was applied to the collector for the first time in Japan in Yazaki SH. Until that time no research had been done on selective surfaces. Since then, progress has been rapid, and a few types of collectors with a selective surface have become commercially available.

The absorber panel of the Yazaki collector is made of a special type of double stainless steel sheets seam welded and finished with a selective coating of interference membrane ($\alpha = 0.93$ and $\epsilon = 0.11$). This selective surface is used in all the solar houses with solar cooling discussed in this chapter. The Toshiba Company developed a selective surface which has been applied to the Toshiba SH collector. Collectors made of copper tube in sheet with copper oxide surface imported from the Beaseley Company in Australia were used earlier in Ohbayashi SH, but the surface seems to become nonselective at excessively high temperatures if the collector becomes dry.

Alminum roll-bond sheeting fabricated by the Showa Alminum Com-

* One ton of refrigeration (RT) equals 3020 kcal/h or 12.7 MJ/h.

pany is quite widely used. The specially treated inside surface is said to make the panel anticorrosive and very durable. Later Showa Company succeeded in making a selective surface on the extruded aluminum sheet with which copper tubings are clamped.

13.7 HEAT MEDIA AND STORAGE

All solar houses except Toshiba SH use a water system. Antifreezing measures are very important in a collector system design. There are a few examples of the use of antifreeze, and many cases in which the water in the tank is circulated during the cold night. Some heat is lost with the latter system, but it was adopted by Kimura SH, Yazaki SH, Ohbayashi SH, Soka SH, Takemura SH, and Mitsubishi Sekisui SH. The drain down system seems simple, but it has not been used in many solar houses except in Uchiyama SH because corrosion is liable to occur inside the pipes and frequent starting and stopping may be harmful to the pump. As these problems are overcome, the drain down system is becoming more widely used in solar houses.

A rock bed underfloor storage combined with an air collector is provided in Toshiba SH. Other solar houses are equipped with water tanks made of steel, FRP,* or concrete. The water tank capacity has been about 60–100 liters per square meter of collector area, irrespective of whether heating or cooling is planned. Recent study, however, recommends a smaller tank. Takemura SH, Tokai University SH, and Konishi SH use rock bed storage combined with a water-type collector. These systems are all unique, but additional research is needed in order to reduce the cost of the storage tank; the collector, on the other hand, could be easily mass producted.

13.8 HOT WATER SUPPLY

Solar water heaters are very widely used along the Pacific coast of Japan, but they are not referred to in this chapter. Since a water supply is needed throughout the year, the use of solar energy to produce hot water is a very effective way of saving energy.

When the solar hot water supply system is incorporated into the solar house design, there are a number of methods to be considered. In some houses, the solar hot water system is set up completely apart from the solar heating or cooling system, which can be considered a simple and fail-safe approach as seen in Toshiba SH.

* Fiber reinforced plastic.

When the hot water supply system is integrated with the heating and cooling system, it is necessary to have the heat exchanged somewhere between the hot water and the heating media, as can be seen in most of the solar houses. Usually city water is preheated by solar energy and backed up by the auxiliary heat.

13.9 CONCLUSIONS

Since Japan has few natural energy resources, the effective use of solar energy must be investigated, and low temperature utilization in solar houses is one of the areas that is receiving great attention. It can be envisaged, therefore, that various types of solar houses will be built at a rapid rate in the years to come.

14

A Low-Energy House in New Zealand

J. STEPHENSON

DEPARTMENT OF MECHANICAL ENGINEERING
UNIVERSITY OF AUCKLAND
AUCKLAND, NEW ZEALAND

14.1 INTRODUCTION

In this chapter the potential for solar heating in New Zealand is discussed, and one approach to the problem of reducing residential electrical energy demand is illustrated with a description of a low-energy house. This house, which was constructed and sold for the benefit of a charitable foundation, was open for public viewing only 10 weeks after the decision to go ahead with design work. After being displayed at several locations around Auckland, the building is now on its permanent site, and a planned monitoring of its thermal performance is underway.

14.2 CLIMATE

New Zealand lies between latitudes 34 and 47° S. The total land area is 27×10^6 hectares and the average width of the two islands is about 175 km. The large sea masses that lie to the east and west have a marked influence on the climate, moderating temperatures and contributing the moisture that brings year-round rainfall.

When seeking the most efficient method of using solar energy, an understanding of the weather pattern is essential. The day-to-day weather pattern is determined by an eastward-moving sequence of anticyclones and low-pressure troughs. With the approach of a trough, freshening northwesterly winds prevail, with increasing cloud and rain. The associated cold front brings a change to cold southerly to southwesterly winds

ISBN 0-12-620860-3

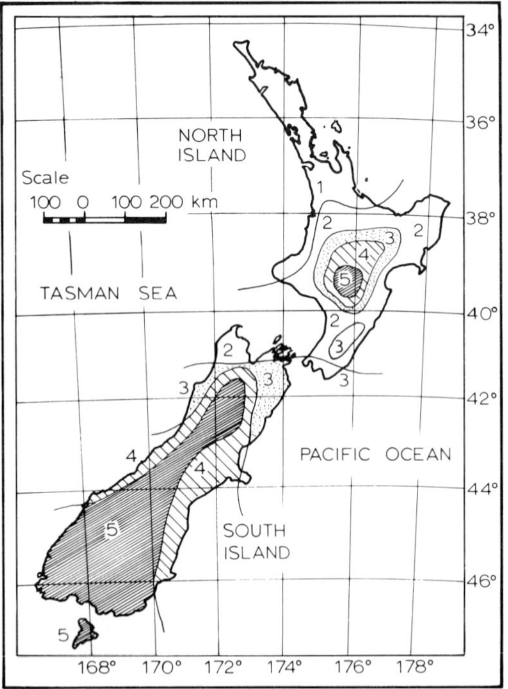

FIG. 1 Climate zones for New Zealand.

and showers. With the approach of the next anticyclone, winds moderate and fair weather prevails. The troughs are unstable systems where depressions readily form, and some develop into vigorous storms that may pass over the islands at any time of the year. When little development occurs, as is common, the pattern is repeated at intervals of about 6 to 7 days. This interval effectively fixes the size of any medium-term solar

TABLE 1

Typical Climate Data for New Zealand

Zone	Annual insolation (h)	% of possible sunshine		Mean temperatures (°C)	Minimum design dry bulb temperature (°C)	Heating degree days (15°C datum)
		Summer	Winter			
1	2150	55	48	15.0	4.5	<1000
2	2050	54	43	13.0	1.5	1000–1300
3	2000	53	40	12.0	1.0	1100–1400
4	1800	47	46	11.0	−1.0	1300–1800
5	1650	41	35	10.0	−2.0	>1800

heat storage that will carry a household through most periods of low-insolation weather.

For a study of residential energy the country can be divided into five climate zones (Fig. 1). Typical insolation levels and other data (from Anonymous, 1976e, and Bastings, 1964) are given for these zones in Table 1. It should be noted that there are marked local variations in climate, particularly in mountainous areas.

14.3 RESIDENTIAL ENERGY USE

A large proportion of New Zealand's population of 3.2 million lives in individual single-story dwellings, typically set on about 820 m^2 of land. The most common form of construction is weatherboard, fastened to a timber framing; it is used for 57% of all houses, while timber framing with brick veneer or asbestos sheathing account for 15 and 6%, respectively. Corrugated or sheet iron is used as a roofing material in 71% of houses. Insulation levels are generally low, since there are no mandatory national standards, although these are expected to be introduced in the near future. Interest-free loans are at present available for retrofitting ceiling insulation.

Partial house heating is still accepted by many people, and even as whole-house heating is adopted, moderate comfort standards are likely to prevail. Typical temperatures are 20°C in living rooms and 15°C in bedrooms. Although there has been some demand for residential air conditioning in the warmer zone 1, there is little real need for it if reasonable design and insulation standards are met. This discussion is restricted to heating applications.

Until a few years ago the expected standards of residential heating were low. A typical house would have been heated by several fixed or portable electric resistance heaters, probably without thermostatic control. This choice of electricity for space heating—and also water heating—was encouraged by the low cost of electricity. Because of the large proportion of hydroelectric generating capacity in the system—about 85% of the electrical energy supply still comes from hydroelectricity—costs have been sheltered from recent oil price rises. In the period 1967–1977 typical residential electricity prices rose by 166% (about 6% in real terms); the average of 2.1¢ (US) per kWh in April 1977 would be regarded as low by world standards.

In 1976 the national residential use of low-grade heat was 32×10^{15} J out of a total residential use of 41×10^{15} J. 70% of the low-grade energy was supplied by electricity, 20% by coal, and 10% by oil. In 1972 electricity was the principal means of heating in 46% of homes, and the average home use for space heating was 5200 kWh per annum.

Manufactured or natural gas is available in only a few localities. In 1975 gas accounted for 5% of the residential and 3% of the total national use. These values will increase as the Maui offshore field comes into production in the late seventies.

In the future, if a greater proportion of electrical growth is supplied by fossil fuel thermal stations, it can be expected that prices will rise rapidly. Improved efficiency, substitution, and even frugality will lower growth rates. There will be greater use of renewable resources, solar energy in particular, for low-grade heat. New construction will take advantage of passive systems.

In order to introduce new energy-conserving technologies successfully, certain preconditions must be fulfilled. Scientific and engineering data that are relevant to local conditions must be available. Production, installation, and servicing facilities must be built up; the economic climate must be favorable. And, most important, the consumers must have an understanding of the need for conservation and of the choices that are open to them. This understanding will become increasingly important as energy scarcities enforce a choice between higher costs and some measure of frugality. The house discussed here represents an early step in the process of informing consumers of some of the alternatives that are available.

14.4 THE LOW-ENERGY HOUSE

In November 1975 a New Zealand television station sought the assistance of a number of experts to design and construct a "stand alone" house that could be displayed at the 1976 Easter Show and afterward sold to provide funds for the Child Health Foundation. After some discussion, the original ambitious plans were modified to provide for a stock house, with low-grade energy for domestic and space heating being supplied by solar heaters used in conjunction with thermal storage.

The specifications given to the designers were that the house should demonstrate some of the techniques and technologies that were likely to be used to conserve energy in New Zealand homes over the next 10 years. Those involved in the design felt that this was an excellent opportunity to test conservation techniques in the residential sector, and it was resolved that the purchaser of the house would be asked to provide reasonable access to any instrumentation needed to measure thermal performance. The house and all necessary equipment and furnishings were to be sought as donations from local manufacturers and suppliers. This factor, and the limited time available, made it necessary to select stock items wherever possible.

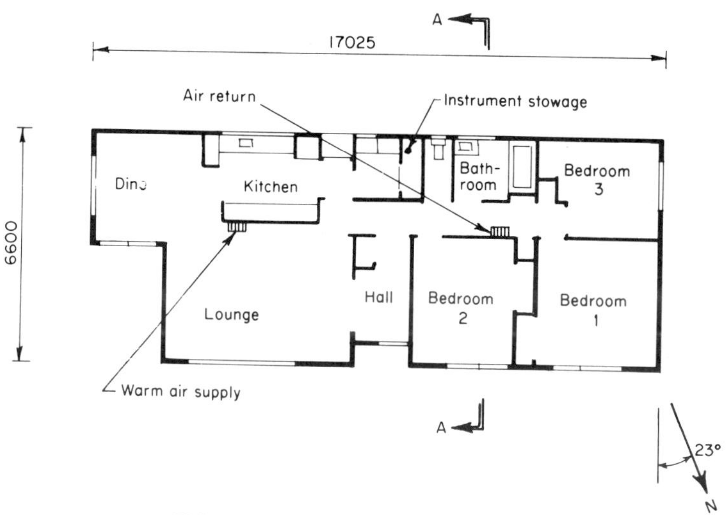

FIG. 2 Floor plan of low-energy house.

The final decision to go ahead with the proposal was taken in mid-January, and the house was opened for inspection by the public on April 2, 1976.

The house selected is typical in floor plan of many New Zealand homes. Based on a standard design by Keith Hay Homes Ltd., it has a floor area of 105 m^2 (Fig. 2). The construction is asbestos cement weatherboard on a timber frame, with bitumen-coated pressed steel tiles on a 15° pitch roof.

The principal structural modification from the standard design was the provision of full height windows along a large fraction of the north (sunward) face. Heavy, close-fitting, aluminum-coated polyester fabric curtains were selected to minimize nighttime losses. Within the existing wall frames, fiberglass insulation thickness was limited to 75 mm; the ceiling insulation was 150 mm. The design this far was representative of good design practice for the milder New Zealand climate zones.

In this particular house the standard space-heating system consists of two fan-assisted electric resistance convective heaters, having a total output of 4.4 kW. This system was not installed.

14.5 THERMAL SYSTEM DESIGN

It was decided that low-grade heat would be supplied by solar water heaters. Since the permanent location of the house was not known at the time of design, space-heating requirements were based on zone 2 climate conditions. In fact, the house was finally located within zone 1. It is worth noting that 53% of the population lives in these two climate zones.

It was clear that by using a standard house design, comfort conditions would not be attainable through a typical bad weather winter cycle without the use of excessively large thermal storage. It was decided to combine thermal storage, in the form of a hot water storage tank, with a water-to-air heat pump.

The basis of the solar installation is an array of "Sunpower" solar water heaters. In this flat plate collector, plates are built up from black chlorinated PVC extruded channels and connected at each end to injection-moulded headers (Fig. 3). The plates are backed by a 25-mm thickness of polyurethane foam insulation. The single thickness of 4-mm glass covering the collector is carried in a galvanized sheet steel housing.

It was determined that an array of 11 collectors could be mounted along the length of the house, at roof-line level (Figs. 4 and 5). The collectors were arranged so as to provide full shading of the north-facing windows in midsummer. Three panels (3.8 m^2 area) were used for domestic water heating, leaving 10.2 m^2 for space-heating needs.

The water storage tank is located under the house, and consists of a PVC plastic liner supported in a timber framework (Fig. 4). The sides and top of the tank are insulated and a vapor barrier is fitted over the top.

FIG. 3 Details of collector plate construction.

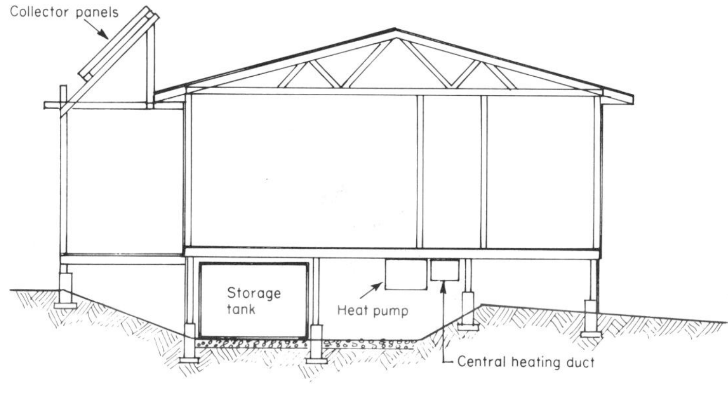

FIG. 4 Cross section of house showing equipment location.

FIG. 5 Northern face of house with solar collectors.

Space heating is supplied by a forced air circulation system. The single floor outlet is located in the lounge, with the return duct in the passage (Fig. 2).

When there is a call for heating, the heat pump fan and the heating coil pump are activated (Fig. 6). With stored water temperatures of 40°C and higher, satisfactory comfort conditions should be attained. When the stored water temperature falls below 40°C, the heat pump is switched on. In this mode of operation energy from the storage tank is supplied to the heat pump evaporator, the duct air being heated as it flows over the condenser coil. The heat pump is a 1.5-kW (input) unit supplied by McAlpine Prestcold Ltd.

In view of the uncertainty at the design stage regarding the house location and the equipment to be used, detailed calculations to optimize the storage system were not possible.

The design was based on the energy loss equation for the house

$$Q = 0.41(15.5 - t_a)$$

where t_a is the ambient temperature and Q the steady state heat loss (kW).

It was taken as a design criterion that over a 7-day period in midwinter (June), the temperature of the water in the storage tank should not fall below 10°C. This provides a reasonable safety margin to prevent freezing

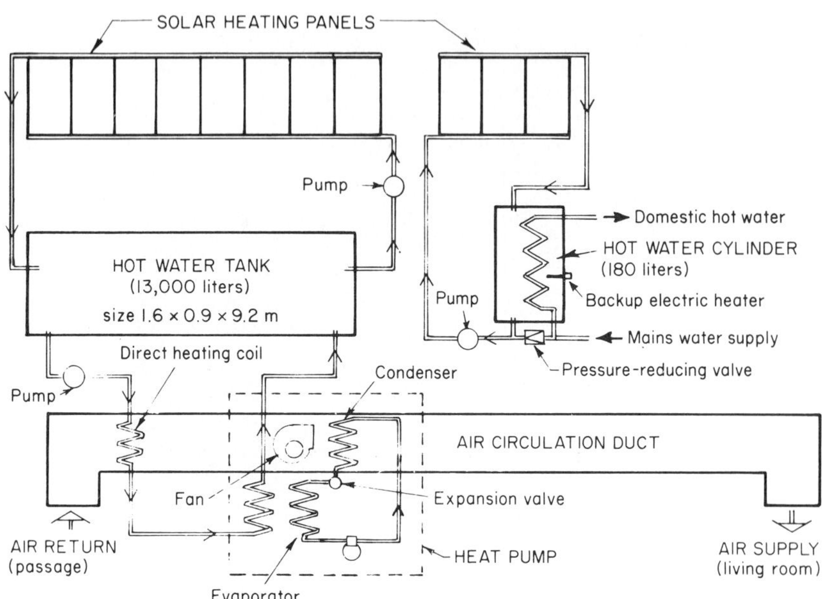

FIG. 6 Schematic layout of heating equipment.

in the heat pump coil and results in a high average heat pump coefficient of performance. It was assumed that during this low-insolation period, the average daily heat supply per panel would be 1 kWh. This takes into account the fact that the water in the solar collectors will be close to, or even below, ambient air temperature.

An energy balance for the design period was carried out, assuming an initial average storage water temperature of 25°C. The required tank size, allowing for losses, was determined as 13,000 litres. In view of the favorable climate at the selected site, some reduction in storage volume could be contemplated.

14.6 FUTURE PROGRAM

A detailed monitoring program has been undertaken through the 1978 heating season; the objectives of the continuing program will be

(a) to refine a computer model for the house and associated thermal equipment

(b) to optimise the configuration and size of various components, in particular the solar heat supply and storage system and the control system

(c) to extend these calculations to other climate zones and housing types.

Preliminary results show that it will be hard to justify the additional cost of a solar and storage system, over and above that of a simple air-to-air heat pump, until there is a marked rise in the cost of electrical energy.

The preceding approach to the design and construction of an experimental house is by no means the ideal one. A longer time spent on the planning and design stages would have been well repaid. However, the project resulted in the provision of an experimental house, embodying some interesting design concepts, in a far shorter time than would have been necessary if funds had been sought through the usual channels.

ACKNOWLEDGMENTS

Many people and organizations were involved in this project, and their contributions to the project, and to this chapter, are gratefully acknowledged.

15

Solar Houses in Britain

J. KEABLE

HELIX MULTIPROFESSIONAL SERVICES
READING, ENGLAND

15.1 INTRODUCTION

Nineteen million solar homes have already been built in Britain. Unfortunately, most of them do not function very well and require a great deal of fossil fuel to bring them up to comfort conditions throughout half the year (typically the heating season lasts from October to March) (see Fig. 1). Nevertheless, the fact remains that very few of our existing 19 million units of housing stock will have been rebuilt during the next 25 years. Current predictions show that during the same period there is likely to be a fossil fuel "energy gap": this may open up by 1995 (in under 20 years, that is to say), while some forecasts put the date considerably earlier than this. It can readily be seen that the capital required for new house building on the scale needed to replace all existing houses over a period of say 25 years would be out of the reach of even a very buoyant economy. So it follows that whatever solutions are proposed for "the solar house in Britain," these solutions will be widely applied only if they can be incorporated into the existing housing stock.

A certain amount of thought and attention had been given in some quarters to the problems of resource depletion, particularly in the field of energy, before 1973, but it was only after the sharp increase in the price of oil that year and the consequent "energy crisis" in Britain that these subjects began attracting widespread interest. Since that time the government has formed a separate Department of Energy; it has created the "Energy Technology Support Unit" (a scientific advisory body to evaluate strategies, trends, and specific proposals); the U.K. section of the International

ISBN 0-12-620860-3

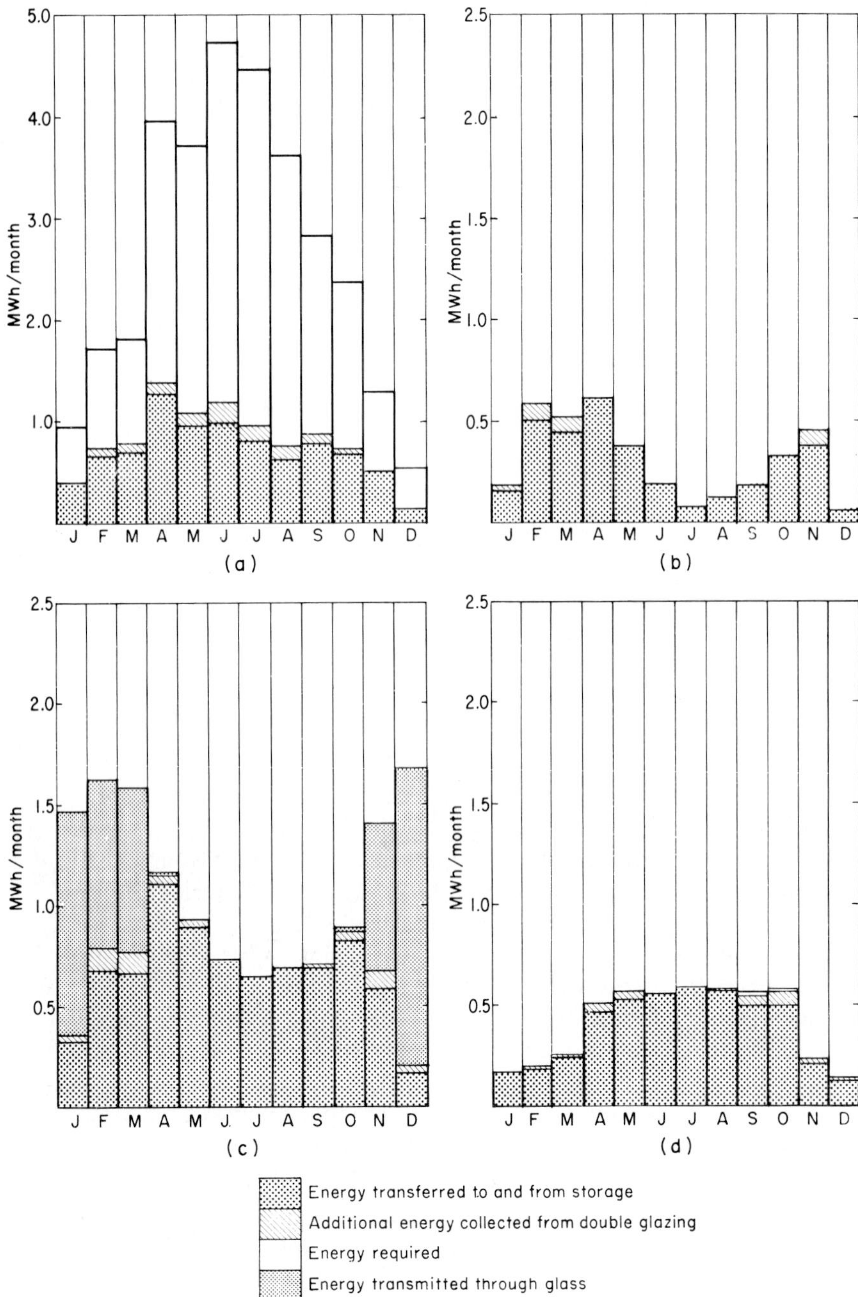

FIG. 1 Solar energy utilization. The figures refer to actual weather data for 1969. (a) Solar energy available; energy collected from total available: 26% from single glazing, 30% from double glazing. (b) Space-heating requirements; solar contribution: 50% of year's total (single glazing), 54% of year's total (double glazing). (c) Total heat requirements; solar contribution: 59% of year's total (single glazing), 63% of year's total (double glazing). (d) Water heating requirements; solar contribution: 68% of year's total (single glazing), 72% of year's total (double glazing).

Solar Energy Society has been formed and has held a series of lectures and seminars covering the field from photovoltaic conversion to solar collection in agriculture. At the same time the practical development of hardware has progressed rapidly, since higher energy costs make it possible to justify higher capital costs for energy collecting devices. Projects by universities have varied from theoretical studies concerned with the levels of energy obtainable in different parts of Britain and the factors for collecting at these levels, through the development of computer programs aimed at assisting design processes, to involvement in practical projects for building experimental houses in conjunction with government or local authorities programs.

It has come as a considerable surprise to many people that levels of solar energy obtainable in Britain are sufficient to justify serious thought and work on the possible ways of using this resource in the future. It is now generally recognized that a flat plate solar collector can make a signicant contribution to the supply of domestic hot water, and the only question remaining in this area is whether the total economics of such installations would justify their widespread use. Hot water is required throughout the year; so the lack of winter sunshine is a disadvantage but not a complete bar to development. However, if we consider space heating, the problems are clearly of another order. In Britain the major climate modification required of a building after it has provided the basic shelter from wind and rain is to offer comfort warmth in the winter. Summer cooling in the sense of a need to reduce inside temperature below that of external ambient conditions is very seldom required, and therefore in considering the needs of the solar house for Britain we shall deal here mainly with house heating and the supply of domestic hot water.

While defining our subject, we should not overlook the fact that wind energy is, strictly speaking, solar powered and as such must be considered a logical and possibly important subsection of this subject, since wind power is more abundant in winter, the very time when space heating is particularly required and when sunshine is particularly lacking (Anonymous, 1976c).

At this point it is necessary to touch on thc qucstion of "life-style." Demands vary very widely, but the general trend has certainly been (until 1973) to demand higher comfort conditions of home heating and larger volumes of domestic hot water for bathing, washing, and washing machines. Domestic electricity consumption between 1970 and 1975 rose by 15% and of gas by 66%. Undoubtedly most of this energy was used in space and water heating, and these trends have to be recognized when considering the solar house.

15.2 CLIMATE

When considering the climatic background to the design of solar housing, we have to recognize wide variations in each of the relevant sectors. Although small, Britain spans a full 9° of latitude from south to north; bright sunshine may vary between 40 h in December and 220 h in June (as at Kew), and for the same location the average intensity of direct solar radiation will vary from 0.41 cal/cm^2/min up to 0.77 in June. Of great significance will be the variation in diffuse radiation: at Kew intensity will fluctuate from 0.05 cal/cm^2/min in December to an average of 0.39 in June. For a great part of the year, only indirect, diffuse radiation will be available, and the number of completely cloudless days in an average year will be very low. Pollution levels will vary widely, in some places resulting in further reductions of energy levels received. The relationship between direct and indirect energy will vary from greater than 4:1 in December to approximately 2:1 in June. These figures explain why the concentration of effort so far has been on the flat plate solar collector and why the largest unsolved problem lies in the area of energy storage (see Figs. 2 and 3).

15.3 SOME EXAMPLES OF SOLAR BUILDINGS

Before attempting to point to the direction in which the solar house of the future might lie, let us look at a few examples of work carried out to date. In 1957 architect Ed Curtis built his own home in Rickmansworth and designed it with a solar wall facing south. The intention was to trap as much heat as possible in the winter (when the solar altitude is low). Heavy curtains improve the nighttime insulation value of the double glazing. The backup system consists of a heat pump, which (after modification) uses domestic cold water as the low-temperature heat source, while internally distribution is through air ducts. The system has been in operation since its construction.

In 1957 the Milton Keynes Development Corporation decided to adapt one out of a standard terrace of houses so that almost all of its monopitched roof should consist of a flat plate solar collector aimed at delivering both domestic hot water and some space heating (see Figs. 4–6). The house is now built and occupied and is being carefully monitored: one of the apparent difficulties here is the problem of dissipating the very large quantities of energy collected in summer heat wave conditions. The hot water collected is stored in relatively large steel tanks within the house, and a gas boiler provides the backup for space heating and hot water where necessary; it would appear the system is producing useful results.

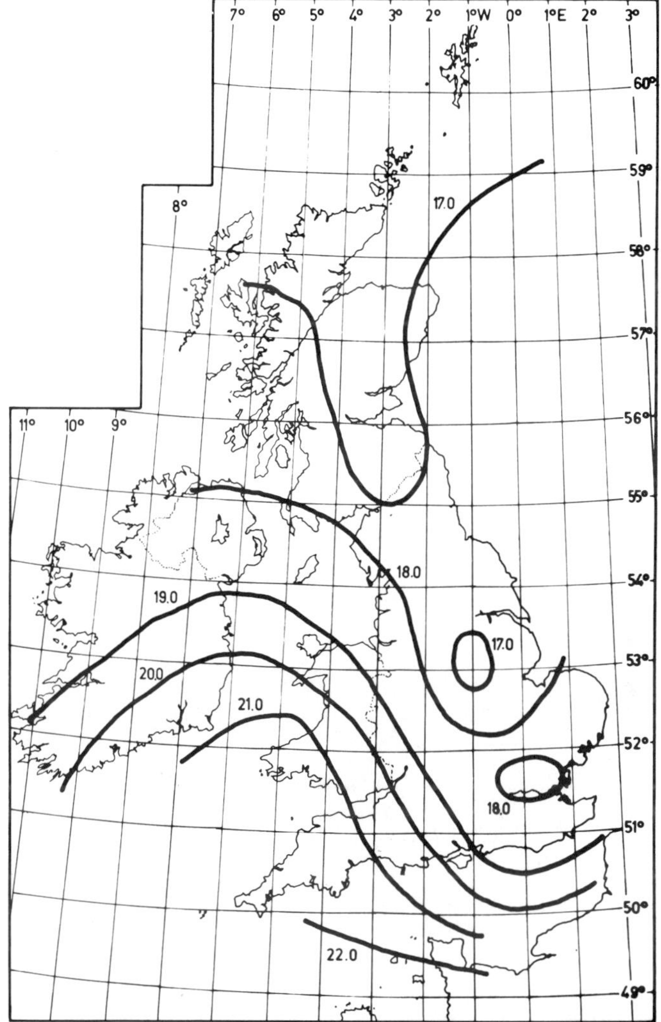

FIG. 2 Global solar radiation; average daily totals for June (MJ/m²).

More radical proposals have been incorporated into the plans for a fully autonomous house designed at Cambridge University by a team under Alexander Pike (Fig. 7). The intention is to demonstrate that it would be possible to build and operate a house completely separated from existing service networks, so that not only energy but also water and sewage are dealt with autonomously. Much of the technology is not only new but as yet unproven, and so the intention is to follow computer studies

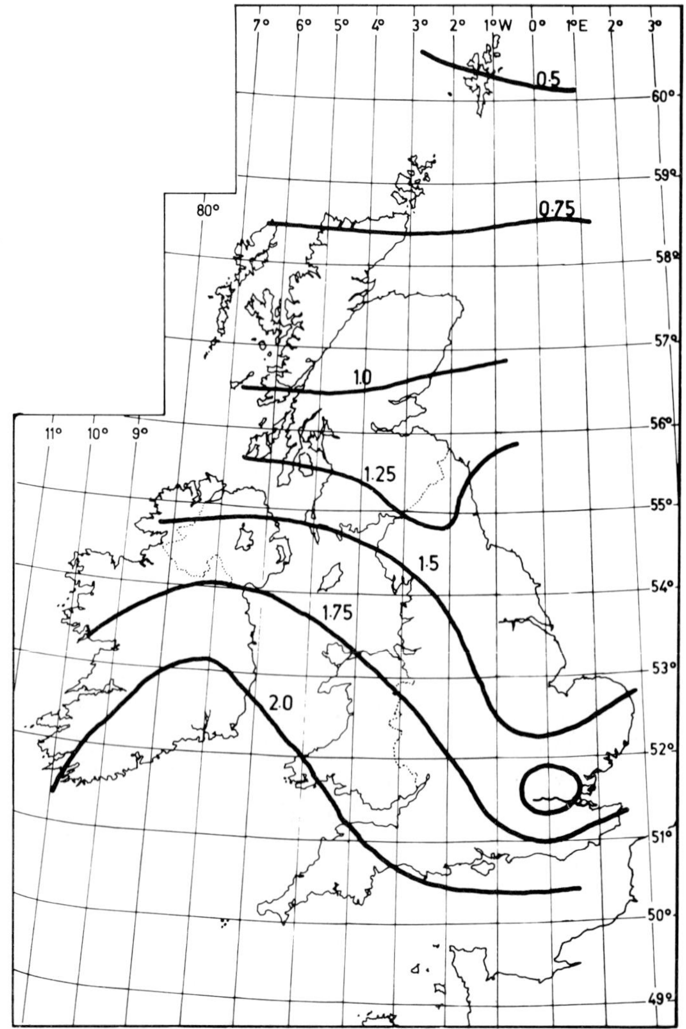

FIG. 3 Global solar radiation; average daily totals for December (MJ/m²).

and research and development work with the construction of an experimental house that would be very rigorously monitored. One of the features of the house is that it divides the living area into a core zone that would be kept at comfort temperature throughout the year and a larger space to be used only when conditions were favorable. This is one way to respond to the problem of low-energy inputs during the winter and would certainly seem to be of wide general relevance. The house would collect

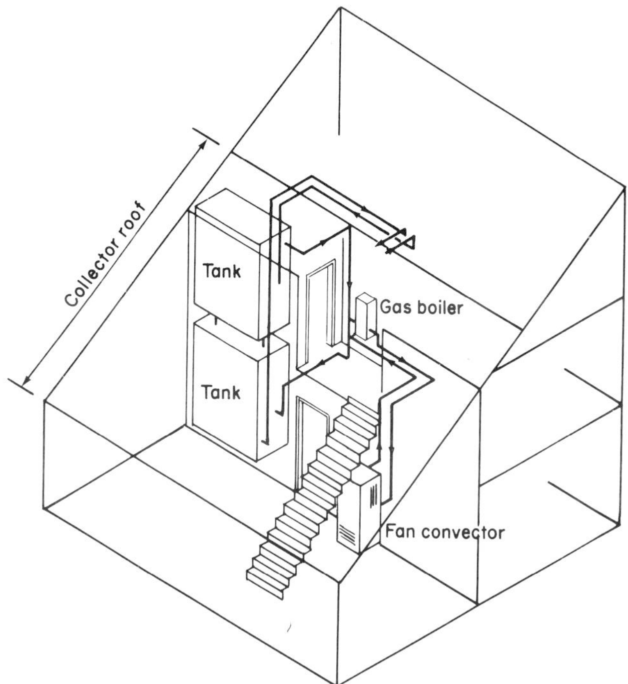

FIG. 4 Milton Keynes solar house showing the major components of the heating system.

FIG. 5 Milton Keynes solar house.

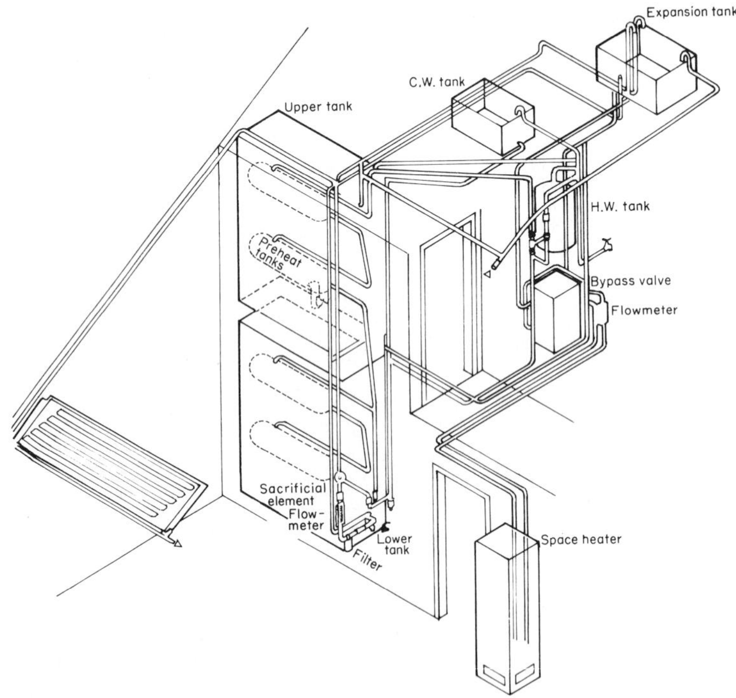

FIG. 6 Layout of the Milton Keynes solar house.

FIG. 7 Cambridge autarkic house.

solar energy, wind energy, and rain and also elaborate methane from waste refuse and sewage besides recycling used water (Vale, 1976).

Alongside these experiments has grown a solar collector industry with an increasing number of small firms offering systems for domestic hot water. There is a growing realization that very rapid response is essential, and typically controls will operate to pump between the flat plate collector and the storage cylinder when temperature differences of 1 to 2°C are reached (see Fig. 8). The value of single or double glazing the collector panels has been widely argued; similarly, the value of making the collector itself of low specific heat has been exploited by some manufacturers. This industry has not, however, attracted any strong support from the government, which so far has failed to be convinced that at current price and tariff levels the systems developed are in fact cost effective.

A number of workers have emphasized in paper studies the value of the passive approach, with the aim of maximizing the benefit of incidental gains and minimizing losses of heat from the house so that the necessity for additional energy inputs is reduced. This approach is of course of value both in conserving fossil fuels and in making it possible to provide the necessary input by means of alternative energy sources. Moreover, ventilation rates have been of increasing relevance with an increase in the insulation values adopted for the structure itself. In the December 1973 study by the Rider and Yates partnership in Newcastle entitled "House under Glass," it was proposed that air change rates should be a maximum of 3 and a minimum of 1, under the direct control of the occupier, and that heat recovery by the use of the Munter's wheel should be adopted. The

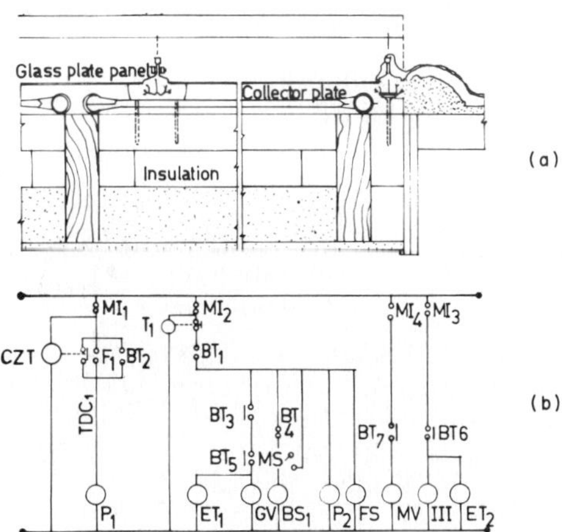

FIG. 8 Control system for pumping and storage of hot water: (a) collector detail; (b) electrical schematic.

basic concept of this study was to erect an outer totally glazed enclosure, inside which the more conventional zones of the house would be planned, the space between the two becoming a sort of buffer. A detailed and careful evaluation produced a final estimate of heat load at 6600 MJ/year based on the inner shell having an average U value of 0.2 $W/m^2/°C$.

Also from Newcastle, Professor A. C. Hardy has been concerned with proposed housing design in connection with Washington New Town. More conventional than "House under Glass" but nevertheless using reasonably high insulation values at 0.4 $W/m^2/°C$, the proposals also foresee a reduction to 1 air change per hour. In this case the ventilation would be by positive pressurization of the house, the fresh air being drawn from the roof space. Since air in the roof is already preheated to a certain extent, an additional reduction in energy demand is provided by this means. The advantage of positive pressurization is that it permits positive emission of air, which would be released from the kitchen and WC. Professor Hardy pointed to the possible risk of overheating in certain areas in the house once criteria of this kind are applied. He also pointed out that if water heating, cooking, and miscellaneous energy requirements are set against space heating in a very well-insulated house with controlled ventilation, then at 19°C internal the space-heating requirement is somewhat less than half the total energy requirement.

FIG. 9 Conservation house built in Wales for National Centre of Alternative Technology (NCAT). (Courtesy Wates Ltd., Norbury.)

This work provides a background for the house constructed in Wales at the National Centre for Alternative Technology according to a design by Peter Bond RIBA. In this case "the passive principle" has been taken to an extreme, and the house has been wrapped in 450 mm of fiberglass within the resultant oversized cavity, while natural light is provided by double windows each double glazed (i.e., four layers of glass). The single external door opening has two doors and the house is fully artificially ventilated (see Figs. 9 & 10). At these levels of insulation and with low air changes, both cooling and heating are required and are provided by a reversible heat pump circuit. The outdoor coil is mounted within the roof space, while the indoor coil and compressor are placed at the heart of the house. The house was finished in 1976. Placed as it is as part of a permanent exhibit to the general public, monitoring is virtually impossible in any strict sense.

The author and his colleague (Keable and Dodson) in 1975 put forward prizewinning proposals in the Copper Development Association solar en-

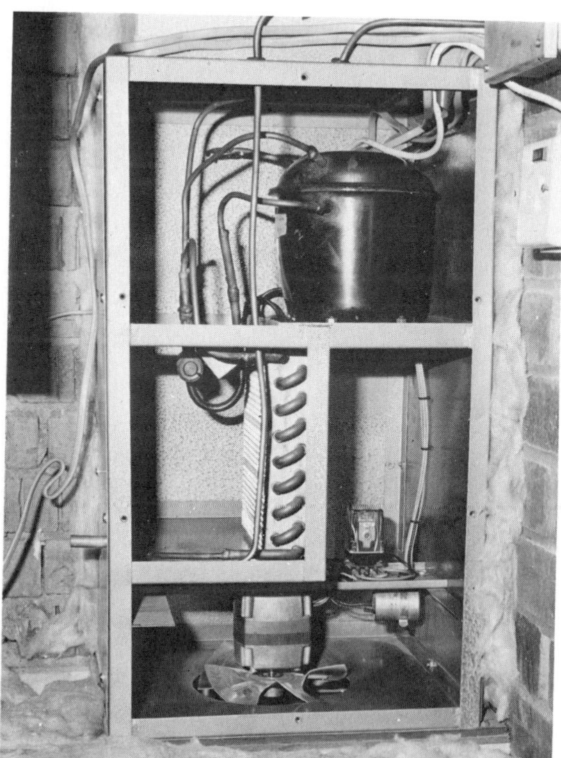

FIG. 10 NCAT conservation house heat pump.

ergy competition for combining the passive approach (high insulation, low ventilation rates, etc.) with a heat pump that would employ as a heat source air that had been preheated using the whole roof area as a solar collector (see Fig. 11).

The roof itself using standard materials would be designed so that air entering the loft space would pass immediately on the underside of slates or tiles and then to the outdoor coil of the heat pump. The heat collected would then be passed to a thermal store arranged below the ground floor of the house using water as the storage medium and sized to give 5 days total requirements to the house at $-7\frac{1}{2}$°C ambient externally. Heat transfer from the store to the house could then be by ducted warm air or directly to radiators. Late in 1975 these proposals were implemented in the existing London house of one of the authors, in this case the roof covering being of slate and the internal heat transfer being via water to radiators. The house was first insulated to a considerably high standard, reducing the total fabric losses from over 20 to under 10 kW, and similarly the ventilation rates were reduced by double glazing, draught stripping, and the installation of door lobbies. The system was operated for the heating season 1975–1976 at an estimated coefficient of performance of the heat pump of 3:1.

At the same time a new house was commenced at Long Sutton in Hampshire, and the same principles were employed, except that in this

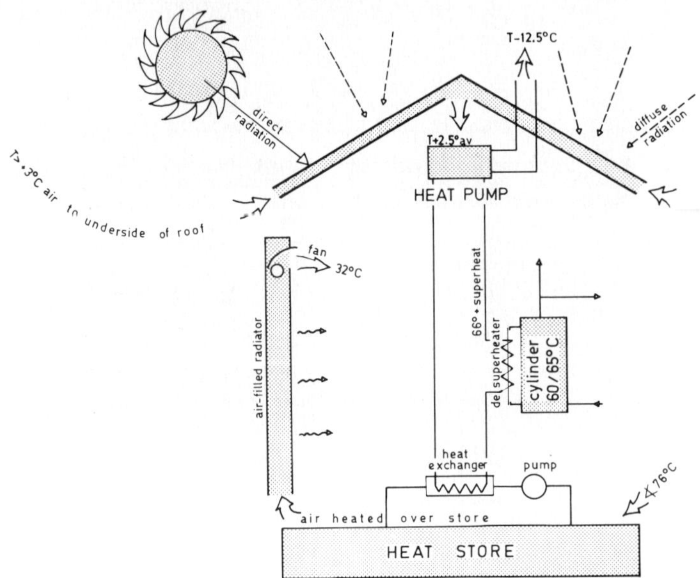

FIG. 11 Keable and Dodson proposed house.

case the roof covering was of concrete tiles and the distribution medium was ducted air. The house was completed in autumn 1976, and the system was operated during the heating season 1976–1977 for the first time (Figs. 12–14).

The use of solar energy in this case is to provide all the heat for the houses concerned via the ambient air. In effect, the air is acting on a continental scale as a heat transfer medium from the thermal stores of land masses and oceans, thus solving the basic problem of seasonal storage, which we have seen is an overriding necessity in the case of U.K. solar housing. For the moment fossil fuel is used via the electrical generation system to provide the energy necessary for the extraction of heat from the ambient air, but in principle this energy could also be supplied by other means. In many parts of Britain it would be feasible to harness wind power via heat-pumping technology to this end, and in other locations solar energy stored as fodder and processed though livestock and methane generators would be fed through gas burners to the same end. In

FIG. 12 Air distribution in house at Long Sutton.

FIG. 13 Standard roof modification in house at Long Sutton.

FIG. 14 Heat pump under the roof of Long Sutton house.

each case the principle involved is that of multiplying the initial energy input by a factor that might vary between 1½ and 4, and in each case the end product is an energy requirement for space heating and hot water of something like one-third that of the available alternatives.

15.4 THE SOLAR HOUSE OF THE FUTURE

This brings us to the question of what the solar house of the future may be like. Let us look first at new housing. Although this is only likely to account for some 100,000 to 300,000 houses a year and therefore only a fairly small percentage of total U.K. energy requirements, nevertheless it offers the opportunity for experiment and thoroughgoing application of research results. The passive approach is likely to be incorporated, and hence careful attention will be paid to orientation permitting a reduction in fabric losses arising from sheltered situations; a maximum gain from solar penetration through windows in winter; the provision of a structure having high thermal mass but low overall U values, probably of the order of 0.2 to 0.3 W/m^2/°C; and almost certainly some form of ventilation control in order to avoid air change rates in excess of 1 to 1½ per hour. There will probably be many cases in which an external south-facing intermediate zone will be designed, a sort of large conservatory adding to the useful space of the house in the intermediate seasons when heat and cold are not the major problems. Experiments with variable insulation may well be seen, perhaps taking the form of well-insulated shutters in place of or in addition to curtains for windows at night.

Domestic hot water is likely to be provided by flat plate solar collectors having an emphasis on rapid response and low thermal mass, and the aim would be to provide for the maximum amount of water required during the summer months but avoiding oversizing the collector in an attempt to provide hot water during the winter. This would tend to provide collector areas in the 1–2 m^2 range, assuming an operating efficiency of say 40% and an average requirement of some 100 kWh per month for a small family. It will be necessary to reduce the total installation price to something of the order of £200 at 1975 prices, but on the assumption of greater standardization and compatibility of the necessary components, this would seem to be a possibility. Given the high insulation values and low ventilation rates for the foreseeable future, there would seem to be every likelihood that small reversible heat pump systems will be developed, capable of supplying up to 3 kW of heating at an input of say 1 kW (or the equivalent), and in many instances the need to absorb, store on a short-term basis, and in certain cases remove an excess of incidental gains from occupancy, cooking loads, and the like will be dealt with in this way.

When it is inherently uneconomical to heat water by flat plate solar collection methods, the heat pump would also provide for this, at an energy ratio (COP_h) of at least 3:1. On this basis only two-thirds of the capital cost allowance that could be attributible to direct solar water heating could be applied to the equipment needed, but this should be more than adequate.

The economic basis is likely to be more closely related to running cost advantages than in the past. The capital cost of equipment or increased insulation costs will increasingly be accepted provided a pay back period of 3 to 5 years can be shown, but it will be necessary to build in assumed inflation rates, increased fossil fuel tariffs, and equipment depreciation rates. Thus, for example, if a flat plate solar collector is built of materials that will require replacement in 10 years, this must also be reflected alongside the fact that the energy bill will be zero as compared with a conventional combustion boiler. With heat pumps, if the life of the equipment is the same, the extra capital cost must show a clear running cost saving long before the replacement becomes necessary.

The general strategies outlined for new housing can in most cases be applied to existing stock. The scope for providing climatic shelter externally will be strictly limited, and the question of orientation will be confined to the opening and closing of window apertures in the course of conversion work. However, the reduction of energy requirements by overall insulation will prove increasingly beneficial. Alternatives to insulation within cavity wall construction are likely to be developed for application to the remaining large stock of solid wall houses. The sealing of windows in the context of controlled ventilation drawn from existing roof space is likely to prove beneficial in many instances, and the techniques of double glazing to existing windows are of course already well developed.

The remarks already made about solar water heating would apply without change to the conversion of existing houses, but in the case of space heating it is unlikely that it will be possible to reduce the total energy requirement below some 6 to 8 kW. In this context, heat pumping in conjunction with thermal storage is likely to prove very beneficial, particularly if the problems surrounding the provision of thermal stores externally to existing houses can be overcome (Anonymous, 1976f, 1976g; Heap, 1975).

It remains true that taken on an interseasonal basis, there is even in the United Kingdom an excess of solar energy in the summer sufficient to provide the entire energy needs of water heating and space heating in the winter. The Danish zero energy house has demonstrated this quite clearly in a similar climatic context, but to date there has not been much evidence to suggest that the cost of providing the very large interseasonal storage required would in general be justified. However, it is likely that even in this context heat pumping will remain a useful tool, since the effective capacity of a heat store can be doubled if the heat can be recovered at much lower temperatures than would be directly usable for space heating. The distribution of heat through normal radiator systems depends on heating water to some 68–70°C. Large surface, low-temperature panels can work effectively in the region of 50°C. Carefully designed air distribution systems may work at even lower temperatures still, but by recovering heat via a heat pump a thermal store may be effective down to say 5°C. The advantage here will be that the recovery of the store will be rapid using ordinary flat plate collector methods in the early stages, since the cooled water will reheat at relatively high efficiencies.

Since it was calculated in the case of the Danish zero energy house that half the total energy requirement for the house in the winter would be dissipated from the store by the unavoidable losses due to conduction (despite the very high insulation values adopted), this latter point becomes of particular relevance. The lower the temperature becomes within the store, the less the waste by heat leakage, until the point is reached where the store drops below ambient and is actually attracting local heat to it.

15.5 CONCLUSIONS

The solar house in Britain is likely to develop along two main lines:

(a) the "passive approach," aimed at the substantial reduction of the need for energy inputs, and

(b) the development of alternative energy sources particularly related to the flat plate solar collector and the collection of ambient energy via heat pumping.

16

Solar One

*M. A. S. MALIK**†

UNIVERSITY OF DELAWARE
NEWARK, DELAWARE

16.1 INTRODUCTION

Solar One is an experimental house that provides a facility for full-scale testing of components and development of an integrated energy system for the fullest exploitation of solar energy to meet the needs of a typical family house in Delaware and to determine the optimum mode of system operation. A combination of energy from the sun and the utility grid has been employed in such a manner as to effect maximum reduction in the daily energy drawn from the utilities, peak shaving, and load leveling.

For a typical sunny day in Delaware, load leveling is achieved by a mix of solar energy utilization and storage, and an efficient mode of running the air conditioners. The valley in the typical demand load curve of a family home during summer can be leveled by running the air conditioners during the night (which is an efficient mode because of low atmospheric temperature) and storing the coolness (as solidification of a salt) to obtain cold air for air conditioning during the day. This availability of cold air during the day, combined with the fact that the useful energy available from the sun (an increasing function of solar insolation) is well correlated with the demand load, leads to load leveling on hot sunny days, when about 40% energy storage is employed. During the winter most of the

* The author was a participant in the work reported here, during his stay at the University of Delaware.

† Permanent address: Engineering Division, Kuwait Institute for Scientific Research, Kuwait, Kuwait.

ISBN 0-12-620860-3

heating of Solar One is provided by solar energy; supplementary heat is provided by resistance heaters or more efficiently by heat pumps.

During 1973–1974, even though more than half the roof was covered by relatively inefficient collectors, about 60% of the winter heating load was met by solar energy; with improved collectors, the corresponding figure has increased to 80%. In summer, the nighttime operation of air conditioners and cooling by stored coolness during the day leads to overall energy savings, load leveling, and peak shaving. Efficient solar collectors using air as the heat transport medium, and commensurate with the operation of solar cells on the plate, have also been developed in the program.

16.2 THE SOLAR HOUSE

A. Thermal System

The house shown in Fig. 1 is constructed in such a manner that it permits maximum flexibility for experimentation (data acquisition, systems and component development) and will be suitable for use, with only minor modification, as a solar house after termination of the experimentation phase. The house is well insulated and contains 1300 ft^2 of floor

FIG. 1 Solar One.

space: living/dining room, two bedrooms, $1\frac{1}{2}$ baths, kitchen, small entrance hall, and washer–dryer corner separated from the main hallway. The wall insulation is 1-in. sprayed-in-place K13 Type T foam (National Cellulose Corporation) plus two layers of $3\frac{1}{2}$-in. fiberglass blankets, one side coated with a heat reflector (aluminum foil), installed back to back between 2- × 4-in. studs. To reduce infiltration, the entire outside wall area is covered with a 4-mil polyethylene film. The exterior has redwood bevel siding on top of an Armstrong sheathing (1 in. thick). Inside the house, the walls are covered with sheetrock, tapejoint spackled, and painted. All windows are fixed double-pane Thermopane. Special care has been taken to insulate with rockwool and by caulking with Thiokol compound the space around each window frame to reduce infiltration. The house has two Andersen sliding glass doors (Thermopane), one in the living room and one in the master bedroom at the opposite end of the house, to provide easy cross ventilation when desired. Above the stove in the kitchen there is a circulating fan with a charcoal filter to eliminate kitchen odors and to reduce infiltration and heat losses. The refrigerator/freezer is located next to an air duct to permit future venting of heat during the summer. The washer/dryer combination is located in a niche facing the garage to permit outside venting (summer) or venting into the garage (winter). The main entrance door opens into a small entrance hall, which is separated by doors from the living area and hallway to minimize infiltration.

The house is heated and cooled by circulating air, with air ducts opening on the floor below most windows. There is currently one main air return located in the closet in the main hallway. The temperature of the house is controlled fully automatically by two thermostats (for night and day).

Adjoining the north side of the bedroom area is the garage, which is currently used as an exhibition area. It is heated by baseboard electric heaters and air conditioned by a separate window unit. The garage walls and ceiling are insulated with 4-in. fiberglass blankets and covered with sheetrock.

The entire house (except garage) has a full cellar. All heat storage, the control system, and the electrical data acquisition equipment are located there. The cellar is thermally insulated from the living area by sprayed-in-place K13T cellulose and sheetrock or, wherever possible, by a suspended ceiling.

The roof has a 45° south slope and a vertical north wall. The 2- × 10-in. roof rafters are spaced 4 ft apart to accommodate the Plexiglas skylight and two 4- × 8-ft panels between each rafter. Roof construction is torsion stabilized by 2- × 6-in. tongue and groove roof decking and wind-braced by 45° rafters on the north wall into the garage. The attic is insulated with 1-in. sprayed-in-place K13T cellulose to prevent excessive temperatures during winter and summer. Thermal data acquisition for collector research and insolation is done in the attic.

B. Electrical System

The house is connected to the power utility grid through a main watt-hour meter with a two wire, 115-V, 200-A connection. From the main distribution panel a separately metered branch feeds the garage. Two other branches are connected to the main house load, insensitive to ac or dc operation and to the mechanical system of the house.

The ac/dc load can be switched by the main transfer switch between the power utility grid and the dc solar energy system powered from solar cells and the connected power processing system.

C. Brief Description of the Present Heating System

An overall schematic diagram of the airflow system in Solar One is shown in Fig. 2. The south-facing roof has 24 solar collectors. Each collector measures about 4 × 8 ft, and the collectors are placed in pairs, 12 pairs in parallel.

An additional 6 collectors are installed in the south-facing wall. The active surface of these collectors measures about 4 × 6 ft. The total overall area of 30 collectors equals about 850 ft^2. The total area of the absorber plates available for intercepting solar energy equals nearly 750 ft^2.

Four of the roof collectors are equipped with CdS/Cu_2S cells of different origin and make, and are connected to the heat-collecting substrate by different means.

The system has three main circuits, which are shown in Fig. 3:

(a) collector circuit, characterized by the solar fan,
(b) house circuit, characterized by the house fan, and
(c) heat pump circuit, characterized by the condenser fan.

At present the airflow through each collector (which can be varied) on the roof is around 225 cfm and that through each collector on the south wall is approximately 100 cfm. The solar fan is, therefore, supplying a total of about 3300 cfm of air against 1.5-in. H_2O overall pressure drop; currently it is coupled with a 2-hp motor. The house fan is powered by a $\frac{1}{2}$-hp motor and supplies air at the rate of 1200 cfm. The overall pressure drop in the house circuit is 0.3-in. H_2O. The condenser fan is powered by a $\frac{3}{4}$-hp motor and is rated at 2400 cfm.

The nonsolar heating and cooling system of Solar One uses a York

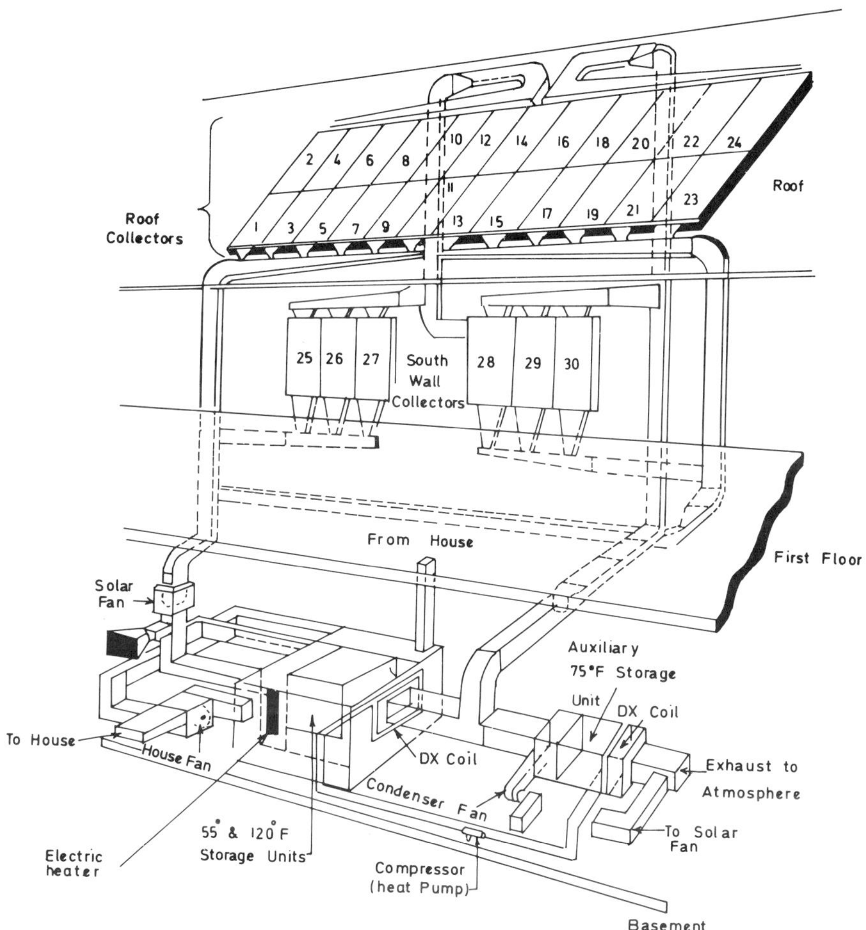

FIG. 2 Overall schematic diagram of airflow system in Solar One.

3-ton heat pump and 9-kW-rated Joule's (resistance) heater as an alternative heat source and for undertaking comparative tests.

D. Automatic Control System

The basic components of the heating–cooling system in Solar One are grouped into three overlapping subsystems: (1) the solar collector–thermal storage loop that contains the solar fan, (2) the thermal storage–house loop with the house fan, and (3) the heat pump–thermal storage house loop with the condenser fan (see Fig. 4); the different modes of operation are represented in Fig. 5.

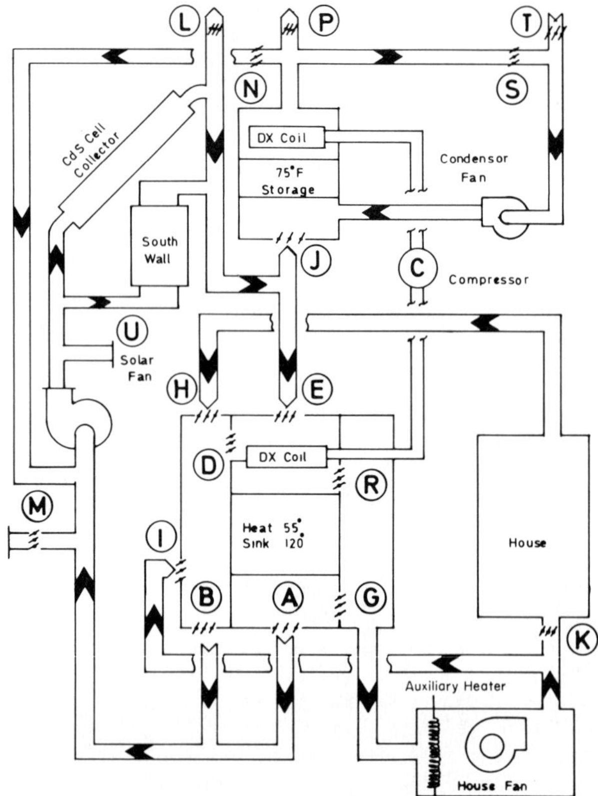

FIG. 3 Solar house thermal system flow diagram.

Temperatures from five locations throughout the heating–cooling system interact through the automatic control system to maintain the house environment at prescribed limits. These temperatures are the solar collector air outlet temperature (T_c), the high thermal storage temperature (T_{sh}), the low thermal storage temperature (T_{sl}), the average room temperature (T_r), and the outdoor temperature (T_o). The criterion for setting the comfort control system into heating or cooling is determined by the outdoor temperature, T_o (Fig. 6).

The normal sequence of control operations for house heating is

(1) direct heating with air from the collectors,

(2) heating from 120°F storage bin,

(3) heating from 120°F storage with heat pump boosting using 75°F storage;

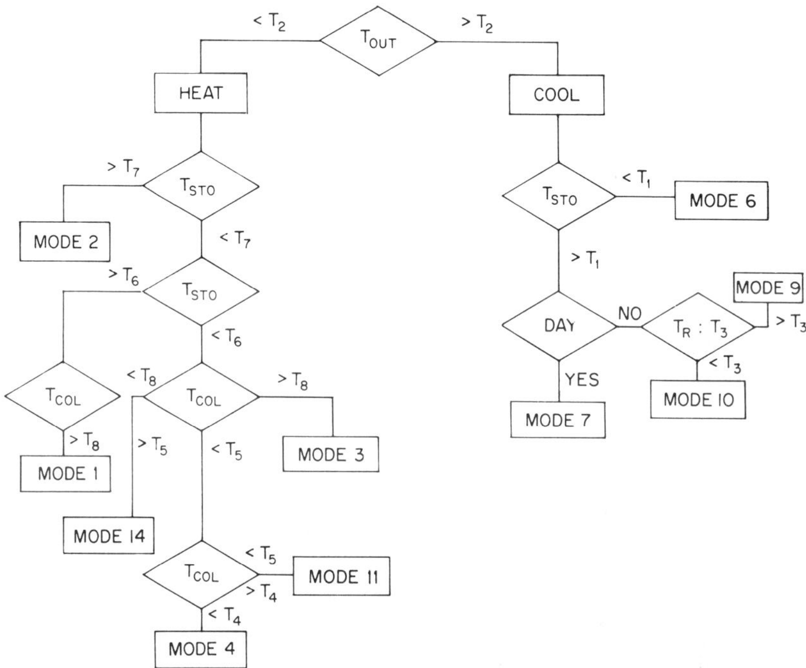

FIG. 4 Automatic control logic. T_{STO}, storage temperature; T_{COL}, collector air out temperature; T_{OUT}, ambient temperature; T_1–T_8, critical temperature for mode selection.

(4) the resistance electric heating, which takes over if (1)–(3) are not available

During cooling in the summer, clock-driven controls operate the heat pump at night to charge the 55°F storage. During the daytime it is this storage that is consumed. When this storage is depleted before 8.30 P.M., the heat pump cools the house directly.

E. Heat Storage

Solar One has three heat reservoirs, using inorganic salt hydrates melting at 55, 75, and 120°F.

Solar heating During the winter heat is stored in sodium thiosulfate pentahydrate $Na_2S_2O_3 \cdot 5H_2O$, melting at 120°F. The material is sealed into pan-shaped containers (measuring 21 × 21 × 1 in.) made of ABS thermoformed material with 0.060-in. wall thickness. One-inch-wide flanges support the pans on their lateral peripheries and space the pans properly to maintain air passages of $\frac{3}{8}$ in. between them for the circulation of air.

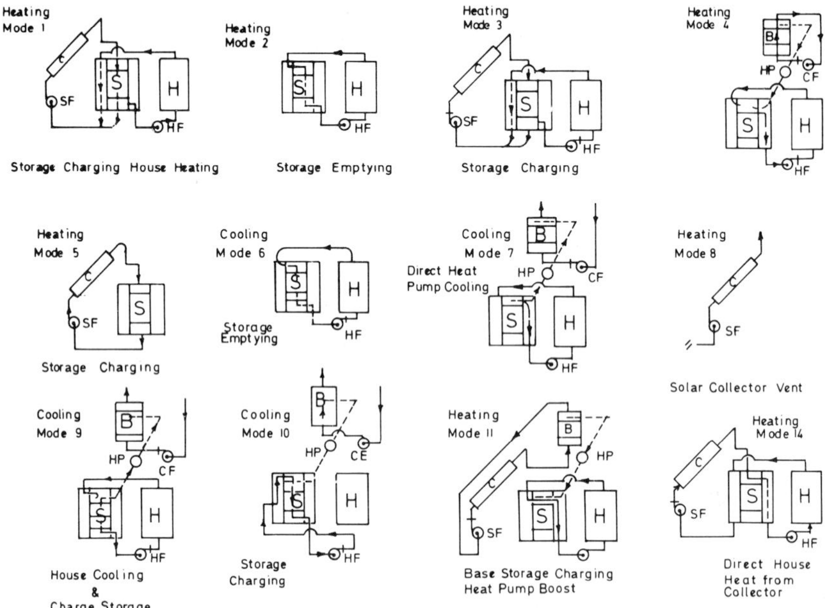

FIG. 5 Twelve heating and cooling modes used in Solar One. HF, house fan; CF, condenser fan; HP, heat pump; S, main storage; B, base storage; H, house; SF, solar fan; C, solar collector. Modes 12 and 13 are not used in the system.

Air conditioning During the summer coolness is stored in a salt eutectic containing $Na_2SO_4 \cdot 10H_2O$, NaCl, and NH_4Cl and melting at 55°F. Borax in small quantities (3–4%) is used as a nucleating agent. An organic thickening agent is added to prevent settling of the borax or major separation of the incongruently melting salt.

Thermal storage material melting at 55°F is sealed into ABS tubes of 1.25-in. diameter, 0.030-in. wall thickness, and 6-ft length. A total of 620 tubes are assembled in an open-crate-type structure within a 2- by 6-ft section of the bin.

Auxiliary storage An auxiliary storage bin has been provided to store heat insufficient to properly heat the house at morning or late afternoon hours or during marginal weather conditions, or to store coolness during summer nights after the heat pump has completely charged the main storage bin. This storage bin acts to improve the operating cycle of the

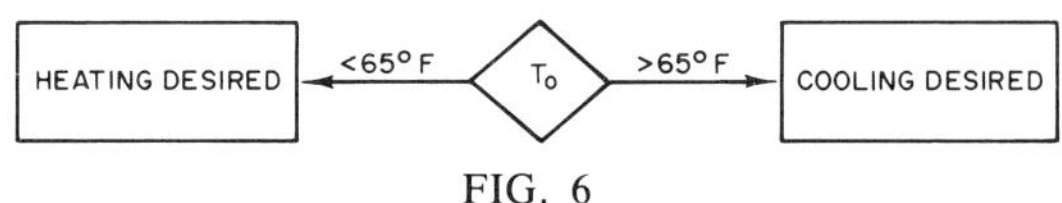

FIG. 6

TABLE 1

Heat loss through:	A, area (ft^2)	U (Btu/h °F ft^2)		$A \times U$ (Btu/h °F)
Walls	1280	0.07		90
Windows	228	0.55		125
Floor	1300	0.06		65
Ceiling	1300	0.08		104
			Subtotal:	384
Infiltration allowance (5%)				20
			Total:	404

heat pump. It contains a salt eutectic containing $Na_2SO_4 \cdot 10H_2O$ and KNO_3, which melts at 75°F.

Again, borax is used for nucleation and an inorganic thickening agent is added to prevent major settling. This heat storage material is sealed into the same ABS tubes as the 55°F material. A total of 310 tubes are assembled in an open crate cubicle structure similar to that of the coolness storage bin.

F. Heat Load

Based on calculations and observations, an estimate of heat load for maintaining a temperature of 65°F inside Solar One on a typical cold day is presented in Table 1. The figure corresponds to 9700 Btu/degree day or 2.9 kWh/degree day. A value of 3.1 kWh/degree day was obtained from the plot of energy used in Solar One for heating purposes (resistance heater) plus lights plus house fan (minus garage operation) during a sequence of cloudy days. However, when considerable electrical energy is used in the basement to run the test equipment, a somewhat lower heat demand for the house is experienced.

16.3 EXPERIMENTS WITH SOLAR COLLECTORS

A. General Description

Since efficient solar collectors are crucial to the performance of the integrated power system of Solar One, a considerable effort was devoted to the testing of solar collectors at Solar One under actual operating conditions and to develop improved collectors. Significant gain in efficiency for tolerable pressure drops was made with a geometrical configuration of collector absorption plates and fins.

The roof of Solar One is equipped to test different kinds of collectors.

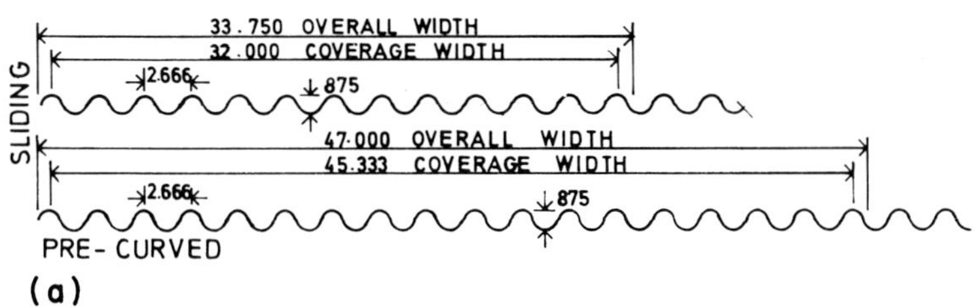

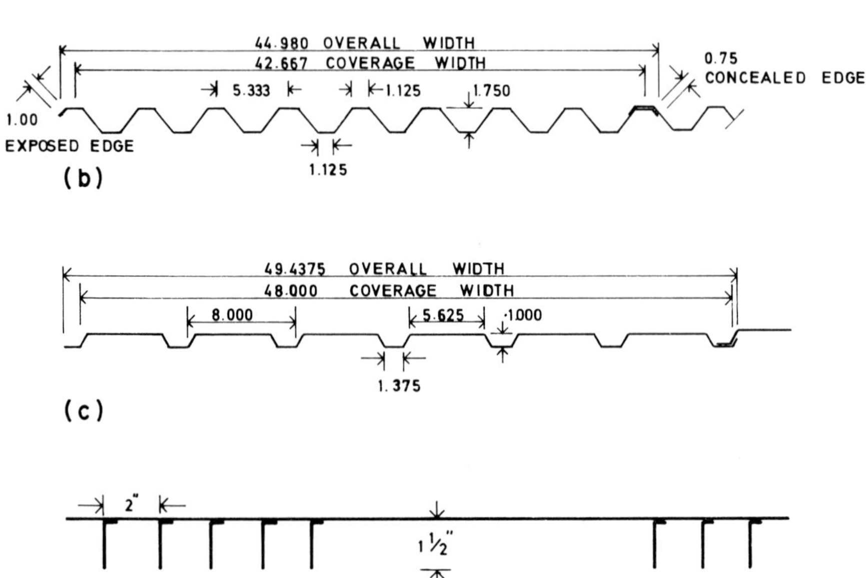

2"
1½"
(d)

FIG. 7 Geometrical configuration of absorber plates: (a) collectors 19–20, (b) collectors 17–18, (c) collectors 15–16, (d) collectors 13–14, 21–22, 23–24. Collectors 13–14 have epoxied and riveted fins; collectors 21–24 have soldered fins.

These collectors are contained in 4- × 8-ft plywood boxes, which can be easily exchanged from the attic. They are pressed against the skylight and held in place by angle irons bolted to the roof joists. During the period February–May 1974, tests were conducted on 12 collectors located on the east bank of the roof to determine their comparative performance. These collectors (which are installed in pairs) are designated as collectors 13–24 and are described in Table 2. Figure 7 gives the geometrical configuration of absorber plates for different collectors. All collectors tested have two cover glazings: an outer $\frac{1}{2}$-in. Plexiglas coated with ABCITE[R] and a double-strength "white" PPG glass.

TABLE 2

Description of Solar Collectors

Absorber metal:	Aluminum	Aluminum	Aluminum	Aluminum	Galvanized iron	Galvanized iron
Coating:	Black no. 4	Black no. 2	Black no. 2	Black no. 2	Black no. 2	Black no. 1
Fin metal:	Aluminum	Aluminum	Aluminum	Aluminum	Galvanized iron	Galvanized iron
Collector no.:	14	16	18	20	22	24
Absorber metal:	Aluminum	Aluminum	Aluminum	Aluminum	Galvanized iron	Galvanized iron
Coating:	Black no. 4	Black no. 1	Black no. 2	Black no. 2	Black no. 2	Black no. 1
Fin metal:	Aluminum	Aluminum	Aluminum	Aluminum	Galvanized iron	Galvanized iron
Collector no.:	13	15	17	19	21	23

As a result of comparative testing, collectors 21 and 22 (which are identical) showed better performance. These collectors were subjected to further testing starting in June 1974. Moreover, collector 17 was replaced in June by a newly designed collector and subjected to a limited amount of testing on the roof of Solar One. This collector is hereafter referred to as collector 17A. The geometrical configuration of the absorber plates and fins for collectors 17A and 21 are shown in Fig. 8. The absorber plate for each collector (except that for collector 17A, which measures 36 × 90 in.) has overall dimensions of $43\frac{1}{2} \times 88\frac{1}{2}$ in.

The distinguishing feature of these collectors is that the fins are staggered. Considerable improvement in heat transfer accompanied by an expected increased pressure drop was observed.

B. Theory

For an air collector in the temperature range of interest, performance hinges primarily on the output of its air duct. Better ducts permit operation at lower plate temperature, hence yield higher efficiency. The duct is rectangular in shape and has uniform solar flux impinging on one side with the remaining three sides being essentially adiabatic. The aspect ratio b/a of the ducts is large (well in excess of 10) and can, therefore, be approximated by a parallel plate channel for theoretical considerations.

The hydrodynamic and thermal boundary layers develop simultaneously in these ducts, and the air velocities are such that the flow lies either in the transition zone or is turbulent. The ducts are characterized by an abrupt entrance and coefficients of friction higher than those of hydraulically smooth ducts. The flow is expected to develop fully (both thermally and hydrodynamically) for most of the duct length. The entrance and intake regions are characterized by higher heat transfer and pressure drops.

The observed different collection efficiencies are due to different losses (mainly convection and conduction) at lower temperatures, with some radiation losses at higher plate temperature. Some leakage is observed at certain collector interconnections; hence the given values should be used only as a first approximation. The following measurements were made:

(1) temperature of the air entering and leaving the collector duct, collector plate, fin or corrugation, glass and Plexiglas cover,
(2) pressure drop across each collector,
(3) airflow through collectors,
(4) humidity ratio, and
(5) total solar radiation.

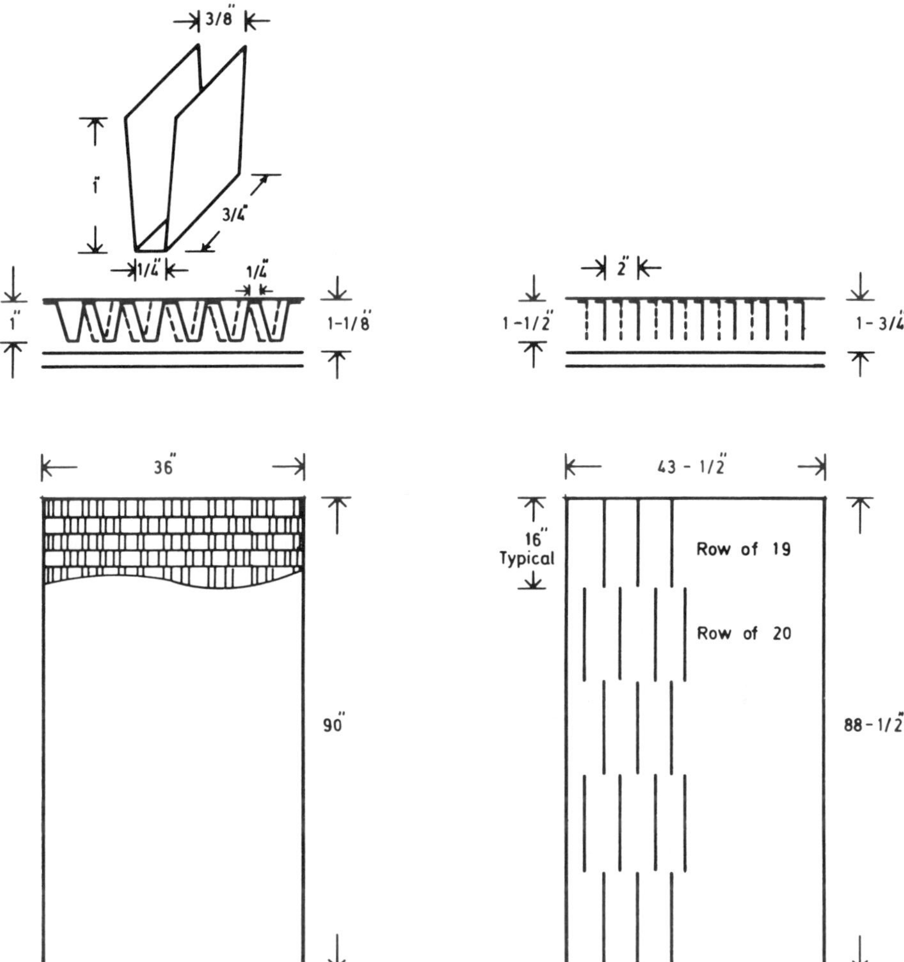

FIG. 8 Geometrical configuration of absorber plates and fins for collectors 17A and 21.

For comparative tests on different collectors a quick disconnect system was installed, which permitted rapid sequential testing of many collectors through a multichannel recorder.

C. Outdoor Testing of Collector 21

Measurements were made at 15-min intervals and in addition to the measurements already listed, the experiments monitored

(1) variations in absorber plate temperature as a function of length along the duct, and

(2) pressure drop as a function of length along the duct.

D. Indoor Testing of Collector 17A Air Duct

Experimental setup The collector and its instrumentation were the same as described earlier; only the outer glazings have been removed. An electric heater was sandwiched between the absorber plate and asbestos sheet, which was then covered on top by urethane foam insulation.

The airflow was monitored in the rectangular section by a hot wire anemometer. Velocities in both the horizontal and vertical directions were measured; the profiles were integrated and the flow was then determined.

Test procedure (a) *Pressure drop in the duct* Data on pressure drop versus length along the duct were recorded under essentially isothermal conditions ($4000 < \mathrm{Re} < 18000$).

(b) *Heat transfer characteristics of the duct* For a given airflow the temperatures were allowed to reach a steady state condition. Temperature measurements pertained to absorber plate, airstream, inlet, and exit air ($4000 < \mathrm{Re} < 17{,}500$).

E. Analysis of Data, Results, and Discussion: Comparative Tests on Collectors 13–24

Some sample results from the comparative solar collector tests are given in Table 3. The finned collector plates (sets 13–14, 21–22, and 23–24) have the higher efficiencies. Pair 13–14 have the fins riveted to the aluminum plate with epoxy filling the remaining gaps. Because of poor thermal contact compared to pairs 21–22 and 23–24 (with fins soldered to the galvanized plate), pair 13–14 has the lower efficiency in this group. Corrugated aluminum plate collectors have the lowest efficiencies (panel sets 15–16, 17–18, and 19–20); the observed differences in collector efficiencies of this group are too small to judge relative advantages of the different kinds of corrugation. Without additional finning the corrugated aluminum collectors are not acceptable as efficient collectors. The slightly higher efficiency of collector pair 21–22 compared to pair 23–24 is probably caused by avoidance of some edge losses as well as by some reduction in radiation losses since paint no. 2 has selective absorption characteristics. The difference between maximum plate and air outlet temperatures is approximately 35°F for collector pair 23–24. This temperature difference is lowest for pair 21–22 (30°F) and increases to 50°F for pair 13–14 and to more than 100°F for pair 15–16 at the given flow rates.

TABLE 3

Comparative Performance of Solar Collectors

Test	Time[a] (EDST)	Collector no.	Flow (cfm)	$t_{out} - t_{in} = \Delta t$ (°F)	Efficiency (η)
1	12:20–12:25	23–24	205	128 − 80 = 48	60
2	12:25–12:30	23–24	205	135 − 81 = 54	66
3	12:33–12:37	21–22	205	135 − 94 = 41	50
4	12:45–12:50	19–20	80	170 − 93 = 77	35
5	12:55–13:00	17–18	205	121 − 97 = 24	30
6	13:05–13:10	15–16	205	121 − 98 = 23	33
7	13:20–13:25	13–14	205	130 − 96 = 34	47
8	13:30–13:35	23–24	205	140 − 88 = 52	71
9	13:40–13:45	21–22	205	140 − 92 = 48	66
10	13:50–14:00	17–18	205	124 − 101 = 23	32

[a] Incident radiation: 12:00–13:00 → 309 Btu/h ft^2, 13:00–14:00 → 275 Btu/h ft^2.

The pressure drop for all corrugated collectors is below 0.01-in. H_2O per panel. The pressure drop of panels with staggered interrupted fins (13–14, 21–22, 23–24) is approximately 0.01-in. H_2O for 300 cfm.

From these measurements the importance of proper finning is evident. The fins must be properly connected to the collector to avoid imperfect bonding and an accompanying low heat transfer. Fins riveted to the absorber plate do experience this problem.

F. Further Tests on Collector 21

Efficiency A total of 14 tests were analyzed; 6 tests pertained to collector 21 and the remaining 8 were related to collector 22. The resulting average value of efficiency for the two collectors is 53 ± 5%.

Temperature profiles A sample data plot for the plate and air temperature shows that the plate temperature maintains a constant gradient for most of the duct length, thereby indicating that the flow is thermally developed.

The data indicate that despite the extensive finning and flow interruption, for practical purposes the pressure drop remains very small (compared to the pressure drop in the rest of the system: 1.5-in. H_2O).

G. Results for Collector 17A

Testing on the roof of Solar One The observations indicate that

(1) absorber plate and bulk air temperature vary linearly along most

of the duct length ($\delta t_B/\delta x = \delta t_p/\delta x = \text{const}$), which is typical of the "constant flux" boundary condition, and

(2) the efficiency of collector 17A is found to be on the order of 75%; plate and bulk air temperature lie within 7°F.

Results from indoor testing of collector 17A (a) *Pressure drop versus duct length* These tests were run under essentially isothermal conditions. The pressure drop is relatively small near the duct entrance and asymptotically approaches a linear profile when the turbulent flow is fully developed. The intake length, defined as the distance from duct entrance to the location where the pressure gradient is constant, increases slightly with increasing flow rate (from $L_e/d_h \simeq 20$ for Re $\simeq 5000$ to $L_e/d_h \simeq 25$ for Re $\simeq 17{,}000$).

For a noncircular duct, the coefficient of friction f for a fully developed flow is given by

$$f = -d_h(dp/dx)/(u_m^2/2g) \tag{1}$$

The coefficient of friction for hydraulically smooth ducts is given by the Blasius law as

$$f_0 = 0.316/\mathrm{Re}^{0.25}$$

The average value for $(f/f_0)^{1/2}$ for all readings was found to equal 4.6 ± 0.2. The values of $(f/f_0)^{1/2}$ were calculated for the following reason. Following the analysis for rough ducts (Kolár, 1965), the heat transfer in rough tubes with symmetric boundary conditions is described by

$$\mathrm{Nu} = 0.04\sqrt{f/8}(\mathrm{Pr})^{0.5}$$

Thus one expects

$$\mathrm{Nu}_{\mathrm{rough}}/\mathrm{Nu}_{\mathrm{smooth}} = \sqrt{f/f_0} \tag{2}$$

An attempt was made to see if this relationship could also successfully correlate the data in asymmetrically heated ducts. (As will be discussed later, a prefactor was found necessary.) The total pressure drop for the duct is given by

$$\Delta p = f(L/d_h)(u_m^2/2g) + \Delta p_0 \tag{3}$$

where for Δp_0, the following value was experimentally determined:

$$\Delta p = 1.95u_m^2/2g \tag{4}$$

(b) *Heat transfer in the air duct* The heat transfer coefficient (h) of the duct was determined for different flow rates corresponding to a range of Reynolds numbers from 4000 to 17,500. The Nusselt numbers of the duct can be expressed as a function of the Reynolds number by the empirical relationship (Tan and Charters, 1970)

$$\mathrm{Nu} = A\,\mathrm{Re}^{0.8} \tag{5}$$

with, however, $A = 0.51$ in contrast to Charters' $A = 0.0156$.

This observation seems to permit an analysis of collector duct 17A in terms of hydraulically smooth ducts with, however, a substantially larger heat transfer coefficient:

$$\mathrm{Nu}_{\mathrm{rough}}/\mathrm{Nu}_{\mathrm{smooth}} = 0.51/0.0156 = 33 = 7.1\sqrt{f/f_0} \tag{6}$$

$(f/f_0)^{1/2}$ was determined from pressure drop curves.

A similar analysis for the duct of collector 21 was carried out with somewhat lower accuracy and gave similar results with

$$\mathrm{Nu}_{\mathrm{rough}} = 3.25 = 2.5\sqrt{f/f_0} \tag{6a}$$

16.4 INSOLATION AND TOTAL SYSTEM PERFORMANCE

A. Solar Radiation

Insolation during winter season 1973–1974 In Solar One the total solar radiation is measured on the 45° roof with the help of a Kipp and Zonan pyranometer, which is coupled with a Lintronics integrator and an Esterline-Angus recorder. The instantaneous and quarter-hourly integrated radiation values are monitored. The monthly average of these radiation intensities are listed in Table 4.

Insolation on bright clear days (a) *Daily total insolation* Less than 10% of the days for the winter season 1973–1974 were found to have a cloud cover of $\frac{1}{10}$ as defined by the weather bureau. The intensity of radiation (in Btu $(\mathrm{ft}^2\ \mathrm{day})^{-1}$) for bright clear days in the month is listed in Table 4 to provide a measure of the variation in radiation received by the 45° roof.

TABLE 4

Average Intensity of Solar Radiation Falling on 45° Roof of Solar One During Winter Season 1973–1974

Month:	October	November	December	January	February	March	April
Monthly average [Btu $(\mathrm{ft}^2\ \mathrm{day})^{-1}$]:	1446	1170	736	667	1301	1255	1247
Monthly maximum [Btu $(\mathrm{ft}^2\ \mathrm{day})^{-1}$]:	2144	2010	1840	1880	2416	2385	2285

(b) *Number of sunshine hours* The hour angle (H) at sunrise or sunset is given as

$$H = \arccos(-\tan l \tan d) \tag{7}$$

where l and d represent latitude and declination, respectively. The number of sunshine hours for clear days was calculated accordingly.

Based on data for bright clear days during the winter season 1973–1974, Fig. 9 was plotted. To a good approximation, the results can be represented by a single intensity curve $I/I_\circ = f(\theta/\theta_0)$, where I represents the instantaneous intensity of solar radiation at time θ, θ_0 equals one-half the number of sunshine hours, and I_0 represents the peak intensity of solar radiation as recorded on the strip chart.

On a 10-year average basis, it is noted that the clear ($C \leqslant 0.3$), partially cloudy ($0.4 \leqslant C \leqslant 0.7$), and cloudy days ($C \geqslant 0.8$) in winter are nearly 25, 25, and 50%, respectively, and that the intensity of solar radiation is not a linear function of cloud cover C.

(c) *Average air temperature* The average daily air temperature at Newark, Delaware, for winter 1973–1974 and the last 20 years is presented in Table 5.

(d) *Thermal system operation* The electrical power for all parts of the thermal system is separately metered and monitored on a day-to-day

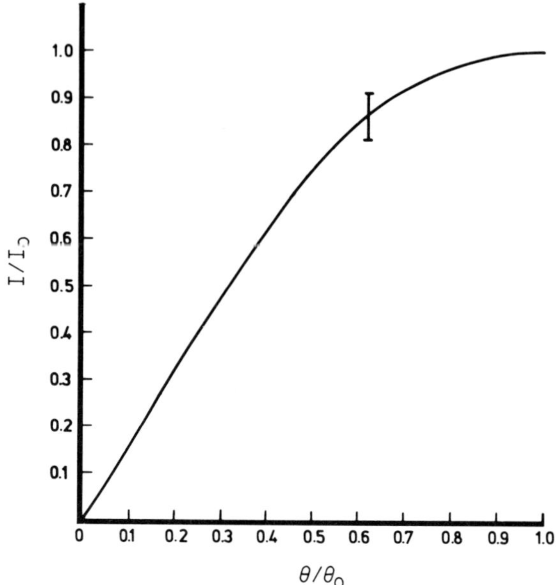

FIG. 9 Dependence of I/I_0 on θ/θ_0.

TABLE 5

Average Air Temperatures (°F) and Degree Days in Newark, Delaware, in Winter for 1973–1974 and a 20-Year Period

Period	October	November	December	January	February	March	April
1973–1974	57.9 (230)	48.0 (509)	38.5 (821)	37.8 (789)	33.8 (873)	43.8 (658)	56.1 (296)
20-Year	56.5	45.1	35.8	31.4	33.1	41.6	53.5

basis, with the exception of the control system, which is connected in parallel to the thermal system. The operation of the thermal system is analyzed with respect to the electrical power consumption of the different components and the fraction of solar energy supplied for heating purposes. During the air-conditioning season, electrical energy harvested through the CdS solar panels will be considered for partially powering the heat pump.

B. Electrical Power Use for Different Subsystems

Heat pump operation/heating The heat pump is used for cooling (summer) and auxiliary heating (winter) operation. In the Delaware climate the pumps are usually oversized for summer operation or need substantial assistance (e.g., resistance heating) during cold winter days.

An attempt is being made to minimize this discrepancy in Solar One by using solar energy for supplemental heating and storing surplus heat properly. Using a heat storage bin with a capacity of more than one cold winter day, one can minimize the use of the heat pump even during a longer sequence of inclement weather days by intermittently using the storage or the heat pump. The heat pump is then activated to heat the rooms directly as soon as the collector temperatures permit efficient heat pump use ($T \leqslant 40°F$) or as long as the base reservoir contains useful heat (again $T \leqslant 40°F$). The base reservoir is charged when the rooms do not call for heat. In many long sequences of cloudy days, there are sufficient cold spell interruptions to permit efficient heat pump operation but not direct solar room heating.

Taking a sequence of 5 cloudy days as the design basis in midwinter with 180 DDs demand, one needs approximately 900,000 Btu to be supplemented by the heat pump during this period. With an operating air temperature at the indoor coil of 85°F (sufficient for room heating through the bin bypass), the heat pump provides (at 4.9-kW input) nearly 35,000 Btu ($COP \simeq 2$) at a base coil temperature of 40°F, for an operating average of 5 h/day. Such operation seems feasible with sufficient base storage capacity.

During the first part of the heating season (November–January 8) 1973–1974, the heat pump was operated without proper match of collectors and heat storage bin (and this in the absence of thermal storage) available for the heat pump operation. During this period the heat pump was able to supply heat satisfactorily to the house to keep the rooms at 65°F at all times.

During the second part of the heating season (January 9–April 27), the heat pump was disconnected and all auxiliary heating was provided by the resistance heater for comparison reasons.

For simplicity and reliability reasons, a periodic switch-off was selected, and during the cooling season 1974 the heat pump ran at a cycle of 45 min on, 15 min off.

Heat pump power consumption The electric power consumption connected with the heat pump consists of use for a compressor, a base exchanger fan (condenser fan), and a storage fan (house fan) when a storage mode is used, while the house air does not circulate.

The average daily consumption for August 1973–June 1974 is listed in Table 6 together with other power consumption. The different measured ratios between the energy consumption of the heat pump compressor and condenser fan are explained by different level loads (COP) of the heat pump and by extensive experimentation.

(a) *Heating cycle* For a performance analysis of the heat pump operation, two 31-day periods were chosen: December 9, 1973–January 8, 1974, for heat pump operation and January 9–February 8, 1974, for resistance-heater-only operation. These periods are chosen so as to provide nearly the same average heating need (DD/day) and approximately the same insolation, hence permitting a simplified comparison. In both periods the same set of solar panels was used. Also, the insolation profile was similar and no heat storage was used.

Therefore, it seemed justified to determine the contribution of solar heating in the second period, when this estimate could be performed easily, and to deduce the solar heating in the first period by multiplication with the insolation ratio of both periods (see below). The contribution of solar heating for the second period can be estimated as the difference between heat needed to keep the house at the set comfort level (65°F) and the heat supplied electrically, i.e., 33.1 kWh resistance heating plus all other contributing factors, such as solar and house fan. With 80% efficiency of such fans, 0.8×7.6 kWh is directly transferred to heat the house air. The difference between the heat needed and that supplied electrically must have been provided from the solar panels; this amounts to 23.6 kWh/day for the period January 9–February 8.

Since the total insolation of useful solar days with insolation (>1000 Btu $(\text{ft}^2\ \text{day})^{-1}$) was 11,586 Btu/ft^2 in the first period and 14,332 Btu/ft^2 in the second period, a ratio reduction of 19% was effected. Hence solar heating supplied an estimated 19.1 kWh/day from December 9, 1973, to January 8, 1974.

This leaves 56.8 kWh/day to be supplied by the heat pump system. With an actual consumption of only 32.5 kWh/day (see column 9 in Table 7), one estimates an overall COP of the combined heat pump auxiliary heating system of approximately 1.7. It is expected that this COP can be considerably improved when heat storage is properly employed in con-

TABLE 6

Average Daily Electrical Energy Consumption (kWh) for 1973–1974

	Aug	Sep	Oct	Nov	Dec	Jan	Feb	Mar	Apr	May	June
(1) Heat pump	54.0	33.7	7.3	3.3	6.3	3.	0	0	0	17.8	26.8
(2) Condensor fan	12.0	9.6	2.4	2.2	6.9	2.3	0 0	0	0	8.3	5.4
(3) Storage fan[a]	—	0	0	—	—	—	—	—	—	0	3.62
(4) House fan	12.9	13.8	12.1	3.2	2.5	2.5	2.5	1.0	2.1[b]	7.9	9.3
(5) Solar fan	13.6	10.7	16.4	9.7	3.6	3.8	11.3	10.0	12.0	3.5	5.4
(6) Garage	10.2	4.0	5.9	3.6	14.3	0.4	0.4	0.5	2.0	0.2	0
(7) ac/dc	12.3	16.6	18.1	14.3	9.0	9.4	9.5	10.7	9.7	9.8	11.1
(8) (1) + (2) + (4) + (5)	92.5	67.8	38.2	18.4	19.2	11.7	13.8	11.0	14.1	37.5	46.9
(9) Mechanical equipment[c]	102.	77.5	47.2	28.0	35.0	54.8	61.0	36.0	52.0	48.6	54.8
(10) Resistance heater	0	0	0	0	5.8	33.1	37.2	15.0	27.9[b]	0	0
(11) ac + experimentation	16.5	17.9	13.8	13.9	14.7	20.8	22.6	13.8	17.3	17.1	10.6
(12) Total electrical consumption	141.0	116.0	85.0	59.8	73.0	85.4	93.5	61.0	81.0	75.7	76.5

[a] Calculated as 0.67 of (2). This energy is included in (4).

[b] An operation period of 79 h of the resistance heater (21.0 kWh/day) and house fan (1.3 kWh/day) is used to charge the heat storage.

[c] Mechanical equipment includes (1), (2), (4), and (5) plus the resistance heater, all automatic control equipment, and the sump pump. During August, September–November, May, and June, the heater was off, leaving ~10 kWh for control equipment and sump pump. Consequently, (10) is calculated as (9) – (8) – 10 kWh.

TABLE 7

Electrical and Solar Thermal Data for Two Similar 31-Day Periods for 1973–1974

	Jan 9–Feb 8	Dec 9–Jan 8
(1) DD	875	906
	Daily averages	
(2) Insolation (Btu/ft^2)	20,938	17,284
(3) DD/day	28.2	29.2
(4) Daily average insolation (Btu/ft^2 day)	675	558
(5) Total heat needed (kWh) at 3 kWh/DD	84.6	87.6
(6) Heat pump + fan (kWh)	—	16.8
(7) Resistance heat (kWh)	41.4	11.9
(8) Solar + house fans (kWh)	7.6	4.7
(9) (6) + (7) + 80% of (8) (kWh)	49.7	32.5
(10) Total house − garage (kWh)	88.7	67.9
(11) Modified total heat need at 2.6 kWh/DD	73.3	75.9
(12) (5) − (11) (kWh)	11.3	11.7
(13) (10) − (9) (kWh)	39.0	35.4
(14) Solar heat (kWh)	23.6	19.1
(15) kWh equivalent of electrical heat needed	49.7	56.8
(16) Mechanical equipment (kWh)	59.0	43.4
(17) (6) + (8) + (10) (kWh)	17.6	31.5
(18) COP = (15):(9)	1.0	1.7

junction with the heat pump, and resistance heating is used on the rare occasions when air temperature at the base exchanger drops too low for the heat pump to provide sufficient heat.

(b) *Cooling cycle* In June 1974 the heat pump operated daily in the condition described in the previous section. With an average daily energy consumption of 30.5 kWh (condenser fan) + 3.7 kWh (charging fan) = 39.8 kWh, charging an average of 200,000 Btu, one obtains an average overall COP of the heat pump–storage system of 1.5. This is substantially below the expected COP of 2.1, which is estimated from an unoptimized system. For an optimized system, a COP probably close to 2.4 is anticipated. The lower COP is attributed to inherent energy losses during intermittent operation and the unnecessary ice formation on the coil, as well as to some heat losses from the bin. Heat losses can also be minimized by starting the heat pump later, so that the time of complete charging of the bin coincides with the first call for cooling.

It should be noted that the average electrical energy consumption in June 1974 for cooling purposes is probably close to the maximum daily consumption for a very warm day (mid 90s), since for reasons of life-cycle testing the coolness storage bin was fully charged and discharged daily.

This occasionally resulted in room temperatures considerably below required comfort levels (65–70°F).

In spite of these shortcomings, the current performance of the systems is sufficient to satisfy the house demand for cooling from storage for most of a hot summer day. The total operation of the cooling system includes utilizing the remainder of the energy of the house fan to circulate air from the rooms through the coolness storage bin (in June, 11.0 − 3.7 = 7.3 kWh); this energy is spent during daytime hours close to peak demand. It is substantially below the heat pump consumption (39.8 kWh).

Summary of thermal auxiliary systems operation It has been shown that the heating and cooling of Solar One is possible using a heat pump and a little additional resistance heating.

Cooling via charging a coolness storage bin during night hours and in turn discharging it during day hours is possible using currently available salt eutectics that melt at 55°F. When the outside temperature is 90°F, a 150,000-Btu storage capacity is sufficient to cool the house to 75°F until the evening hours. The current overall COP of the heat pump storage/cooling cycle is 1.5, and with design improvement may increase to 2.5.

Heating can also be easily achieved with the 3-ton heat pump and only a little additional resistance heat even without any solar energy storage in the 120°F storage bin. An overall COP of 1.7 for the total heating system (including resistance heat) is observed during an average winter month (December 9–January 8) with an average of 29.3 DD/day. With solar heat storage and improved solar collectors, a marked increase of the COP to at least 2.5 is expected.

The performance of solar collectors has been discussed in Section 16.3. Four roof collectors were covered with CdS/Cu_2S solar cells. These cells produce electrical energy and have a reduced life expectancy with the maximum design temperature of 150°F. With much improved finning, a 7°F difference between plate and air temperature at 100 cfm with a conversion efficiency of approximately 75% at a temperature difference of 50°F between maximum plate and the ambient temperature was attained. This set of panels has a maximum plate temperature of 150°F, and the air leaves the panel at 143°F, with losses of 3°F through the duct work. The temperature at the bin inlet is thus 140°F, sufficient to melt the heat storage salt (melting point 120°F).

Thermal energy storage bin; heat storage The heat storage bin of Solar One has been designed to match the proper solar collector. With a heat transfer surface of 940 ft^2 and an estimated heat transfer coefficient of 4 Btu/h ft^2 °F, one estimates a heat transfer of 78,000 Btu/h for the heat storage at $\Delta T = 20°F$. To fully charge the storage, one needs approximately 12 h of sunshine at 300 Btu/h ft^2 solar radiation. This seems to be reasonable for charging within 2 sunny days.

Taking into consideration a loss of up to 3°F in each branch of the ducts connecting the solar panels to the bin, a ΔT of 20°F should be available (30°F is available across the collector). With an estimated 4000 cfm, the required heat (1 MBtu) can therefore be supplied in the given time.

The observed high value of the heat output from the bin shows that all nucleation devices are working properly and indicates the feasibility of using $Na_2S_2O_3 \cdot 5H_2O$ as a storage material.

16.5 OVERALL PERFORMANCE

The house has been used most of the time as a solar laboratory with continuous changes in operational modes and of several components, which necessitated temporary shutdowns. Thermal storage was used only during the very end of the heating period. Thus, an estimation of overall performance will present only a lower limit. However, it may be of interest to observe how much solar energy was used during the heating season 1973–1974 under such adverse conditions.

A summary of all relevant input data indicates that in spite of the experimentation and comparative collector tests with more than 50% of the roof surface still covered with collectors below 40% conversion efficiency, the overall contribution of solar heat to the energy balance has been almost 60%.

It is expected that with current improvements, approximately 80% of the heat demand of the house can be satisfied by solar energy during an average heating season.

ACKNOWLEDGMENTS

The author gratefully acknowledges the contribution of all scientists, engineers, and technicians involved with the Solar One project at University of Delaware. Special thanks are due to Dr. K. Böer, Dr. M. Telkes, Dr. M. K. Selcuk, Dr. T. Kuzay, Dr. F. Costello, Mr. K. O'Connor, Mr. Steve Ridenour, and Mr. Jim Higgins.

17

Solar Heating of Greenhouses

M. IQBAL

DEPARTMENT OF MECHANICAL ENGINEERING
THE UNIVERSITY OF BRITISH COLUMBIA
VANCOUVER, BRITISH COLUMBIA, CANADA

Greenhouses of one form or another have been built for more than a century. Like an ordinary home for human beings, the basic purpose of a greenhouse is to provide shelter to plants from the rigors of climate. Greenhouses are used to protect plants against mechanical damage from winds, frostbite from low temperatures, and excessive heat from the sun. In some instances they are also used to conserve water resources.

Greenhouse architecture varies widely around the world. In poor countries with excessive sunshine, it may consist of a simple flat straw roof supported by wooden poles. On the other hand, in rich areas with excessive insolation, as, for example, in Israel and some oil-rich Arab countries, highly sophisticated structures exist with mechanical cooling devices. Plastic-covered greenhouses are quite popular in temperate areas such as the Mediterranean region, and glasshouses are employed almost exclusively in northwestern Europe.

Plastic- or glass-covered greenhouses are natural solar collectors. Numerous studies exist on the design analysis and climate control of such houses (Businger, 1966; Nisen, 1969; Selçuk, 1971). In this chapter discussion will be limited to the use of additional devices for receiving solar energy for the heating of greenhouses.

Since the fall of 1973 there has been a concentration of effort on devising technological means of capturing solar energy for the heating of greenhouses, just as there have been similar studies for the solar heating of homes. A unique development in France will be mentioned first (Damagnez *et al.*, 1975).

ISBN 0-12-620860-3

For a long time it has been known that a water solution containing small amounts of copper chloride strongly absorbs electromagnetic waves in the infrared region and is transparent to the visible wavelengths. The photosynthetic wavelengths are situated in the visible region, and the infrared band only helps to increase greenhouse temperature. This means that in regions in which there is an excess of solar energy during the day and a heating requirement at night, one can employ the above principle to capture the infrared portion of solar energy during the day, store it, and utilize it during the night. Damagnez *et al*. (1975) employed this principle in the recent construction of a greenhouse, sometimes called a selective greenhouse. Following the same principle, Deminet (1976) at the Boeing Company has proposed rather complicated glass collectors for greenhouse heating. The work of Damagnez *et al*. will be described here, however, since it is more advanced.

The greenhouse of Damagnez *et al*. consists of a solar collector with a double sheet of rigid plastic covering most of the south-facing surface of its structure (Fig. 1). A water solution containing 2.5% $CuCl_2$ circulates through the double sheet. During the daytime the plastic sheet absorbs the infrared portion of solar energy and stores it in a storage tank underneath the greenhouse. During the night, this solar energy is recirculated through the same passage to provide heat to the system. The spectral transmission characteristics of the solution are given in Fig. 2. A typical reduction of the maximum and minimum temperatures in the greenhouse during a 24-hour period is given qualitatively in Fig. 3. It is evident from this figure that the system encompasses not only the solar heating of greenhouses but also the solar cooling.

The two main advantages of the system just described are the collectors do not occupy additional valuable real estate, nor do they appreciably reduce the quantity of visible light incident on the plants. However, there is always a serious risk of damage to the plants in the event of leakage of the solution to the soil and plants.

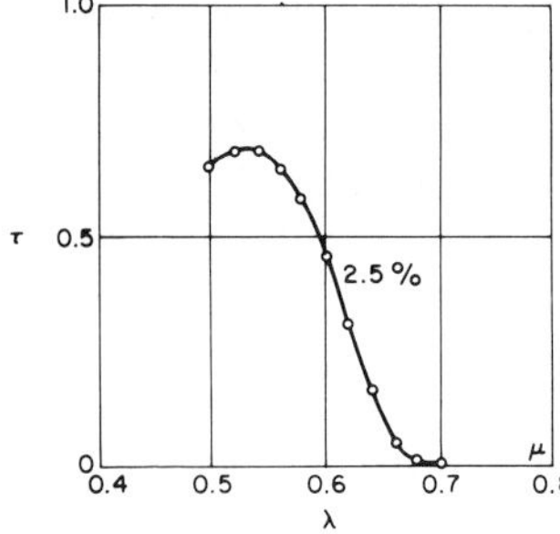

FIG. 1 Coefficient of transmission of 2.5% copper chloride solution.

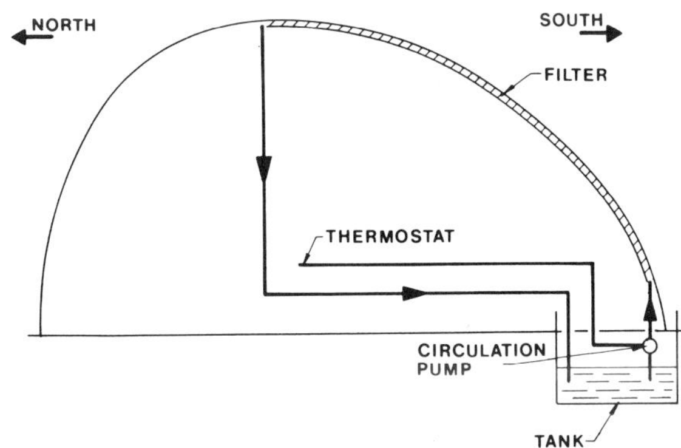

FIG. 2 Shape of greenhouse and copper chloride solution circulation loop. (Damagnez *et al.*, 1975).

In the following paragraphs, a number of solar heating systems for greenhouses, most of them presented in a recent workshop, are described (Jensen, 1976).

The principles of solar ponds are well known. On an experimental basis they have been studied for the space heating of residences by Rabl and Nielsen (1975) and for power production by Tabor (1963). Their application in greenhouse heating is given in a preliminary report by Short *et al.* (1976).

A number of attempts have been made to use nonintegral solar collectors of the conventional type. McCormick (1976) at the Lockheed Company has attempted to heat a hobby-size greenhouse with water-heating collectors installed on the wall of a separate shed that also served

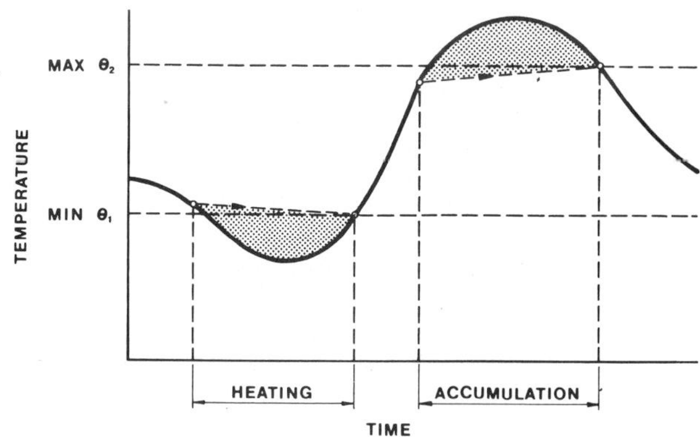

FIG. 3 Principle of temperature regulation. (Damagnez *et al.*, 1975).

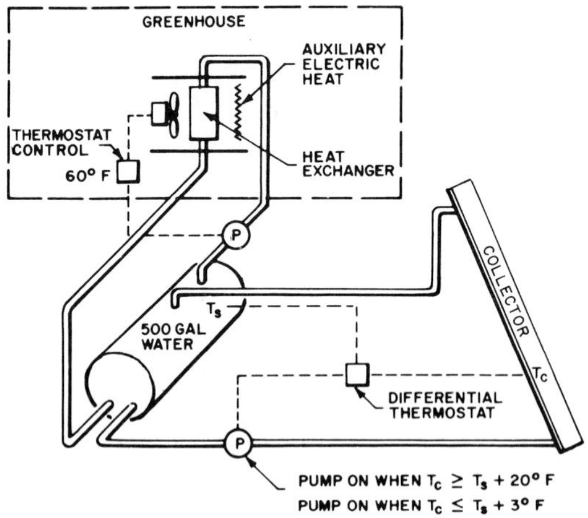

FIG. 4 Schematic of greenhouse heating system (Jensen, 1976).

as a heat storage area. A general schematic of McCormick's system is given in Fig. 4. Using conventional-type solar water heaters, Baird and Mears (1976) have heated a full-size greenhouse.

The main disadvantage of using ponds and nonintegral collectors is that they take up expensive real estate. On the other hand, locating conventional collectors on a greenhouse roof cuts off a portion of the essential photosynthetic light. Liu and Carlson (1976) have presented a conceptual design for installing collectors integral to the greenhouse by changing

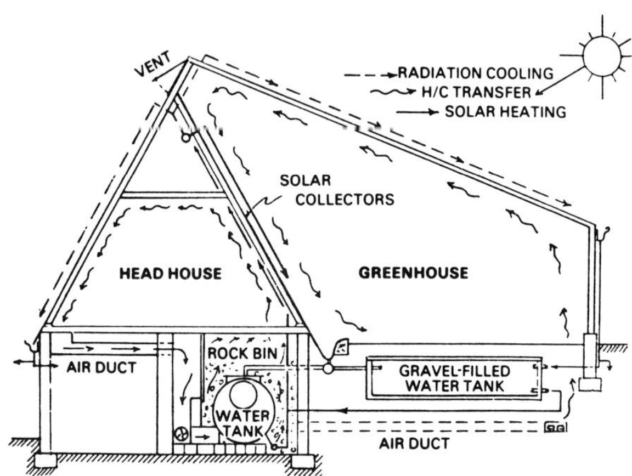

FIG. 5 Sketch of a proposed concept for maximizing solar energy use in a greenhouse (Jensen, 1976).

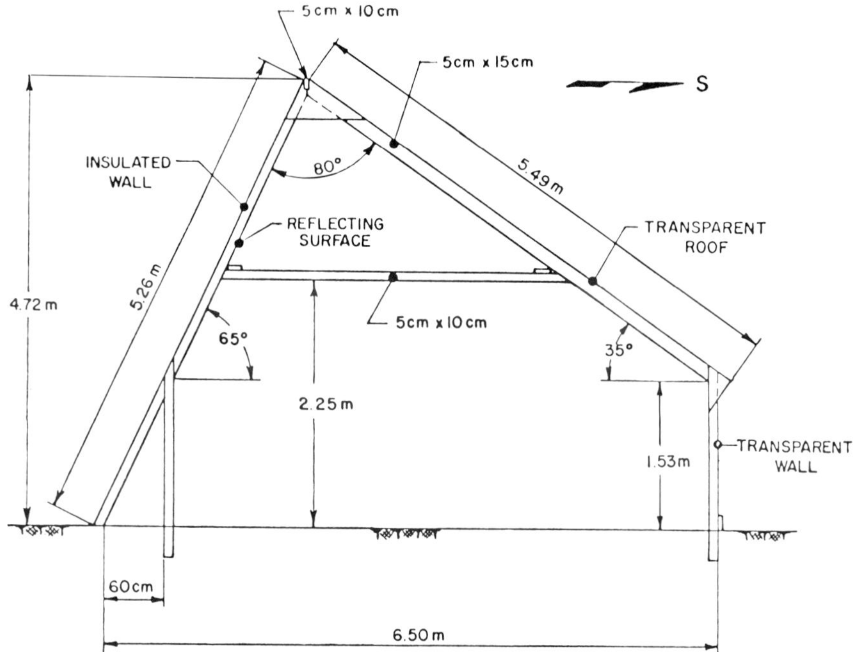

FIG. 6 Section view of brace experimental greenhouse.

its conventional architecture. Their proposed configuration of the greenhouse and collector system is shown in Fig. 5.

In the Northern Hemisphere a transparent northern wall contributes very little to the greenhouse insolation (the reverse is true in the Southern Hemisphere). In fact, the north side contributes equally to the energy loss at nighttime. Lawand *et al.* (1974) have built a greenhouse that has an insulated north wall with a reflective coating on its inner surface. Such a wall has the dual effect of increasing daytime insolation and reducing nighttime energy loss. A schematic of the system is given in Fig. 6.

A Swiss manufacturer (Gabler, 1976) has recently presented a greenhouse design with a type of movable vane that is open during the daytime to let in solar radiation. Along the center of each vane runs a pipe that collects reflected beam radiation (Fig. 7a). At night when the vanes are closed, the stored energy is radiated on to the plants through the same pipes (Fig. 7b).

In many instances a greenhouse has an excess of solar thermal energy during the daytime. Price *et al.* (1976) have attempted to use this excess heat by storing it in the greenhouse soil during the day. The stored energy is then utilized for heating the greenhouse at night.

It may be added here that some attempts have also been made to de-

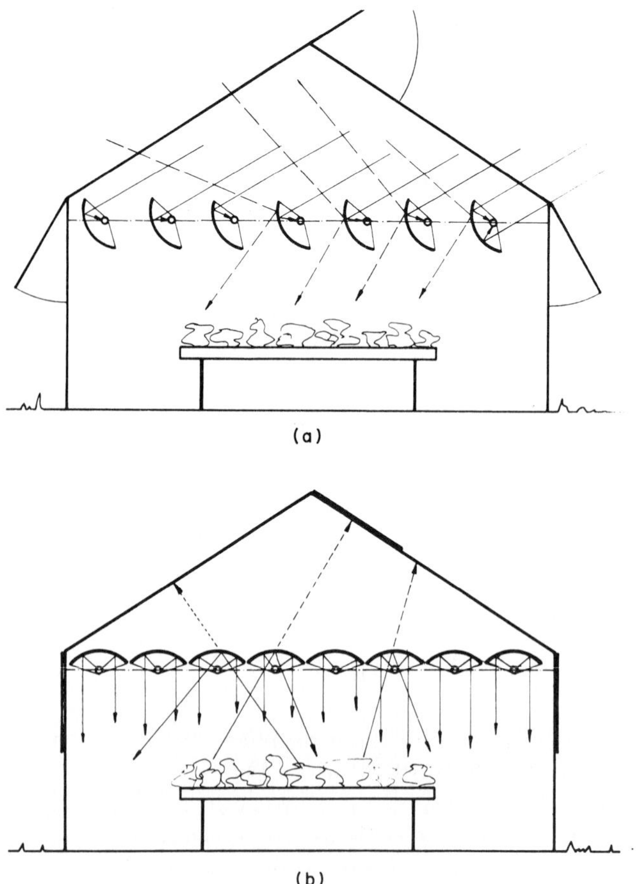

FIG. 7 Wilhelm Gabler greenhouse with (a) shutter vanes open and (b) shutter vanes closed.

sign combination greenhouse–residences. Zornig *et al*. (1976) have presented design criteria for such a combined system. Jensen and Hodges (1976) have actually built a combination greenhouse–office area.

The technology of greenhouse solar heating is very much in the infancy stage. Considerable effort will be required to bring it to a level at which different systems could be designed, as for the solar heating of residences (Klein *et al.*, 1976).

18

Solar Housing in India

M. C. GUPTA

SOLAR ENERGY DIVISION
DEPARTMENT OF MECHANICAL ENGINEERING
INDIAN INSTITUTE OF TECHNOLOGY
MADRAS, INDIA

D. HARIHARAN

BUILDING TECHNOLOGY DIVISION
DEPARTMENT OF CIVIL ENGINEERING
INDIAN INSTITUTE OF TECHNOLOGY
MADRAS, INDIA

18.1 HISTORICAL PERSPECTIVES

A. Introduction

India is a vast subcontinent, where the new and the old are seen in stark contrast. Her checkered history has left a legacy of priceless monuments of a very diverse nature. Even today, many buildings and monuments are old, modern, and Indian at the same time. The builders were keenly aware of India's climate, people, and traditions, and very sensitive to its ecology. The intuition of the architects was as much aesthetic as metaphysical. These aspects have found a fine expression in many buildings in India. Water, curves, textured surfaces, and orientation to the sun give the buildings a characteristic look and a sense of openness in spite of their modest proportions. Quiet elegance integrated with the values of economy, functional adequacy, nearness to nature, and beauty are characteristics of some of these buildings. Religious, aesthetic, and metaphysical aspects are characteristics of others. Some world-renowned temples

ISBN 0-12-620860-3

have exterior erotic sculpture, but they evoke the most sublime and divine thoughts as one steps into the interior.

This chapter presents some thoughts relating to solar energy utilization in the historical monuments, temples, museums, palaces, etc., and attempts to project present and future trends in this vast subcontinent.

B. Veneration of Sun, Synthesis of Astrology, and Astronomy in Post-Vedic Times

The sun has been held in veneration in India from time immemorial. This has been proved by various scholars who have traced the development of sun worship and the temples and monuments dedicated to the sun. In fact, sun worship has survived the passage of time, as have the religious values attached to it. In light of the developments in space science and technology and interplanetary movement, sun worship may be considered archaic today. As the logic and scientific thinking behind this symbolic manifeststion have been overshadowed by its ritualistic manifestation, it should be emphasized that there has always been a scientific orientation in this practice, even in Vedic and pre-Vedic times (Pandey, 1963).

During the Vedic period, as is evidenced from references made in the "Kathopanishad" (1950) attributing importance to the sun, the sun appears to be the central theme in the mystery of time—the fascination of day following night, of sunset following sunrise, and of the periodicity of the seasons. In the post-Vedic period, the sun is identified with "Prana," the vital force in man and in the cosmos. Long after the Vedic period, astrology influenced man's thinking, and human destiny was believed to be influenced by the position of the sun and the planets, which resulted in the synthesis of two age-old sciences, namely astrology and astronomy. Even today, we see the remains of massive and fascinating masonry structures that were built in the pursuit of astronomy. Figures 1 and 2 represent structures that were built at Delhi and Jaipur, respectively. They are used to fix the position of the sun, the diurnal time, and the latitude of the place. The Samrath Yantra at Delhi (Fig. 1) consists of a gnomon and two quadrants, both of masonry construction. The gnomon is triangular and stands in the meridional plane in the north–south direction, with the longer axis on the ground and the inclination of the hypotenuse as the latitude of its location (28°0′39″). The quadrants are arcs of circles that lie in the plane of the equator with their centers at the edge of the gnomon; the center which describes the upper arc of the quadrant is at the upper end of the hypotenuse, and that which describes the lower arc is at the lower end of the hypotenuse. Thus, each edge of the gnomon carries two centers for each quadrant, constructed on either side of the gnomon. The gnomon

FIG. 1 Masonry structures used to study the solar system. They are massive in size, large in scale, and look more like rocket launching pads! Depicted here is the Samrath Yantra.

and the quadrants are graduated to measure the declination of the sun and the hour angle, respectively. The morning shadow of the gnomon falls on the upper end of the western quadrant; the shadow descends as the sun ascends, till noon, when it leaves this quadrant. Soon after, the shadow appears on the eastern quadrant just at the point where the gnomon and the quadrant join; the shadow gradually ascends as the sun descends. The space on the quadrants is divided into hours, minutes, and seconds, and

FIG. 2 Masonry structures such as this one, the Misra Yantra, were used to find the time and declination, zenith distance and altitude, and the zodiacal sign (Rasi) of the sun. Their scale in their setting forms abstract sculpture of unsurpassed nature.

the time is read off directly against the edge of the shadow. The graduations, which are the tangents of the declination angles, are marked on both faces of the gnomon. The hour angle and the declination of the sun are observed by holding a thread connecting the graduation on the gnomon and that on the quadrant when the sun is in line with the thread.

Figure 2 shows the Misra Yantra, so called because it combines the functions of Samrath Yantra (time and declination), Dakshinobhitti (zenith and altitude), Niyatchakra Yantra (declination of the sun at sunrise and sunset), and Kark Rashivala (Rasi or zodiacal sign of sun).

Figure 3 shows one of the wheels of the sun temple at Konarak, which was built between A.D. 1238 and 1264. The Konarak temple could be considered purely as a manifestation of the adoration of the sun, as we do not see any practical utilization of the sun in the construction of this beautiful and exotic structure. However, a careful scientific investigation may lead to some of the thoughts of the architect who had the concept of solar energy utilization incorporated in the temple.

FIG. 3 The wheel of the sun chariot, Konarak Temple. The temple is built in the form of a wheeled car, the chariot of the sun having 24 wheels. The chariot is said to be drawn by seven richly caparisoned horses. Konarak is only one of many temples built to venerate the sun.

C. Sun Science of the Early Days

In the early days the sun was recognized not only as the vital force for sustaining life on this planet but also as a benefactor for all human activities. Hence, ways and means of utilizing sunshine by semiempirical precepts and rules of thumb supported by astronomy and by knowledge communicated orally from one generation to another were established. The therapeutic value of solar energy was also recognized. Sun basking and chromotherapy were advocated for maintaining health. This also indicates the extensive knowledge prevailing at that time regarding physiology and the selective use of solar radiation for treating common ailments.

D. Building Technology of the Early Days

The technology of housing and town planning had advanced to a very fine degree in the early days, and this fact has been corroborated by the archeological excavations made at Harappa and Mohenjo-Daro (now in Pakistan), and the recent excavations at Kurukshetra, Nalanda, and other places. Archeologists report that buildings revealed in the excavations belong to *ca.* 3000 B.C. These buildings were found to have mud brick walls, which were quite massive, no openings on the northern side, and only nominal walls on the south and southeastern sides. The design of the buildings indicates the popularity of the courtyard-type design, which has now been accepted as an ideal design in regions of climatic extremes and large diurnal temperature variations. In winter, the massive mud brick walls would store incident solar energy, providing cooler temperatures in the rooms during the day; this energy was then used for providing warmth during the cold nights, thereby maintaining comfort conditions during both the day and night. During the summer people would stay indoors where it would be comfortably cool during the days, and they would sleep on the flat rooftops during the nights; the cattle would be herded into the central courtyard to secure them from marauders and to protect them from the environment.

18.2 SOLAR AND METEOROLOGICAL CHARACTERISTICS OF INDIA

A. Meteorological Data

A wealth of data has been collected by the Meteorological Department of India since its establishment in 1880, and these data are available for most parts of India. Information published periodically by the Meteorological Department includes temperature, humidity, wind speed, rainfall, and solar insolation. Abstracted information of meteorological data relating to the building industry is published in "Climatological and Solar Data for India" (Seshadri *et al.*, 1959).

B. Climatic Zones

A large part of India lies in the tropical belt, and a relatively small area lies in the temperate zone. However, because of its vastness and geographical diversity, there is great variation in temperature, humidity, and rainfall. Monsoons, which are very characteristic of India, affect the climatic conditions.

The climatic zones, mean temperatures, and effective temperatures are shown in Figs. 4–6 for winter, premonsoon, summer, and postmonsoon periods. Effective temperature is an accepted index of physiological comfort, and it is defined as that temperature of saturated still air wherein the same degree of physical comfort is felt as in the existing conditions of temperature, wind speed, and humidity. The monsoons and the winds that accompany them have an impact on the discomfort caused by dry-bulb temperatures, and this can be seen on the maps of mean and effective temperatures (Figs. 5 and 6). The climatic zones indicated on the maps are based on the effective temperature.

Table 1 presents solar insolation data for both summer and winter. A high incidence of solar radiation occurs even during winter, indicating the

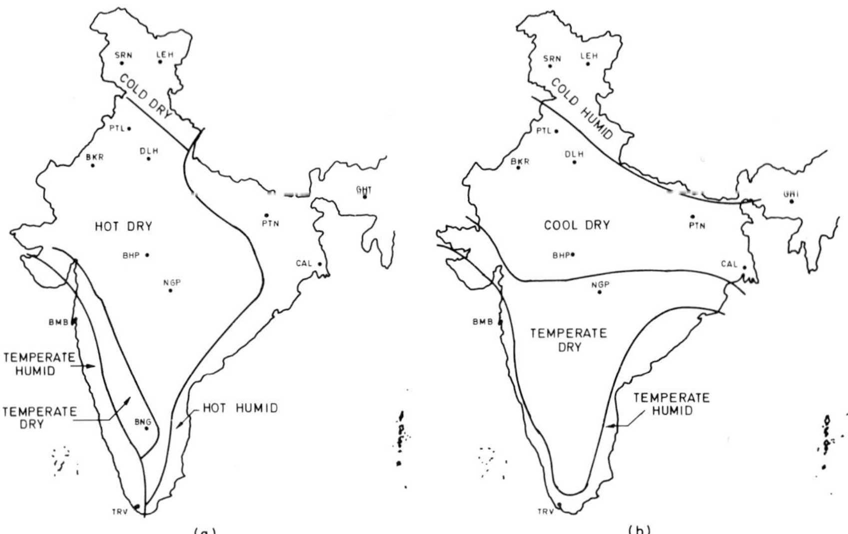

FIG. 4 Climatic zones during (a) May and (b) January.

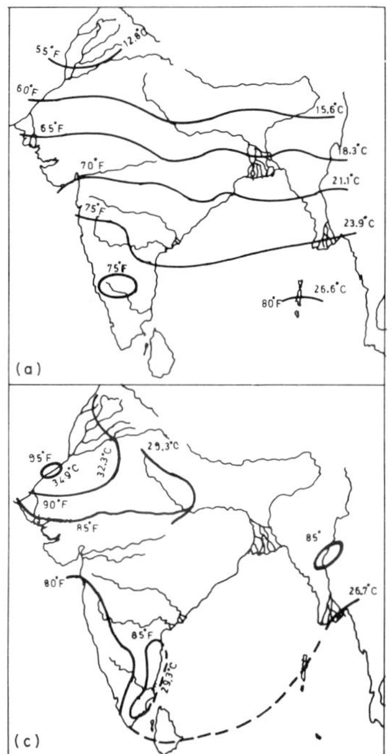

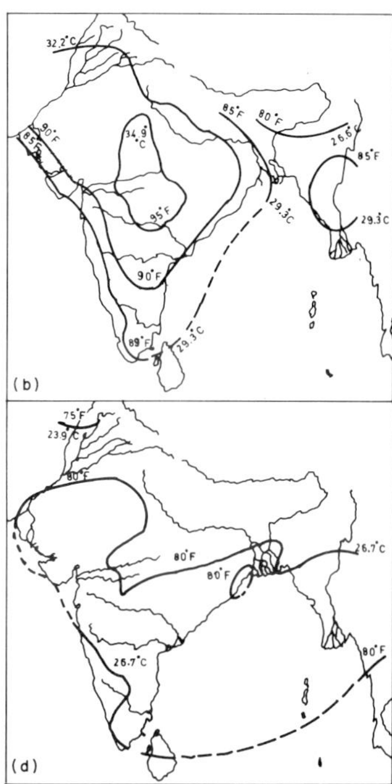

FIG. 5 Mean temperature: (a) January (winter), (b) May (premonsoon), (c) July (summer), (d) October (postmonsoon). Based on meteorological data for India.

potential for solar energy utilization all year round both for cooling and heating.

C. Traditional Solutions in Building Design

The bulk of the Indian population lives in rural areas, and a high percentage of this population exists below the poverty level. While the people may be economically poor, it has not deterred their innovative spirit, and many of them make use of novel ideas for comfort with the help of local materials and rural craftsmanship. The common man is not able to invest, even to a modest extent, in any device to utilize solar energy; hence any such device must involve local resource materials and expertise and not entail much financial investment. This means that the investment should be limited to what would normally have been spent with no extra expenditure. Some ways in which innovation has contributed to comfort without additional expense are described in the following paragraphs.

A typical rural house in India is built of mud walls or walls of sun-dried

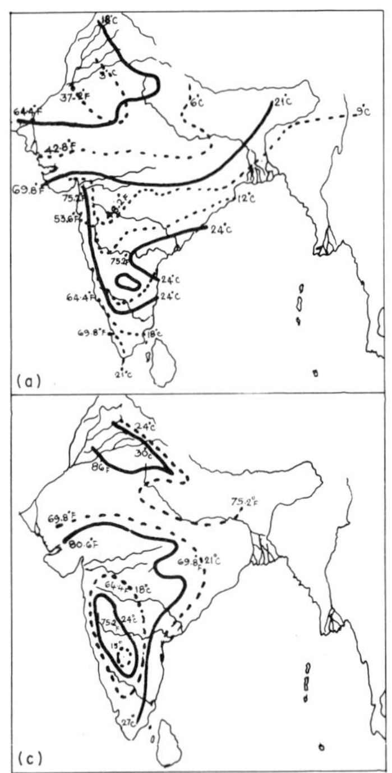

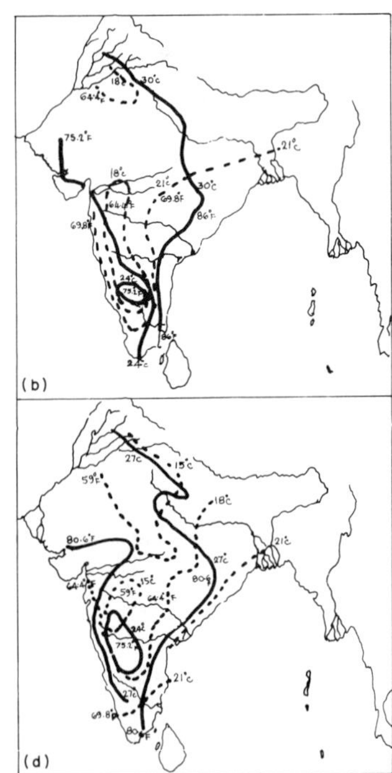

FIG. 6 Effective temperature: (a) January (winter), (b) May (premonsoon), (c) July (summer), (d) October (postmonsoon). Based on meteorological data for India.

or kiln-baked bricks. In some parts of India such as Assam and Western Ghats, where rainfall is very heavy, people build houses with mud walls reinforced with bamboo or timber framework. Buildings are properly oriented for utilizing solar energy to provide comfortable conditions in both winter and summer. In addition to proper orientation, windows are positioned so as to provide natural ventilation to alleviate solar discomfort. Some buildings, though they appear to be somewhat primitive with restricted openings in the walls, judiciously combine natural ventilation and a sense of security. With regard to construction materials, the walls, though they appear to be made of earth, consist of mud mixed with paddy husk or dried stalks of wheat or paddy, which improves the insulating property of the walls and adds to their strength. Also, the exterior walls are treated with lime wash, which reflects sunshine and provides a hygienic surface.

A large proportion of the houses in India are primitive in nature, with mud walls and thatched roofing. In the past few decades, with the ravages of time and changes in the sociopolitical structure, all that was good in the

TABLE 1

Solar Insolation Data[a]

Latitude (deg N)	Orientation of surface:	Horizontal		Southwest/ southeast		South	
		Dec	July	Dec	July	Dec	July
34		220	631				
33		230	630	280	197	396	74
31		247	630	295	186	397	57
29		267	630	298	176	402	43
27		285	627	298	165	402	31
25		305	626	298	153	399	16
23		324	621	298	145	398	7
21		344	618	297	136	394	2
19		359	612	297	126	392	
17		377	603	296	118	385	
15		396	597	294	109	377	
13		414	589	290	102	371	
11		426	580	288	94	359	
9		442	571	285	86	347	

[a] Measured in g cal/cm^2/day.

traditional ways was lost sight of, and the newer styles of houses survived to perpetuate irrelevant construction. The building technology that is now being developed in most countries should take an objective view of past practices and incorporate the good elements into new methods for a resulting homogeneous blend.

18.3 SOME THOUGHTS ON SOLAR ENERGY TECHNOLOGY APPLICABLE TO HOUSING IN INDIA

A. Building Technology in India

Though rapid strides have been made in India with regard to town planning, building construction, and rural housing, no systematic efforts have been made to incorporate solar technology in housing. However, considerable research and development activities in the area of building construction are being pursued by various institutions and national laboratories, such as the pioneering work done by the Central Building Research Institute at Roorkee, the National Buildings Organization, and the National Physical Laboratory at Delhi. The Department of Science and Technology of the Government of India has been making concerted efforts to coordinate all solar energy research and development activities in the country.

In a developing country such as India, where financial resources are meager, every effort should be made to keep down the cost of a building so as to make the additional cost of solar energy devices acceptable. The cost of a building is a function of materials (quality and quantity), design, and construction procedures. Materials account for about 70% of the total cost, and hence economies should be effected in this area first. One means is the utilization of second-grade hardwood, treated against vermin attack, in place of teakwood, which is very expensive. Criteria for materials specifications should be based on function rather than outmoded arbitrary standards. Building design should aim to achieve efficiency as well as beauty. Construction practices should be altered to use a judicious combination of mechanized construction and the plentiful labor supply available. The use of standardized construction materials throughout the country would enhance efficiency, streamline construction procedures, and reduce costs.

B. Building Design and Incorporation of Solar Thermal Devices

Buildings with large windows facing south and arranged to admit solar radiation when the sun is low in the winter sky can be termed "solar houses." The gains to be realized from properly oriented windows are significant. However, in the northern parts of India, where winters are severe, losses during periods of low radiation, nights, and cloudy weather, must be controlled so that net gains can be realized.

Figure 7 illustrates a method of controlled exposure to the sun through a window. Deciduous trees that shed their leaves during winter should be planted near the window so that more sunshine is admitted during the

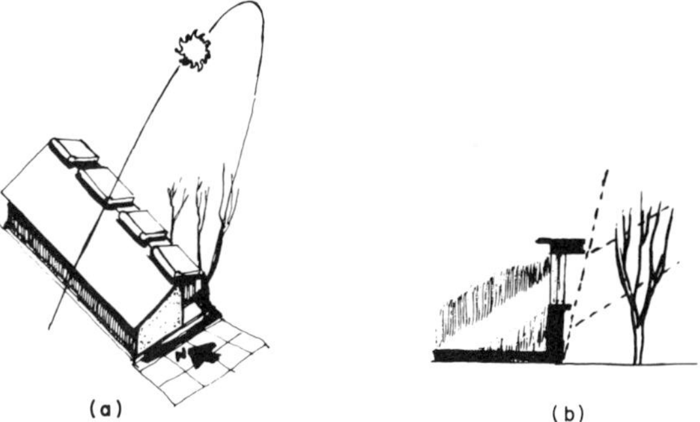

FIG. 7 (a) Planning considerations. (b) Section, controlled exposure to the sun.

winter and shade is provided during the summer. Also, row houses should be designed with the long axis in the north–south direction so that there is ample scope for solar energy utilization for all the houses.

Figure 8 illustrates the installation of flat plate collectors on sunshades, applicable even to multistory buildings, with insulated storage tanks integrated into the interior decor of the building and used for the domestic hot water supply all through the year and for space heating during winter. The space adjacent to the storage tank in the loft can be conve-

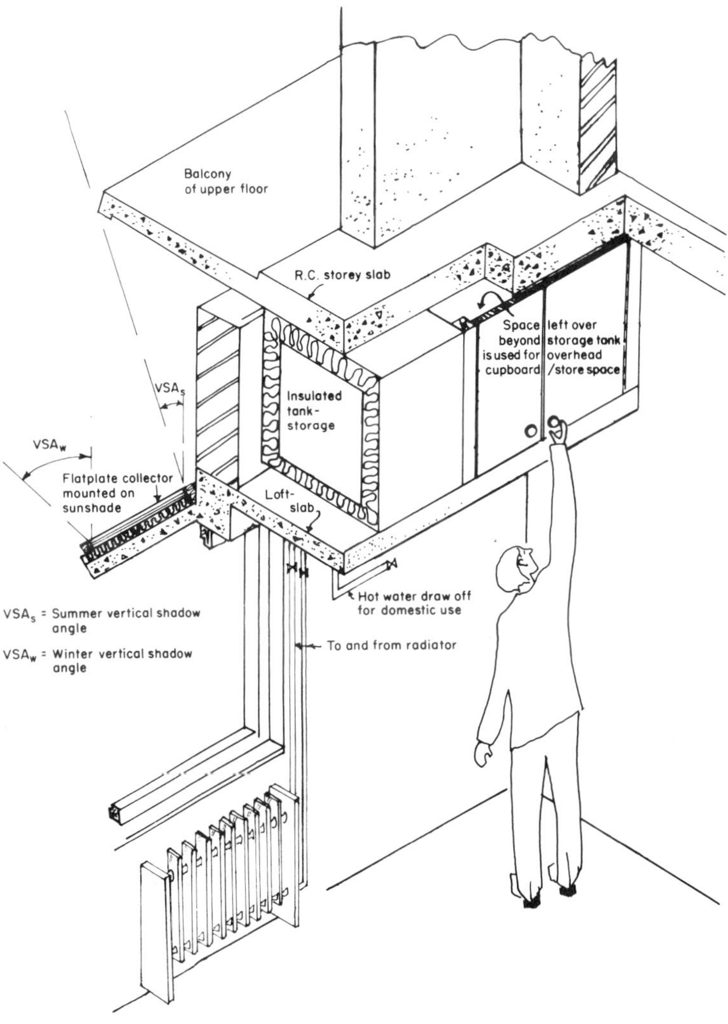

FIG. 8

niently used for a cupboard. The projection of the balcony is restricted by the vertical shadow angle during summer, as shown.

In dry tropical weather, evaporative cooling is quite effective. Air is drawn into the building through mats suspended in front of the window and sprayed with water. Cooled air provides comfort during summer. The mats are made of the roots of a sweet-scented grass called "khas." This adds a pleasant fragrance to the cool air. Large units working on this principle, called desert coolers, are available commercially. This system is not applicable to humid climates.

One space-cooling method is solar refrigeration. An intermittent ammonia–water system using a flat plate collector has been designed and developed at the Solar Energy Laboratory of the Indian Institute of Technology, Madras. This system seems to be an attractive and technically viable solar device for providing cooling in buildings and also for food preservation in remote rural areas without access to the electrical supply or for which long-distance transmission lines are too expensive.

Animal husbandry is an important aspect of the country's economy. Cattle, poultry, and swine must be housed properly and the building temperature regulated so as to obtain maximum output. When cattle are exposed to chilly weather conditions during the nights of hot summer days, milk production is reduced. Similarly, poultry production also suffers. Hence solar housing technology when properly deployed for animal husbandry would yield profitable results.

C. Conclusion

The foregoing is a brief overview of past and present building technology in relation to solar energy utilization. Currently, the wide-scale application of solar energy is saddled with difficulties, such as the initial cost of solar energy devices, public apathy and lack of information regarding solar energy potential, lack of concerted effort, and a pragmatic approach. However, the authors are very optimistic regarding future solar energy utilization, as it is the only alternative energy source that is abundant, pollution free, and devoid of political control.

19

Integration of Solar Systems in Architectural and Urban Design

ERTUGRUL BILGEN

ECOLE POLYTECHNIQUE
MONTREAL, CANADA

JACQUES MICHEL

PARIS, FRANCE

19.1 INTRODUCTION

With the present rate of increase of energy demand, no nation will be able to provide the required energy from nonrenewable energy sources in the next century. All predictions are that the only possible way to overcome an energy shortage is to conserve energy as much as possible and to develop before the end of this century the such renewable energy sources as solar energy, geothermal energy, wind energy, energy from fusion. To conserve energy is a matter of education, and it may take several decades to achieve substantial progress with the present design, construction, and utilization trends. Only with increasing energy shortages may this progress be accelerated. On the other hand, the utilization of renewable types of energy is a matter of economy. Solar, geothermal, and wind energy have already been shown to be technologically feasible and they are being used in a small way in various parts of the world. Energy from fusion, however, requires scientific as well as technological breakthroughs.

It is well known that in Western countries a considerable part of the total energy is spent for heating, ventilating, and cooling residential and office buildings and for hot water requirements. The temperatures required for air conditioning are usually moderate and can easily be pro-

ISBN 9-12-620860-3

vided by solar energy. Although solar energy is free and available where it is needed, without distribution and transportation problems, its intermittent and diffuse nature requires large and expensive collector areas and thermal storage systems. Recent studies have shown that solar heating and cooling of buildings is not economically feasible for two major reasons: First, collector and thermal energy storage subsystems are too expensive and constitute 50 to 80% of the total solar heating and cooling installation cost; second, architectural and mechanical features with respect to heating and cooling requirements of the buildings are not usually optimized. Another decade of development is apparently required for these complicated solar heating and cooling systems to become economically feasible.

"Natural air conditioning" or heating and ventilating of buildings can be accomplished by good energy management through the use of an integrated solar collector–heat storage system. The buildings in winter countries are usually well insulated to reduce the outward heat losses during the winter season. The insulation is so good that the solar energy received on the building is practically lost in the wintertime, while in the summer the high sun, considerable sunshine hours, and high humidity create uncomfortable conditions.

It is possible that by using the vertical and other suitable surfaces of present types of buildings as solar collectors and heat storage units, a considerable portion of the solar energy received on these surfaces could be used for winter heating and summer ventilation, and energy conservation would be implemented.

"Natural air conditioning" can be accomplished through the use of an integrated solar collector–heat storage system in the following manner: By using the greenhouse effect the solar energy received on various surfaces of a building is collected to the greatest extent possible, and a part of it is used to heat the space as required while the rest is stored in a concrete (or brick, stone, etc.) structure for later use through radiation and natural convection. The heat storage structure acts as a large thermal time constant to average diurnal temperature variations and as the heat supply for nocturnal heating and ventilating.

19.2 SOLAR SYSTEM

A. Collector System

The principle of the greenhouse effect is well known: the solar radiation, which has a spectral distribution in the visible and the near-infrared wavelength range of 0.3 to 3 μm, passes through the glass or other suitable transparent material and is absorbed by an opaque surface situated behind. The ideal absorption surface is the blackbody of the physicist, but most comparatively dark, nonreflective surfaces, for example, rough concrete, provide excellent absorption results.

The heated collector surface emits energy in the infrared wavelength range between 4 and 30 μm. This radiation is almost completely stopped by the glass partition, which is substantially opaque at this wavelength range. Consequently, the glass heats up and radiates toward the emitting surface as well as toward the exterior. If a second glazing is interposed 1 or 2 cm in front of the first one, the cover system, which is always transparent to solar radiation, will more effectively block the long-wave radiation from the receiving surface with considerable improvement in solar energy collection. Analysis of the transmittance–absorbance including reflection as a function of the angle of incidence shows that the energy ultimately absorbed by a receiving surface can quite accurately be calculated and that two or at the most three covers would be adequate for the collecting surfaces in winter countries.

The simplest application of this concept was an experimental house built near Chicago (at the Illinois Institute of Technology) in which large south-facing double-glass windows with a concrete floor as the collector–heat storage unit were used. On sunny days, solar energy transmitted through the large windows furnished heat, part of which was collected and stored in the concrete floor for later use. The fuel saving was considerable but not ascertainable (Anonymous, 1945a,b). Further full-scale experiments were carried out at Purdue University with two identical experimental houses, one with large south- and east-facing windows, the other with conventional windows. The Solar-heated house had 80% more window area on the south- and east-facing walls than the orthodox house; otherwise they were the same with the same kind of double-glass windows and insulation (Hutchinson 1945, 1946). The results showed that the number of degree-hours required for heating the solar house was reduced by 9% when the houses were both unheated; however the solar house showed a heating requirement 9.5% greater than the orthodox house when both were electrically heated (Hutchinson, 1947).

B. Heat Recovery System

It is quite evident that if such a collector system is placed on a roof, the trapped warm air, which is lighter than the air in the house, must be conducted mechanically toward the rooms to be heated. This may be the solution and in certain cases cannot be avoided, but the defect is that the system consumes electric energy and ceases to function as soon as the

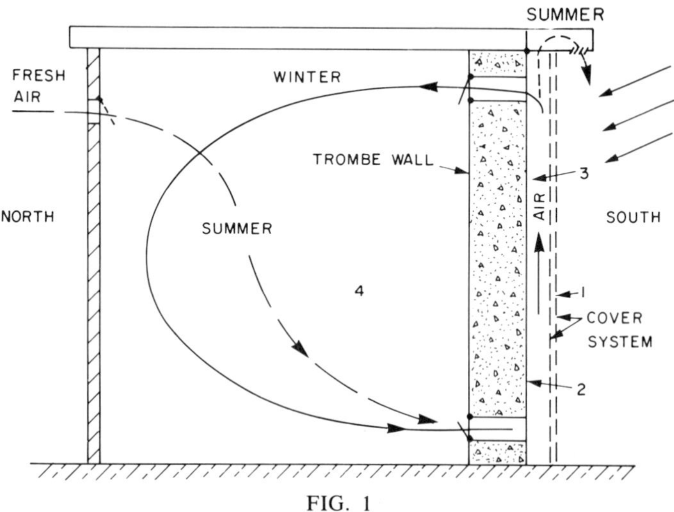

FIG. 1

electric current is cut off. In natural air conditioning collectors are positioned on the vertical facades. The possibility then exists of permanent circulation of warm air behind the glazing; conveniently placed apertures permit the air to pass behind the receiving surface. Figure 1 illustrates the process. Solar radiation passes through glazing 1 and is absorbed by collecting surface 2, which is a wall or a mass of water having a certain thermal mass. The heated air in the greenhouse 3 constitutes a column the density of which is less than that of the air in the room 4. The result is a permanent circulation following the direction of the arrows, provided the air of the collector is warmer than the air in the room. When this is not the case, for example at the end of the night, the lower aperture can be closed to avoid a reverse circulation.

Such a system functions without any mechanical intervention and can continue to function for a comparatively long time after sunset because of the partial storage of received solar energy.

C. Thermal Storage System

The heated surface in an individual dwelling may be a concrete wall. The received solar energy on the exterior surface of this wall is transmitted partly to the circulating air and partly to the wall that warms up throughout its depth. This stored energy serves to sustain thermocirculation well after sunset. It must also be emphasized that, in this system, the thermal storage unit is part of the house structure as a load-bearing wall; thus no additional investment is required.

In the case of light metallic or wood constructions, built for example, without using concrete, thermal storage can be ensured by using a certain mass of concrete block or water in metallic or plastic reservoirs.

D. Choice of Vertical Collector Surfaces

The particular interest in south-facing vertical surfaces for capturing solar energy in the Northern Hemisphere is well known in the literature (Trombe, 1971, 1974a; Michel, 1974). East- and west-facing vertical surfaces also receive far from negligible solar energy in winter. These surfaces represent, to some extent, natural rheostats adapted to the energy requirements of dwellings in a large range of latitudes from the Polar Circle to the Tropic of Cancer. Moreover, at certain times they introduce excess heat into the dwellings, which can be used to air condition them. Figure 2 represents solar energy in kWh/m²/day for various surfaces for different months of the year in Montreal; $\phi = 45.5°$ (Bilgen *et al.*, 1977).

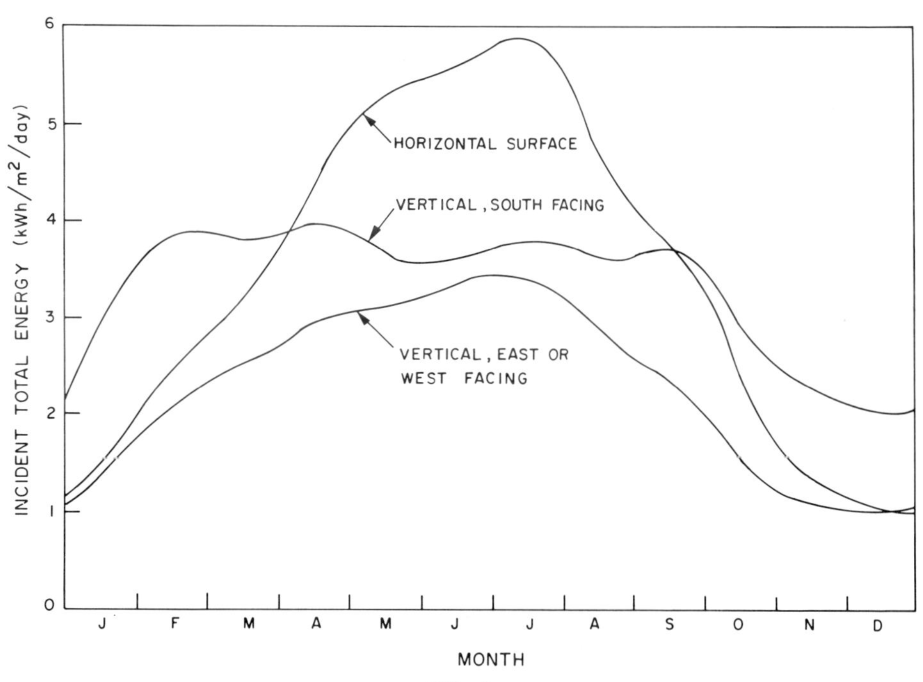

FIG. 2

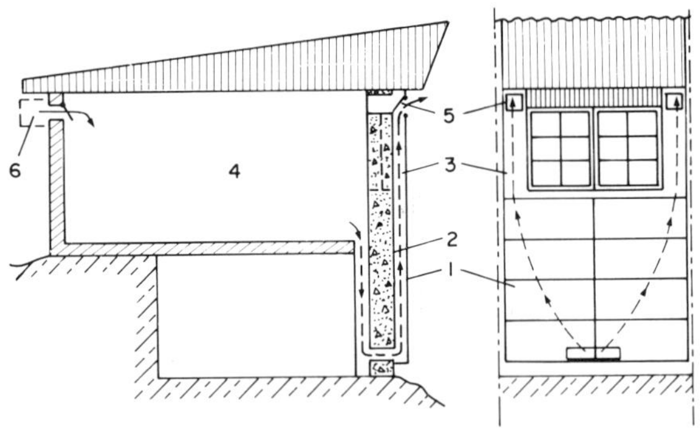

FIG. 3

E. Natural Ventilation

The design represented in Fig. 1 is extremely efficient. But numerous cases exist, particularly for east- and west-facing facades, where the quantity of heat delivered is excessive for the needs of the building; it must therefore be dispersed. Moreover, in many climates, the struggle against heat goes on for many months of the year, just as in winter important deliveries of energy are necessary.

Figure 3 represents the ventilating mode where facade 2 serves as a ventilator and the air is expelled above collector (5). A depression is created in the dwelling, which permits cold air coming from a northern facade or from air-conditioning equipment (6) to enter the living space where it circulates by natural convection and is exhausted by thermal syphon action through the space between the cover system and collector wall (3) and via a damper at the top (5). With this possibility, the integrated solar collectors have much more universal application than when they serve to provide heat only. It should be noticed in Fig. 3 that the receptive part of the collector is surmounted by lateral vents on either side of the window. These vents facilitate an increase in the movement of the air circulation performing the same role vis-a-vis the heating collector as the gas flue of the traditional fireplace that directs smoke upward (Trombe *et al.*, 1970; Trombe and Michel, 1975).

19.3 ANALYSIS OF THE SOLAR SYSTEM

A. Empirical Analysis

The concept of natural air conditioning was first used in the CNRS solar houses in Odeillo, France (Trombe, 1971). By referring to southern facades and to the latitude of Odeillo, the results of the calculations of mean solar energy collection lead to a definition of the ratio between the volume V of the dwellings and the surface A_c of the collector situated on the southern facade.

For a house with relatively poor thermal insulation and an average thermal resistance of $R \leqq 1$ m² °C/W, the ratio A_c/V is equal to 0.16. For a fairly well-insulated house of the E.D.F. (Electricite- de France) all-electric type with an average thermal resistance of $R \geqq 2$ m² °C/W, the ratio $A_c/V < 0.1$, for which the contribution of solar heat may represent between one-third and three-quarters of the total energy required to heat a dwelling. The remainder is supplied by electric current, mainly at off-peak periods.

Instead of concrete, the first solar house at Odeillo had for its heated surface black-painted water reservoirs; water circulated through the collector and was stored in the area over the outside wall above the ceiling. In 1974 this house was modified with better insulation and a brick wall installed behind the cover system as a receiving surface (Fig. 4). In 1967, two identical houses were built with massive concrete walls of 0.60 m thick serving as a receiving surface and thermal storage. All three houses have been occupied by CNRS staff.

The main characteristics of the 1967 prototype house together with those of three solar houses completed in 1974 are shown in Table 1.

The main dimensions of the solar collector of the 1967 prototype house are: concrete wall 0.60 m thick, painted with an acrylic black paint

FIG. 4

TABLE 1

Main Characteristics of Solar Houses at Odeillo, France

	Prototype 1967	Solar houses 1974		
		No. 1	No. 2	No. 3
No. of story	1	2	2	3
No. of principal rooms	4	6	6	7
Living area (m^2)	76	210	180	250
Volume (m^3)	300	525	450	650
Insulation, $\bar{R}$ (m^2 °C/W)	0.75	2.5	2.5	2.5
Insulation, $\bar{G}$ (W/m^3 °C)	1.63	1.0	1.0	1.0
Collector area (m^2)	48	55	45	65
Ratio, A_c/V	0.16	0.1	0.1	0.1
Auxiliary heating	Electric	Electric	Electric	Electric

(ρ = 2200 kg/m^3, k = 1.75 W/m °C, C_p = 0.25 Ws/kg °C); cover system of double glazing with 3-mm-thick glass, mounted in a steel frame, 0.12 m from the receiving surface; collectors, each 1.27 × 4.38 m; distance between the ventilating passages, 3.5 m; dimensions of the vents, 0.565 × 0.110 m or 0.0622 m^2. This prototype was thoroughly evaluated in 1974/1975. (Cabanat and Sesolis, 1976; Trombe *et al.*, 1976). It was reported that during the winter heating season daily total efficiency was 32–40%, that on a clear day 30–35% of the total solar heat was used in thermocirculation, and that the daily total efficiencies were strongly dependent on the climatic conditions of the previous days. The monthly total efficiency was found to be about 37% in the winter time, with decreasing efficiency from April to October due to lower incident solar energy on the southern vertical collector. This lower total efficiency corresponds in fact to lower energy demand in the April–October period, hence the Trombe–Michel system is well justified in its utilization of solar energy for passive heating and natural ventilation of dwellings (Fig. 5).

B. Analytical Study

In the course of the development of the Trombe–Michel system for heating and ventilating dwellings, it was noticed that the 0.60-m-thick walls in the 1967 prototype houses were too thick to allow the solar heat to propagate in time to the interior of the wall. In fact the phase difference was from 14 to 16 h. For this reason, the wall thickness in the 1974 solar houses was reduced to 0.37 m by using empirical relations. However, it was seen that the phase difference was a bit too small, namely, from 9 to

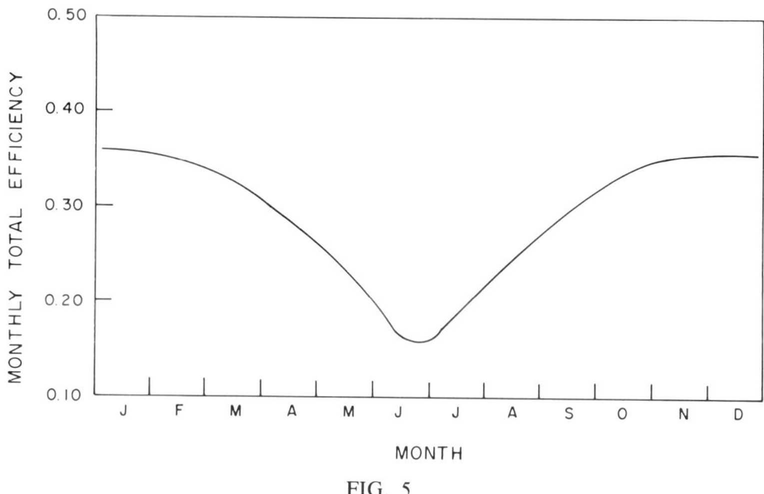

FIG. 5

10 h. It was clear that a satisfactory thermal simulation of this system was necessary.

A mathematical study of this problem has been carried out recently and compared to the experimental results obtained in the 1967 prototype house (Jeldres *et al.*, 1977; Jeldres, 1978; Bilgen and Jeldres, 1978). Two types of operation have been considered:

(1) The ambient temperature is maintained constant at 20°C, by the auxiliary heating system when necessary.

(2) The ambient temperature is variable; at the outset, the inside ambient temperature is equal to that outside.

An analysis will be presented for both types using a double-glazing cover system. For details, see Jeldres *et al.* (1977).

Constant ambient temperature of 20°C The energy balance equations for the Trombe wall collector shown in Figs. 1 and 6 are for the second cover glass:

$$(h_3 + h_{r_3})(T_2 - T_\infty) + (h_{r_2} + h_2)(T_2 - T_1) = 0 \tag{1}$$

for the first cover glass:

$$(h_{r_2} + h_2)(T_1 - T_2) + h_{r_1}[T_1 - T(0, t)] + h_1(T_1 - T_f) = 0 \tag{2}$$

for the air between the cover system and the receiving surface:

$$h_1'(T_1 - T_f) + h_1[T(0, t) - T_f] - Q_t = 0 \tag{3}$$

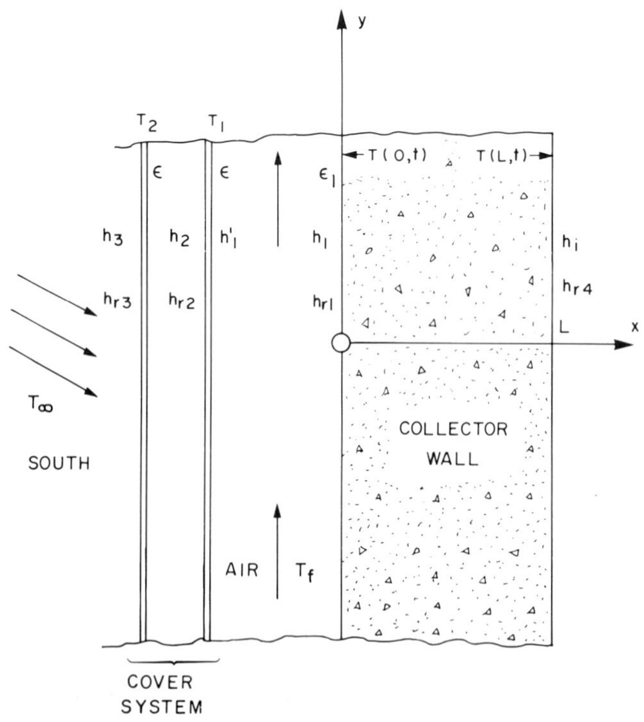

FIG. 6

and by considering one-dimensional heat transfer with uniform temperature distribution in the wall,

$$\partial T/\partial t = \alpha(\partial^2 T/\partial x^2), \qquad 0 < x < L \quad \text{and} \quad t > 0 \tag{4}$$

where α is the thermal diffusivity, h the coefficients for convection and radiation heat transfer, Q_t the energy carried by air, and T the temperature.

The boundary condition at the receiving surface with time-dependent solar energy input $S(t)$ and with convection and radiation exchange between the receiving surface and the cover system is

$$-k \left.\frac{\partial T(x, t)}{\partial x}\right|_{x=0} = S(t) - \sigma F_1[T(x, t)^4 - T_1^4] - h_1[T(x, t) - T_f] \tag{5}$$

$$x = 0 \quad \text{and} \quad t > 0$$

where F_1, F_2 are the emissivity factors.

The boundary condition at the interior surface with convection and radiation exchange in the dwelling is

$$-k \left. \frac{\partial T(x, t)}{\partial x} \right|_{x=L} = \sigma F_2[T(x, t)^4 - T_e^4] + h_i[T(x, t) - T_i] \qquad x = L \quad \text{and} \quad t > 0 \tag{6}$$

where T_e is the equivalent temperature inside the dwelling and is related empirically to the exterior temperature T_∞:

$$T_e = 17.67 + 0.184\ T_\infty(t) \tag{7}$$

The initial conditions of the wall collector are assumed to be quadratic for $0 \leqq x \leqq L$ and $t = 0$:

$$T(x, t) = a + bx + cx^2 \tag{8}$$

where a, b, and c are constants.

Equations (1)–(3) are linear and can be solved easily; however, Eq. (4) together with boundary conditions equations (5) and (6) represents a nonlinear system that can be solved by using an approximate method. The solution of this system of equations is obtained by using a semi-implicit method together with an implicit scheme to express the boundary conditions.

Temperature variation between the wall and cover system The energy balance equation for air is

$$Q_t = (\dot{m}/L)C_p(\partial T_f/\partial y) \tag{9}$$

where m is the mass flow rate and C_p is the specific heat. Using Eqs. (1)–(3) and (9), a differential equation is obtained for the air temperature:

$$(\dot{m}/L)C_p(\partial T_f/\partial y) = A_1(T - T_f) + A_2(T_\infty - T_f) \tag{10}$$

The solution gives a logarithmic temperature distribution

$$T_f(y) = [T_{fi} - (A/B)]e^{-Y} + (A/B) \tag{11}$$

where

$$A = A_1T(x, t) + A_2T_\infty, \qquad B = A_1 + A_2, \qquad Y = A_cB/\dot{m}C_p$$

The convection and radiation heat transfer coefficients in Eqs. (1)–(3) are combined in coefficients A_1, A_2. These coefficients are estimated using the published relationships in the literature. See, for example, Duffie and Beckman (1974).

Heat carried by the thermocirculating air The heat captured by the absorbing surface is transferred partly by conduction to the wall, partly by radiation to the cover system, and partly by convection to the circulating air between the absorbing surface and the cover system.

The portion transferred by convection is important for heating the dwelling by thermocirculation and can be expressed by

$$Q_t = h_1[T(0, t) - T_1] \tag{12}$$

where h_1 is calculated by using an empirical relation determined at the conditions considered for the vertical wall collectors (Jagadish and Peres, 1977):

$$\mathrm{Nu} = 0.0018\mathrm{Gr}^{0.56} \tag{13}$$

where Gr is calculated by using the temperature difference and the distance between the wall surface and the first cover.

Ambient temperature is variable This is the case for various projects described in Sections 19.6, D and H. For this case Eqs. (1)–(5) are valid; however, the boundary condition at the inside surface of the wall is expressed in terms of a time-dependent ambient temperature T_p as

$$-k \left. \frac{\partial T(x, t)}{\partial x} \right|_{x=L} = \sigma F_2[T(x, t)^4 - T_e^4] + h_i[T(x, t) - T_p(t)] \tag{14}$$

where $T_p(t)$ is a function of $T_\infty(t)$, $S(t)$, the thermal resistances of the dwelling, etc.

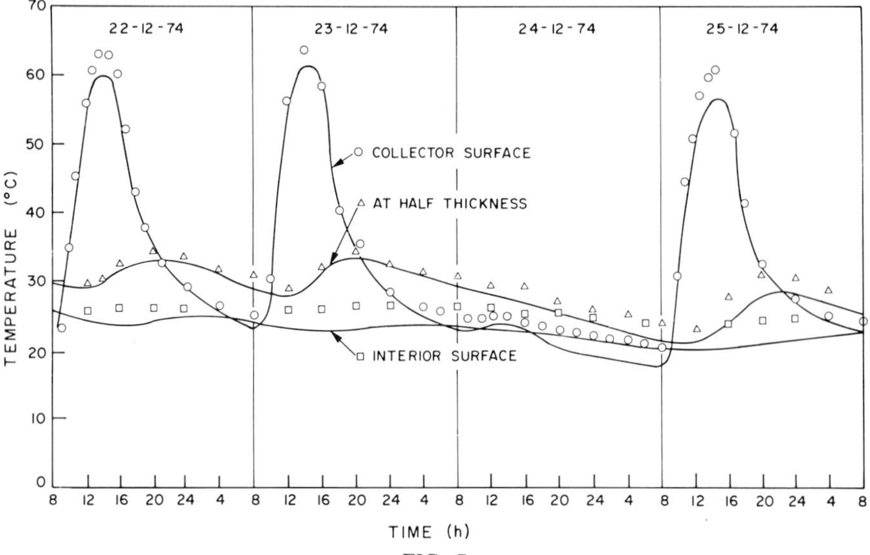

FIG. 7

In Fig. 7 the thermal simulation results are compared with the experimental results obtained at the prototype solar house at Odeillo (Cabanat and Sesolis, 1976; Trombe *et al.*, 1976) for the period December 22–25, 1974. The temperature variations on the outside surface, at the midpoint, and on the inside surface of the 0.60-m wall are compared in this figure. The temperature variation in the wall as a function of time is shown for December 23, 1974, in Fig. 8. The air temperature at the outlet of the collector as a function of time is shown for the same day in Fig. 9. The air temperature along the height of the collector as a function of time is shown for the same day in Fig. 10.

It can be seen that the theoretical predictions are in reasonable agreement with the experimental results. On the other hand, the thermal simulation study carried out for various wall thicknesses has shown that for the experimental conditions of the 1967 prototype house, the optimum thickness should be 0.40 m.

C. Experimental Research House "Solab" at Ecole Polytechnique, Montreal

It appears that the passive heating and natural ventilation of dwellings can be achieved satisfactorily by using vertical solar collectors with or

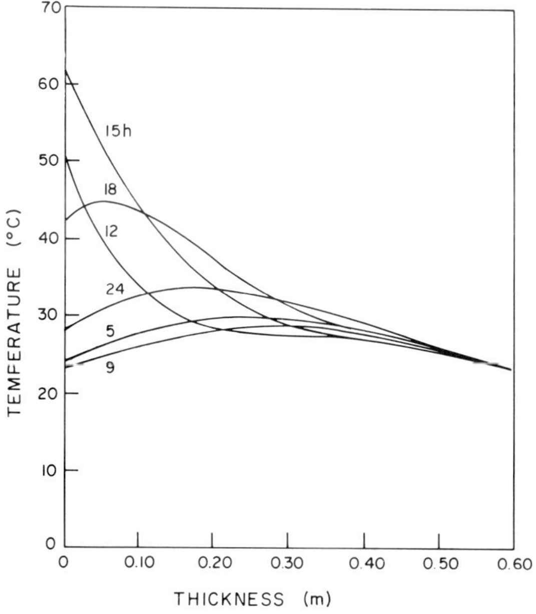

FIG. 8

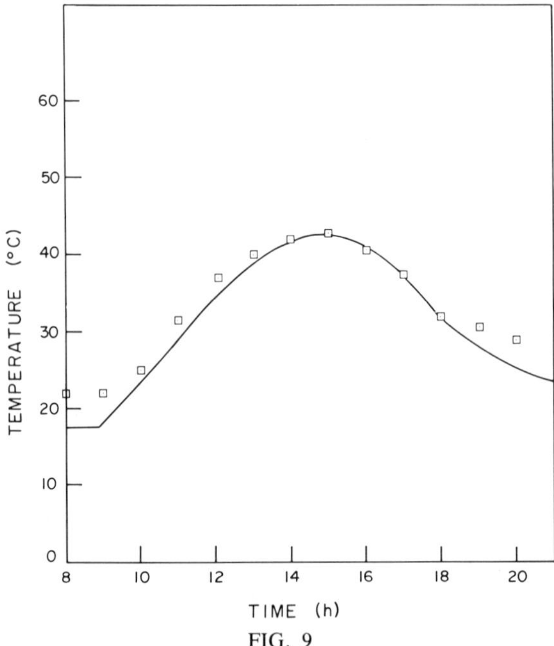

FIG. 9

without thermal mass. In order to study experimentally the variations of this system under the climatic conditions of Canada and to develop the most effective means of capturing, storing, and utilizing solar energy, an experimental solar house was built on the campus of the University of Montreal in 1977 (Fig. 11).

The main features of this house are as follows: The living space floor area is 70 m^2 with a volume $V = 175\ m^3$. In addition, there is a 70 m^2 basement area. The construction is the traditional canadian with a wooden structure. The thermal resistances R (m^2 °C/W) are

walls	3.52
ceiling	5.28
floor	0.26
windows	3.29
doors	1.25

The house has eight modular sections, each with a 1.2 × 2.5-m south-facing double-glazed window. In addition, there are two windows at the eastern facade and one at the western. Each window can be used as a solar collector when equipped with an absorbing surface either with or without thermal mass.

4.0
3.0
2.0
1.0
0
HEIGHT OF THE COLLECTOR (m)
10h
21
11
19
12
17
14
15
15
20
25
30
35
40
TEMPERATURE (°C)

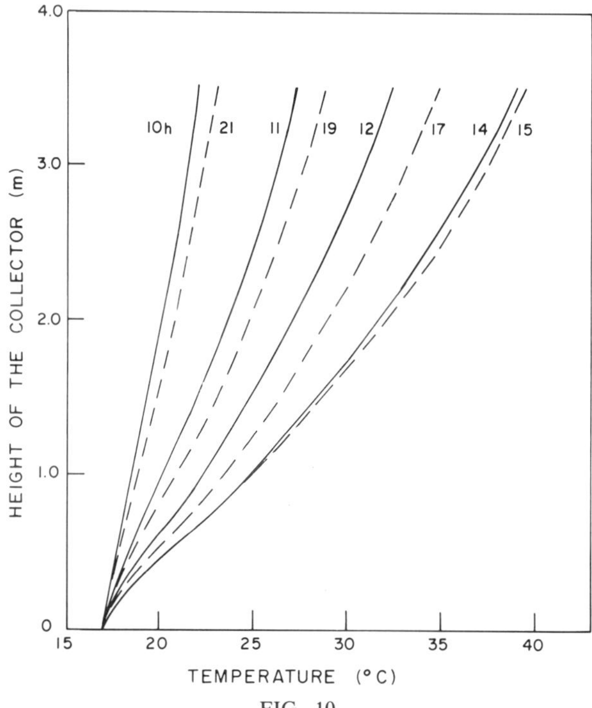

FIG. 10

FIG. 11

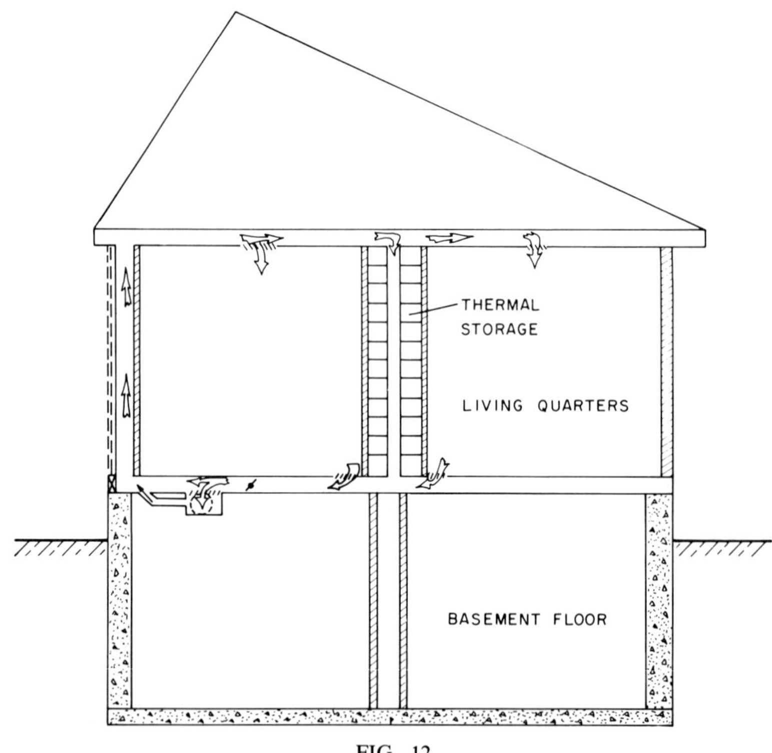

FIG. 12

The present experimental program includes the following:

(a) 1 Trombe wall collector;
(b) 1 one-half Trombe wall collector;
(c) 4 collectors without thermal mass, and thermal storage in the dividing wall inside the house (Fig. 12);
(d) 3 one-half passive collectors without thermal mass at the eastern and western facades;
(e) various types of cover system (Superseal);
(f) natural ventilation during the summer;
(g) economic evaluation of various systems.

The thermal simulation of this house shows that the solar contribution during the heating season will be on the order of 50%; the rest of the heating will be provided by electricity (Bilgen and Jeldres, 1978).

19.4 ECONOMIC ASPECTS

As can be seen, passive solar heating and natural ventilation can improve the internal climate of dwellings. However, if it is desirable to obtain a fairly constant temperature at all times, a flexible auxiliary system should be available without bringing into play additional thermal masses and thermal storage systems. Electric heating, particularly at off-peak periods, and wood-burning stoves are the ideal complements for solar heating.

The actual cost of a collector with double glazing, without taking into account the cost of the load-bearing wall, is on the order of $60/m^2. In the Montreal region a south-facing vertical surface receives on the order of 1100 kWh/m^2/year, and the wall retains and transmits about 35% of it to the air, hence to the dwelling. The cost of the solar kilowatt hours can be determined by considering only the investment for the collector, since solar energy is free. Hence, assuming a 20-year life for the system and a 9% interest rate, by using the cash flow method the cost of solar energy can be calculated to be 1.71¢/kWh, which is about 10% cheaper than the 1978 rates of Hydro-Quebec.

Comparable figures for France are 300F/m^2 (1975) and 1600 kWh/m^2/year. For a 12% interest rate, the cost of solar energy is 7.17 centimes, which is two to three times less expensive than electric energy.

Of course, the cost of solar collectors could be considerably reduced if a simpler wooden structure for the cover system were to be used. Thus, a worthwhile improvement in living conditions could be brought to isolated cold countries that lack fuel by rudimentary capture of solar energy, which would be carried to the inside by thermocirculation, conduction, and radiation.

19.5 ARCHITECTURAL ASPECTS

The first houses built by CNRS had fairly large southern facades and roofs with single slopes; they were simple in appearance and perhaps a little poor architecturally. Moreover, given the relatively poor thermal insulation of these houses, the fenestration area in the southfacing side was somewhat restricted in relation to the total surface. Because these first structures were too functionally oriented, the solar house was looked upon unfavorably by virtue of its architectural aspect. It was believed that the solar house was of necessity associated with one style of architecture, where as this is not so. Moreover, a study of the thermal insulation of the all-electric-type house has shown that a ratio of $A_c/V < 0.1$ could be adapted. Thus, it is possible to arrange large lighting surfaces on the southern facades, with warm air inlet vents located laterally to these surfaces, as shown in Fig. 3. In fact, several solar houses built later show clearly that one can build architecturally harmonious blocks, while retaining the ability to capture and utilize solar radiation.

In order to appreciate the integration of solar systems into architectural design, it is appropriate to discuss briefly the climatic architecture of Le Corbusier. One good example is the project of Chandigarh and Amedabad in India, where climatic architecture enabled the conception of a basic architecture in harmony with tradition while respecting the local way of life. This resulted in buildings that harmoniously combined solar and climatic elements in the construction of sleeping terraces and living spaces. Unfortunately, this kind of architectural design that was closely adapted to climatic influences was poorly accepted. Consequently architecture was arbitrarily left open to designs and styles that permitted thermal engineers to heat and/or cool buildings by expensive means.

10.6 PRACTICAL APPLICATIONS

A. Climatic House

Figure 13 shows the climatic house in Lissey, eastern France, built in 1962 in accordance with the principles of climatic architecture. This circular house has living quarters that are constantly exposed to the sun from east to west, with a resulting saving in heating costs of about 20%.

FIG. 13

B. Solar House at Chauvency-Le-Château

Figure 14 shows the solar house built in Lorraine at Chauvency-le-Château, in eastern France, in 1969. The objective was to build either private houses or apartment buildings with modular and extensible living quarter units, each equipped with a Trombe–Michel wall fitted for an adequate surface–volume combination. The insolation in this area is about 1750 h/year, and the heating degree days are 2900. The characteristics of the house are: 106 m^2 of living space in 5 principal rooms, 275 m^3 total volume, average thermal resistance $\overline{R} = 1.1$ (m^2 °C/W), collector area $A_c = 45$ m^2 (hence $A_c/V = 0.16$), and electric heating as the auxiliary heating system. The collector system is a Trombe wall of 0.60-m-thick concrete situated exclusively on the southern facade of the house. The thermal resistance R (m^2 °C/W) in various parts of the house is: walls, 2.5; Trombe wall, 0.6; cover system, 0.3; ceiling, 2.5; floor, 1.0. The volume loss coefficient is $\overline{G} = 1.30$ W/m^3 °C.

The experimental balance sheet of this house, which has been occupied for 2 years, is as follows when the rooms are kept at a temperature of

FIG. 14

18 to 20°C: solar energy delivered annually about 18,000 kWh gross, solar kWh actually utilized about 10,000 kWh. The energy saving is estimated at about 46%.

Since demand in off-peak hours is of the same order of magnitude as the demand in peak hours, the balance sheet for 2 years shows an annual mean cost of auxiliary electric heating of approximately $170.00

Parallel to solar energy research is the attempt to construct dwellings (individual houses in a terrace or collective buildings), starting with a "module of habitable space" that can be juxtaposed and used in different types of accommodation; this module is constructed of self-contained standard elements and can lead to an economy and speed in construction that would be difficult to attain with traditional processes.

The house has a metallic tubular framework of steel, supporting latticed trusses separated by a habitable half-module and perpendicular to the solar wall; the roofing is of steel with thermal insulation and multilayered felt. The eastern, western and northern facades are covered externally by precoated steel panels with high thermal insulation and internal linings erected dry. The glazings of the openings are composed of the same standard elements as the glazing of the collectors. The suspended ceiling has the same thermal insulation. The central core incorporates sanitary arrangements and the kitchen, and groups all the ducts and various conduits; it is made of plastic. The internal partitions are erected dry with the cupboards built to modular sizes.

C. Solar Houses at Odeillo–Font Romeu

Figure 15 shows the block of three solar houses built in 1974 on a rocky peak situated not far from the 1000-kW solar furnace. In this project, the architecture takes account of the environment, climatic conditions, and direction; the facades incorporating solar collectors face south, east, and west; the openings form one unit with the solar collectors. The thermal system is based on the Trombe–Michel system shown in Fig. 3, with a concrete wall 0.37-m thick. These three dwellings are owned by M. Ducarroir (no. 1), B. Armas (no. 2), and F. Trombe (no. 3), all from CNRS Laboratories. Personalized architecture, taking account of different programs, demonstrates the flexibility of application of vertical collectors in a free plan. The houses are built using traditional materials. The main characteristics of these houses are given in Table 1. The collector is a concrete wall with the same characteristics as that of the 1967 prototype. It has a coating of vinylic brown paint on the external receiving surfaces and a double-cover system two of 4-mm-thick glass sheets installed

FIG. 15

the same way as in the 1967 prototype. Each collector is 1.55 × 2.50 m. The distance between the vents is 2.20 m and the dimensions of the vents are 0.84 × 0.095 m or 0.0798 m².

These houses were evaluated during the 1974/1975 season; however, the results were not conclusive because of the residual humidity in the concrete walls (Trombe *et al.*, 1976). Later in 1977, the thermal performance of these solar houses as evaluated by the owners was similar to that of the prototype.

D. AFPA Center, Beziers

Figures 16 and 17 show the center for the professional education of adults in Beziers, southern France, in the construction stage. The building has 11 workshops with a total living area of 4000 m². For heating and ventilating the Trombe–Michel system with a concrete block wall situated at the southern facade is used to ensure a minimum temperature of 14°C during the winter and natural ventilation by thermocirculation during the summer, following the principle illustrated Fig. 3. For this location the ratio $A_c/V = 0.06$. Moreover, the administration offices, foyer, and restaurant are heated by a solar system utilizing laminar-type water collectors installed on the roof and a thermal storage connected to the heating system. The 1500-m² south-facing solar collector makes the building completely independent with regards interior environmental control.

FIG. 16

FIG. 17

E. Solar House near Dourdan, Paris Area

Figures 18 and 19 show the Dourdan solar house in the construction stage (1977). Figure 20 provides architectural views of the house. The building has a solar air heating system with a pebble bed thermal storage. Auxiliary heat and summer air conditioning are provided by a heat pump. The collector system is installed at the basement level in order to free the facades of the living quarters.

The technical characteristics of the solar house are as follows:

Collector system The 66-m² system installed is at the southern facade and is 25-m long. Air distribution is by means of collectors at the bottom and top of the receiving surfaces.

Air distribution The distribution of air is provided by means of a fan also serving the heat pump through the fiberglass distribution system.

Thermal storage The storage is a 6-m-long × 2.5-m-wide × 2.5-high pebble bed storage system of about 40 m³ and 60 tons.

Auxiliary heating Auxiliary heating is provided by a heat pump (Technibel) but can operate with air temperatures as low as 5°C, below which electric strip heaters are used to prehat the air.

Operation modes. Depending on the availability of solar energy, there may be

(1) direct solar heating;

FIG. 18

FIG. 19

(2) storage of solar heat in the thermal storage system;

(3) heating from the thermal storage system;

(4) heating by means of a heat pump using direct solar heat, heat from the thermal storage system, or exterior air with or without direct electric heating depending on the exterior air temperature.

The volume of the house $V = 500\ \mathrm{m}^3$ with $\overline{G} = 1.15\ \mathrm{W/m^3\ °C}$. At Saint-Cheron, heating requirements are 2500 degree days (Celsius). Hence, the seasonal heating requirements are estimated at 25,877 kWh. Table 2 shows the detailed estimates for the heating season.

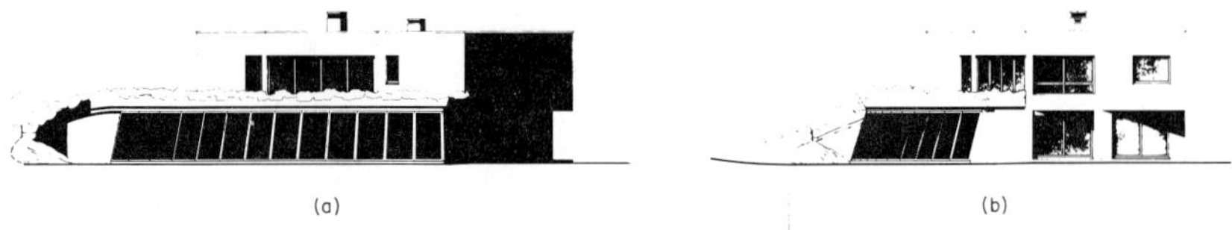

(a) (b)

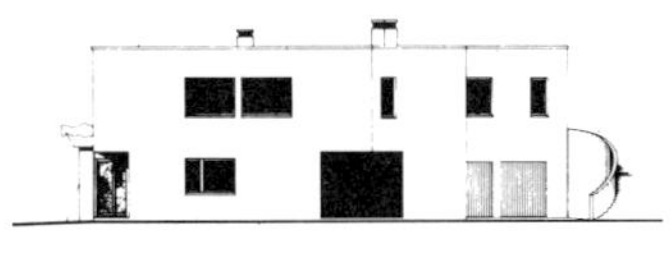

(c)

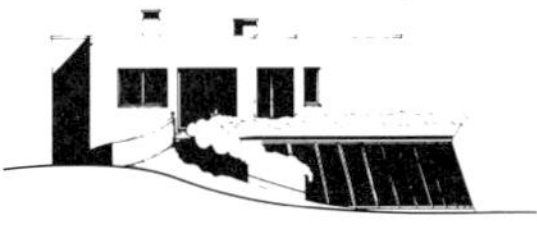

(d)

FIG. 20

TABLE 2

	Degree days Celsius	kWh	Solar energy[a] (kWh/m²)	Absorbed energy[b] (kWh)	Q_u (kWh)	Q_{aux} (kWh)
October	170	1760	94.5	2495	1760	—
November	347	3592	53.6	1415	1415	2177
December	442	4575	40.5	1069	1069	3506
January	466	4823	53.6	1415	1415	3408
February	391	4047	71.5	1888	1888	2159
March	335	3467	111.2	2936	2936	531
April	222	2298	116.9	3086	2298	—
May	127	1315	119.4	3152	1315	—
Total	2500	25877		17456	14096	11781

[a] Tricaud (1976).
[b] For an average efficiency of 40%.

It can be seen that 55% of the total demand is provided by solar energy. The rest, about 12,000 kWh, is provided by the heat pump. For a COP = 2, this would represent about 6000 kWh of electric energy for the heating season.

F. Solar House of Dr. Grand at Orsay

Figure 21 shows an architectural perspective of the house at Orsay. This house, as in the case the Dourdan house, is heated by a solar-assisted heat pump without, however, a thermal storage system. It has $V = 550$ m³ living space with $\overline{G} = 1$ W/m³ °C and a 2445 degree days (Celsius) heating requirement. The seasonal energy requirement is 24,205 kWh. The solar

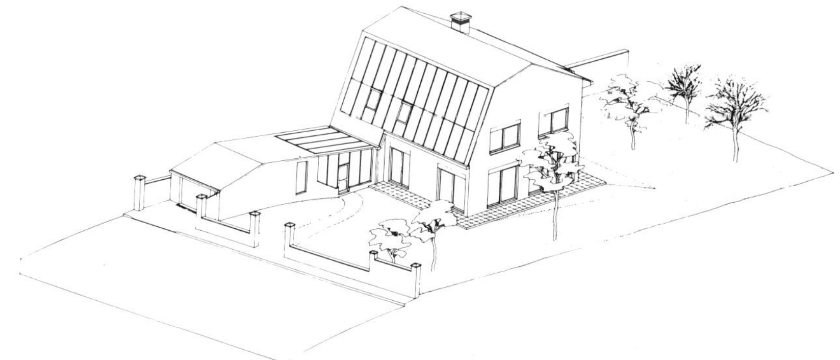

FIG. 21

heat contribution is about 35% or 8472 kWh, the rest is provided by the heat pump with a total electricity consumption of about 7850 kWh.

The solar air heaters are the economical laminar "Rexotoit Leroy" collectors with an installation cost of about $50/m^2. They are installed at the southern facade at a 70° inclination.

There is also a solar greenhouse built at the principal entrance of the house.

G. Office Complex and Social Center of the Agricultural Cooperative at Port-sur-Saone

An overall view of the complex is shown in Fig. 22. This two-story building has 64 offices and meeting rooms, which are connected to the laboratories in the basement. Diurnal illumination of the laboratories is ensured by the northwestern facade and the pyramid-shaped skylight.

The offices are connected at the back to the social center, which is built according to the same bioclimatic concept. The building has large solar facades oriented toward the southeast, south, and southwest.

This complex is a good example of the bioclimatic concept and natural utilization of solar energy.

The wide double-glazed facade is equipped with ventilated galleries where the solar energy is received. The concept of the ensemble is to favor the transition from building to nature by introducing a green area in

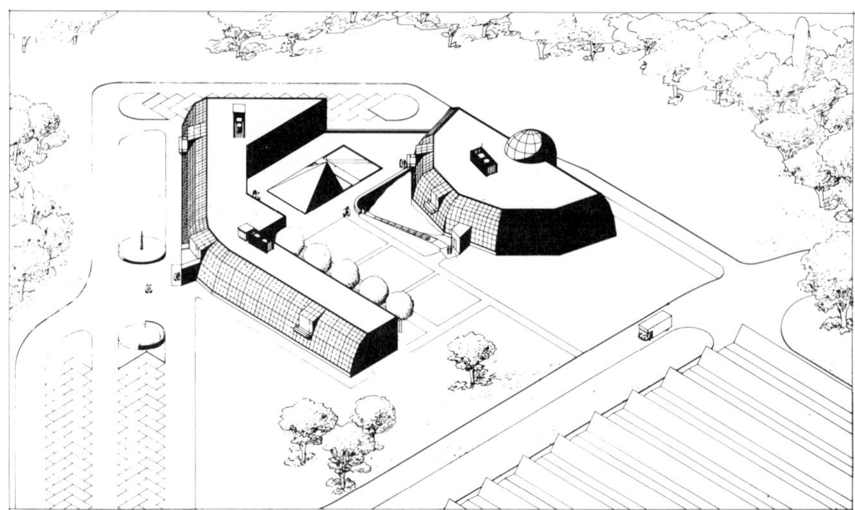

FIG. 22

front of the offices in the bioclimatic space at the southeastern and southwestern facades.

Auxiliary heating is provided by the heat pumps installed on the roof, which extract heat from thermal storage system or, when necessary, from the exterior air. The offices are partially heated by the thermal storage system. The heat pumps serve also to air condition the space in summer season.

The building is designed to have good internal thermal inertia favoring natural air conditioning. The southeastern, southern, and southwestern facades are made of steel and aluminum structures with selective type double windows and "Solomatic" mobile blinds that permit the minimizing of heat loss during the night or heat gain during the summer.

Sanitary water is provided by solar collectors installed at the top of the galleries at the southeastern, and southwestern facades.

H. Solar Gymnasium at St-Peray, Ardèche

The general view of the solar sport center is shown in Fig. 23. The center is $40 \times 20 \times 7$ m with a total volume of 6000 m^3.

The vertical solar collector system at the southern facade of the building has a total area of 322 m^2. Each collector element is installed within the walls, forming a thermal enclosure like a honeycomb that reduces convection and radiation losses (Patent Michel-Diamant-Durafour).

A minimum temperature for the gymnasium of 12°C during the winter is provided entirely by the solar system.

Sanitary water as well as the heating of the administration sections is supplied by 96 m^2 water collector and thermal storage systems.

The preliminary estimates show that the additional investment for the solar system will be amortized in 6 years.

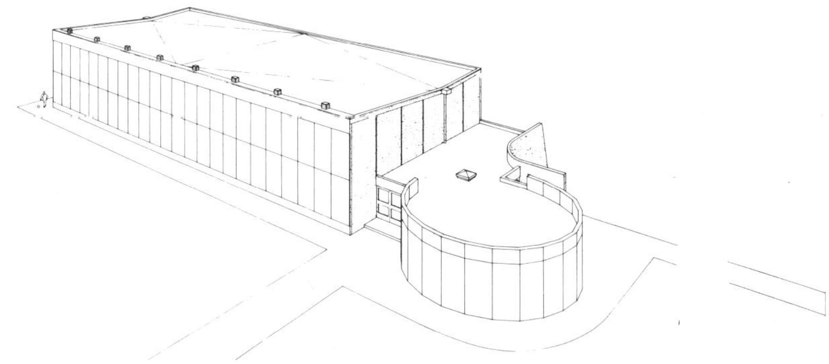

FIG. 23

I. Prototype Solar Houses at Paris Fairs

The 1973 prototype solar house shown in Fig. 24 incorporated new lightweight collectors with liquid storage, both static and with transfer after solar heating in a nonconducting accumulator arrangement; the latter, when situated in a room not exposed to solar radiation, permits heating of the room by forced ventilation.

With each volume corresponding to a module of 3.60 × 3.60 m of tubular framework was associated a facade comprised of two 1.8-m panels. One of these panels consisted of a pane of glass ensuring lighting; the other comprised the glazing of the vertical collector, whose collecting surface was calculated to heat the volume enclosed in the module of 3.60 × 3.60-m.

This prototype house had four principal rooms, 72-m^2 living area, 210-m^3 volume. It was designed for 2406 degree days (Celsius) with $G = 1.16$ (W/m^3 °C) and had the following thermal resistances R (m^2 °C/W):

walls	2.50
collectors	2.50
windows	0.30
ceiling	2.58
floors	0.57

FIG. 24

The estimated solar contribution was 50%. The rest was provided by auxiliary electric heating.

The prototype solar house at the 1975 Paris Fair is shown in Fig. 25. This house was a wooden structure. The solar system was a water-circulating system. The water solar collectors, 45 m^2 in area, were installed at an angle equal to the latitude, on the roof of the house. Two water circuits were used: In the primary circuit the water circulating through the collectors transported the solar heat to the 3000-liter-capacity thermal storage system. In a second circuit the water was circulated through the radiators in the house. The sanitary water as well as the heating of an adjacent swimming pool, was also provided by the solar system.

The prototype solar house at the 1976 Paris Fair had 5 rooms, 100-m^2 living area, and 250 m^3 volume. The solar system was a passive system using Trombe–Michel walls for collecting and thermal storage of solar energy. This prototype was developed for various latitudes and used classical construction materials, aluminum or steel frames, with ordinary or selective glass, simple or sandwich-type concrete walls for a better insulation and thermal storage. It could easily be adapted for prefabricated solar

FIG. 25

TABLE 3

Design Adaptations of 1976 Solar House for Three Climatic Zones in France

Region	ϕ	t_{design} (°C)	$\bar{G}$ (W/m³ °C)	Q_{tot} (kWh/year)	A_c (m²)	% solar heat
Paris	49° 30′	−5	1.18	12390	30	50
Strasbourg	48° 40′	−12	1.12	13175	45	50
Marseille	43° 20′	−3	1.39	11680	26	64

houses with three to six principal rooms. This house was designed for three climatic zones in France (Table 3).

19.7 METHODOLOGY FOR URBAN PLANNING

The urban planning and architecture for a given region and site with particular climatic and environmental conditions should be carried out on the basis of scientifically obtained data of various parameters.

The methodology considered here is based on the utilization and coordination of the three following programs:

1. The first program deals with climatic data obtained from meteorological stations for a period of several years using a computer program. For a given site, these data are analyzed to obtain hourly and daily values, and the deviations with respect to average yearly values. These data are used to ensure comfortable conditions in the home and to improve the urban environment. In this respect, renewable energy sources, such as solar and wind energy can be utilized fully to create a bioclimatic urban atmosphere and architecture, and to provide acceptable comfort conditions.

2. The second program deals with site evaluation based on low-altitude aerial photography. The data obtained on the sites, already urbanized or not, can be converted to numerical values and stored on magnetic computer tapes. With this information, and using an appropriate computer program, it is possible to simulate any urbanization project in three dimensions and to study various aspects at any angle, including the information obtained from the first computer program on climatic conditions. Hence, the position of the sun with respect to the buildings, the effects of shading, and possible convection currents for a given hour, day, or season can be simulated.

3. The third program encompasses experimental studies with a model built using the information from programs 1 and 2 in a wind tunnel.

This methodology can be utilized for regional planning and urbanization as well as for urban rehabilitation. The general program can be used effectively and objectively to study the above problems, to make the preliminary design, and then to improve the design by making necessary modifications for a given site and input data. In this way, the quality of life can be improved and the conservation of energy can be implemented.

19.8 CONCLUSION*

André Malraux has once said that the twenty-first century will be a century of metaphysics. Starting from this statement, let us go back and examine the rules of aesthetics in architecture in past centuries.

The western Middle Ages excellent in the dynamism of the asymmetry of forms juxtaposed freely with cubic, circular, and parallelepiped elements with a touch of lookout turret or machicolation in the necessary places.

The golden number was used by the masters of the art.

From ancient times until the utilization of the metric system, the golden number upset the tempo of utilization of such other systems as the inch, foot, and cubit; the 8-in. long brick, which is about half a cubit, represented an element of measurement at the work site on the human scale.

Architecture took into account not only the cosmic relation of the golden number but also climatological data; dwellings were open to the sun and turned their backs to the strong winds.

Villages and cities used the experience of centuries to their own benefit. The end result was a perfect integration with the environment; hence, the old cities offered architectural variety in which each occurring form was the result of a particular intention that blended in perfect harmony with the whole.

At the end of the Renaissance, Cartesian thinking was introduced, and architecture became more rigid and orderly. The magnificent ensembles of the seventeenth and eighteenth centuries brought about architectural points of view that were majestic and sometimes a bit abstract, in which the human scale was often ignored. The concept of climate was also forgotten. Nevertheless, residential architecture followed fundamental traditions, particularly in the country. In Europe the successful utilization of materials gave the designer a creative freedom that continued until the industrialization of construction in the 1950s.

At the beginning of the nineteenth century, engineers became king; one still admires the steel constructions of that epoch by Baltard and

* From Jacques Michel, *Conference on Solar Energy Fundamentals and Applications, Maracaibo, Venezuela, March 10, 1978.*

Eiffel. At the same time the fundamental choice of energy means, which derived from the options of steam engines and, later, internal combustion engines, made fossil fuel, at first coal and later petroleum and natural gas, an important commodity. Starting with this era, the thermal application of solar energy could possibly have influenced the quality of life and given it a more human direction. Unfortunately, the work of Buffon and Lavoisier in the eighteenth century and that of Muchot in the nineteenth century are now considered only works of scientific curiosity without any future likelihood of application.

By the end of the nineteenth century, the easy path of thermal engineering obliterated the meaning of natural elements in architectural concept and the psychology. In the name of rationalism, sites and landscapes have been destroyed, first by the railways and public works, later by the super highways.

Because of the systematic exploitation of natural resources, nature has become an enemy to be destroyed; in fact, the literature of the nineteenth century and the colonial conquests well illustrate this trend.

In 1925, the new spirit of Le Corbusier combined material with human proportion, utilizing natural light and polychromy, in structures built in the Mediterranean tradition. The Bauhaus school in Germany has revolutionized the forms of architecture with Mondrian-like figures, and the cubic influence directly introduced the architecture of the "curtain wall" facade. The students of Bauhaus who immigrated to the United States rationalized architecture in "orthogonal symbolism."

The real climatic architecture in the United States is in fact expressed by Frank L. Wright in an "indian language," in which color and cosmic space are blended with the landscape.

The academic architects of Europe, by demolishing the historical places, have started building towers without thermal inertia or solar protection, towers that are too expensive to heat during the winter months and to air condition during the summer months.

The reconstruction of Europe during the 1950s was architecturally catastrophic, insofar as prefabricated elements were widely used, often for economic reasons, in many reconstruction projects. Presently, in Europe, in the face of the dehumanizing effect of urban spaces, there is a return to the older architecture and urbanism; however, the real problems concerning the possible future of society and the fundamental choices regarding the quality of life have not been identified and tackled as required. Therefore, in order to prevent pollution and the total destruction of natural resources, the concerned ecologists have reacted by utilizing renewable resources, such as solar energy at low and high temperatures,

FIG. 26

wind energy, and hydraulic energy, as well as by recycling materials and engaging in biological agriculture.

In conclusion, following the thoughts of Malraux, the application of solar energy in architecture will not start if only technical aspects or economics are considered; it is obvious that the present studies and the Cartesian conclusions, reported in terms of the percentage of the energy needs of a country, are based on the trends of the consumer society of the last decade.

Following Le Corbusier, one should start from "*La tringle à rideau*" and design solar houses and buildings within human proportion and in accordance with the needs of people. The sun should be fully utilized to heat, ventilate, and illuminate the home in as ingenious a way as, for example, in the solar–bioclimatic house shown in Fig. 26. The design incorporates Trombe–Michel walls and a patio to utilize the solar energy, day–night heat exchange, and bioclimate, as a result of which year-round comfortable living conditions are created.

ACKNOWLEDGMENT

The architect of the solar houses described in the text and shown in Figs. 13–26 is Jacques Michel, D.U.H. The financial support of the National Research Council of Canada (to E. Bilgen) is acknowledged.

References

Allen, G. (1975). The Gananoque Solar House, *Proc. SESCI Conf., Ottawa, June* pp. III 65–69.

Anderson, B. (1976). "The Solar Home Book." Cheshire Books, Harrisville, New Hampshire.

Anonymous (1976). Bull. TD-5A, Teflon (R), E. I. duPont de Nemours Co., Wilmington, Delaware.

Anonymous (1976). Bulletin TD-5, Tedlar, (R), E. I. duPont de Nemours Co., Wilmington, Delaware.

Anonymous (1945a). *Pencil Points* 112.

Anonymous (1945–1946). *The Archit. Forum* **82,** 125–144.

Anonymous (1933). "British Building Research Board, Report for 1923," p. 87. H. M. Stationery Office, London, 1932.

Anonymous (1960). The Climate of Canada, Meteorological Branch, Dept. of Transport, Toronto, Ontario.

Anonymous (1968). Controlled Humidity Method with Hydrol Liquid, Section 14, Niagara Blower Company, March.

Anonymous (1971). Monthly Radiation Summary, Environment Canada, Government of Canada, U.D.C. 551.506.1 (71).

Anonymous (1974a). Can the Sun be Utilized to Heat Canadian Residences in the Winter? J. Wadsworth Development Group (Central Mortgage and Housing Corporation), May.

Anonymous (1974b). The Potential of Solar Energy in Meeting Canadian Energy Needs. Brace Research Institute, R.86, January.

Anonymous (1975a). *Proc. Workshop Solar Energy Storage Subsyst. Heating and Cooling of Buildings* American Society of Heating, Refrigeration and Air-Conditioning Engineers, New York.

Anonymous (1975b). Appropriate Building and Energy Systems for Quebec Indian Communities, Shelter Systems Group, School of Architecture, McGill University, December.

Anonymous (1976). Solar-heat utilizing experimental housing for practical use (Nangano Solar House), *TEMPTROL* **3,** No. 9.

Anonymous (1976a). Middleton Associates. Canada's Renewable Energy Resources, An Assessment of Potential, report commissioned by the Canadian Ministry of Energy, Mines and Resources, April.

Anonymous (1976b). Solar Heating and Cooling in Canada's Homes, Solar Energy Society of Canada Incorporated Newsletter, SOL 4, pp. 1–11, March.

Anonymous (1976c). U.K. Workshop on Heat Pumps, Lincoln College June/July Report published by Energy Technology Support Unit at Harwell.

Anonymous (1976d). All Weather Solar House Catering Cooling, Nikkei Architecture, Issue of May 31, 1976 (in Japanese).

Anonymous (1976e). New Zealand Official Yearbook. Dept. of Statistics, Wellington, New Zealand.

Anonymous (1976f). Wind, sun and waves, *One Day Symp. Lorch Foundat., October.*

Anonymous (1976g). BRE Current Paper, "Heat Pumps for Use in Buildings" Reference CP1976.

Anonymous (1976h). Solar Energy: A UK Assessment, UK-ISES, pp. (375).

Anonymous (1977a). Solar Activated Cooling Projects for Solar Heating and Cooling Applications, Energy Research and Development Administration, Request for Proposals, EG-77-R-03-1439.

Anonymous (1977b). Controls and Passive Systems for Solar Heating and Cooling Applications, Energy Research and Development Administration, Request for Proposals, EG-77-R-03-1443.

Anonymous (1977c). A house in Kokubunji, *Nikkei Architecture,* January (in Japanese).

Anonymous (1977d). "HUD Intermediate Minimum Property Standards Supplement 1977-Solar Heating and Domestic Hot Water Systems." U.S. Government Printing Office, Washington, D.C.

Argue, R., and McCallum, B. (1976). Environmentally appropriate housing, alternatives **5,** No. 3-4, 6–17.

Armet, A., and Nardini, A. (1976). Performance of a Passive Solar Wall, Brace Research Institute Publi. No. E.P. 22, April.

Arner, W. J. (1950). Personal communication, Tech. Dept., Libbey-Owens-Ford Glass Co., Toledo, Ohio, July 11.

Ashrae (1977e). "Handbook of Fundamentals." American Society of Heating, Refrigeration and Air-Conditioning Engineers, New York.

Bahadori, M. N. (1973). A feasibility study of solar heating in Iran, *Solar Energy J.* **15,** 3–26.

Bahadori, M. N. (1976a). Thermal energy storage, *Iran. J. Sci. Technol.* **5,** 159–171.

Bahadori, M. N. (1976b). Thermal Performance of Old Buildings in Central Iran Employing Air Towers, presented at *Passive Solar Heat, Cool Conf., Los Alamos Scientific Laboratory, Albuquerque, New Mexico, May.*

Bahadori, M. N. (1978). Passive cooling systems in Iranian architecture, *Sci. Am.* **238,** No. 2, 144–152.

Bahadori, M. N., and A. Kosari, (1977). Performance of the natural ice makers, to be presented at *Int. Solar Energy Conf., New Delhi, India, January.*

Baird, C. D., and Mears, D. R. (1976). Performance of a hydronic solar greenhouse heating system in Florida, *Proc. Solar Energy Food-Fuel Workshop, Tucson, Arizona.*

Balcomb, J. D., Hedstrom, J. C., and McFarland, R. D. (1977). (Simulation Analysis of Passive Solar Heated Buildings: Preliminary Results.) Los Alamos Internal Rep. LA-UR-1719, ERDA Contract W 7405-Eng 36.

Bannon, J. H., and Steele, L. P. (1960). Average water vapor content of the air, Meteorological Office, Geographical Memoir 13, No. 102, H. M. Stationary Office, London.

Ballantyne, E. R. (1973). Building design and solar energy, *Build. Int.* **6,** No. 5, 471–494.

Barkmann, H. G., and Wessling, F. C. (1975). Use of building structural components for thermal storage, *Proc. Workshop Solar Energy Storage Subsyst. Heat. Cool. Build., Charlottesville, Virginia* N SF-RA-N-041, pp. 136–140.

Bartoli, B. *et al.* (1976). Natural cooling: results and problems, *in* "Heliotechnique and Development" (M. A. Kettani and J. E. Sousson, eds.), Vol. 2. Cambridge, Massachusetts.

Bastings, L. (1974). Insulation and Heating of Buildings, Information Ser. No. 18, 2nd ed., NZ DSIR, Wellington, New Zealand.

Bilgen, E. A. (1976). Solar cooling, applications of solar energy, *Can. Plains Proc.* (*Regina*)' **3.**

Bilgen, E., and Jeldres, R. (1978). "Computer Modelling of the Solar House "Solab" Proc. Solar Energy Seminar." Maracaibo, Venezuela.

Bilgen, E., and Jeldres, R. (1978a). On the Optimisation of Trombe Wall Solar Collectors, AS.M.E. 78-WA, Solar Rep. 13.

Bilgen, E., Camous, R., and Trombe, F. (1977). "Chauffage Solaire et Climatisation des Bâtiments." Ed. Ecole Polytechnique, Montreal.

Bilton, T. *et al.* (1969). Atmospheric turbidity with the dual-wavelength sunphotometer, *J. Appl. Meterol.* **8,** 955–962.

Bird, R. B., Stewart, W. E., and Lingfoot, E. N. (1960). "Transport Phenomena," 1st ed. Wiley, New York.

Blackwell, M. J. (1954). Five years continuous recording of total and diffuse solar radiation at Kew Observatory, Meterol. Res. Publ. 985, Meterol. Office, London.

Bliss, R. W. (1964). The performance of an experimental system using solar energy for heating and night radiation for cooling, *Proc. UN Conf. New Sources Energy* **5,** 148.

Boldrine S., Lazzarin, R., and Sovrano, M. (1974). A Study of Infrared Radiation from the Atmosphere, Rep. No. 55, Fiscia Tecnica, Ingeneria, Padova.

Bolton, J. R. (1975). Solar energy—An important energy resource for Canada, *Chem. Can.* **27,** No. 8, 29–32.

Brinkworth, B. J. (1977). Refrigeration and air conditioning, *in* "Solar Energy Engineering" (A.A.M. Sayigh, ed.), Chapter 16. Academic Press, New York.

Brunt, D., (1932). Notes on radiation in the atmosphere, *I.Q.J.R. Meteorol. Soc.* **58,** 389–420.

Bultot, F. (1971). Atlas climatique du Bassin Congolais, Inst. Nat. pour l'Etude Agronomique du Congo, Democratic Republic of the Congo (I.N.E.A.C.).

Businger, J. A. (1966). The glasshouse (greenhouse) climate, *in* "Physics of Plant Environment" (W. R. Van Wijk, ed.), 2nd ed. North-Holland Publ., Amsterdam.

Cabanat, M., and Sesolis, B. (1976). Chauffage de l'Habitat par l'Energie Solaire. Emperimentations sur les Maisons C.N.R.S. d'Odeillo, Thesis, Univ. Paris VII.

Carman, P. C. (1956). "Flow of Gases through Porous Media," 1st ed. Academic Press, New York.

Catalanotti, S., Cuomo, V., Piro, G., Ruggi, D., Silvestrini, V., and Troise, G. (1975). The radiative cooling of selective surfaces, *Solar Energy* **17,** No. 2, 83–89.

Close, D. J. (1965). Rock pile thermal storage for comfort air conditioning, *Mech. Chem. Eng. Trans. Inst. Eng. Aust.* **MCl,** 11.

Close, D. J., and Dunkle, R. V. (1970). Energy storage using dessicant beds, *Int. Solar Energy Conf., Melbourne, Australia* Paper. 7/24.

Close, D. J., and Pryor, T. L. (1975). Energy storage in absorbent beds, *Int. Solar Energy Conf., Los Angeles, California* Paper No. 31/2.

Cramer, R. D., and Neubauer, L. W. (1965). Diurnal radiant exchange with the sky dome, *Solar Energy* **9,** No. 2, 95–103.

Croome, D. J., and Roberts, B. M. (1975). "Air-Conditioning and Ventilation of Buildings," 1st ed. Pergamon, Oxford.

Damagnez, J., Chiapale, J. P., Denis, P., and Jourdan, P. (1975). Solar greenhouse: New Process for heating cooling and water economy under greenhouses, *ISES Conf., Los Angles, California.*

Daniels, F. (1964, 1975). "Direct Use of the Sun's Energy." Yale Univ. Press, 1964 (Presently published as a paper-back by Ballantine Books, New York).

Dannies, J. H. (1959). Solar air conditioning and solar refrigeration, *J. Solar Energy* **3,** 34–39.

Deminet, C. (1976). Glass solar collectors for greenhouses and integrated greenhouse—Residential systems, *Proc. Solar Energy Food-Fuel Workshop,* Univ. of Arizona, *Tucson, Arizona* (M. H. Jensen, ed).

Dietz, A. G. H. (1954). Weather and heat storage factors, space heating with solar energy, *Proc. Course-Symp. M.I.T. August 21–26, 1950.* M.I.T., Cambridge, Massachusetts.

Dietz, A. G. H. (1963). Diathermanous materials and properties of surfaces, *in* "Introduction to the Utilization of Solar Energy" (A. M. Zarem, ed.). McGraw-Hill, New York.

Dietz, A. G. H. (1969). "Plastics for Architects and Builders," p. 71. M.I.T. Press, Cambridge, Massachusetts. From material supplied by Rohm and Haas Co., Philadelphia, Pennsylvania.

Drummond, A. J. (1973). "The Extraterrestrial Solar Spectrum" (A. J. Drummond and M. P. Thekaekara, eds.), p. 1. Institute of Environmental Science, Mt. Prospect.

Drummond, A. J., Hickey, J. R., Scholes, W. J., and Laue, E. G. (1968). *Nature* (*London*) **218,** 259–261.

Duffie, J. A., and Beckman, W. A. (1974). "Solar Energy Thermal Processes." Wiley, New York.

Duffie, J. A., and Beckman, W. A. (1976). A review of solar cooling, *Proc. Joint Conf. "Sharing the Sun 76,"* Winnipeg, Canada.

Dunkel, R. V. (1965). A method of solar air conditioning, *Trans. Mech. Chem. Eng. Inst. Eng., Aust.* **MCl,** 73–78.

El-Salam, E. M. A., and Sayigh, A. A. M. (1976). Estimation of diffuse solar radiation in the Arabian Penninsula, *IAMAP/WMO Symp. Radiat. Atmos., Munich, West Germany, August 19–28.*

Elterman, L. (1968). UV, Visible and IR attenuation for Altitude to 50 km, AFCR-68-0153. Office of Aerospace Research, U.S. Air Force.

Esbensen, T. V., and Korsgaard, V. (1976). Dimensioning of the solar heating system in the Zero Energy House in Denmark, *Proc. Conf. European Solar Houses, April,* UK-ISES.

Faber, O., and Kell, J. R. (1958). "Refrigeration for Air-Conditioning," 1st ed. Architectural Press, London.

French, P. W. (1950). Personal communication, Research Laboratories, Pittsburg Plate Glass Co., Creighton, Pennsylvania, July 13.

Frohlich, C. (1976). The solar constant: a critical revue, *IAMAP/WMO Symp. Radiat. Atmos. Munich, West Germany, August 19–28.*

Fukuo, A. *et al.* (1961). *UN Conference on New Sources of Energy, Rome, May.* Gabler, W. (1976). Neue sonnen energie nutzungsanlage, Private communication.

Gates D. M., and Harrop, W. J. (1963). Infrared transmission of the atmospheric to solar radiation, *Appl. Opt.* **2,** 887.

Gilman, S. F. (1976). Solar Energy Heat Pump System for Heating and Cooling Buildings, ERDA DOC COO-2560-1.

Gutierrez, G. *et al.* (1974). Simulation of forced circulation water heaters: Effects of auxiliary energy supply; Load type and storage capacity, *Solar Energy* **15,** No. 4, 287–298.

Hamilton, B. (1976). Preliminary Evaluation of an Experimental House at Manitou College, McGill School of Architecture, October.

Hamilton, B., and McConnell, R. (1976). Experimental evaluation of a solar house heating system in Quebec, *Proc. ISES/SESCI Conf., Winnipeg, August* **3,** 120–135.

Hardy, A. C., and Perrin, F. H. (1932). "Principles of Optics." McGraw-Hill, New York.

Hay, H. R. (1973a). Roof, ceiling, and thermal ponds, *Paris Conf., July* ISES Paper 41/16.
Hay, H. R. (1973b). The California solarchitecture house, *Paris Conf., July* ISES Paper EH 73.
Hay, H. R., (1973c). Housing, technology, and solarchitecture, *Am. Soc. Mech. Eng. Mech. Eng.* **11,** 95.
Hay, H. R. (1975). Roof ceiling and thermal ponds, *Int. Solar Energy Congr. Exposit., Los Angeles, California* I.S.E.S.
Hay, J. E. (1975). Solar energy utilization in Canada: A climatological perspective, *Proc. Solar Energy Soc. Canada (SESCI) Conf., Ottawa, June* pp. I-25–I-40.
Hay, J. E. (1976a). Climatological constraints of the development of solar energy in Canada, *Proc. Joint Am. Sect.* **4,** 258–270. International Solar Energy Society and Solar Energy Society of Canada, Inc. (ISES/SESCI) Conference, Winnipeg, August.
Hay, J. E. (1976b). The climatology of available solar energy for Canada, *Proc. ISES/SESCI Conf., Winnipeg, August* **1,** 211–225.
Hay, H. R., and Yellot, J. I. (1970). A naturally air conditioned building, *Mech. Eng.* January, 19–25.
Heap, R. D. A. (1975). Domestic Heat Pump Operation, a paper presented at the Seminar on Heating and Air Conditioning at Imperial College in June.
Heywood, H., (1965), The computation of solar radiation intensities, Part I, *Solar Energy* **9,** No. 4.
Heywood, H. (1966). The computation of solar radiation intensities, Part II, *Solar Energy* **10,** No. 1.
Hickey, J. R. (1976). *Proc. Joint Conf. "Sharing the Sun," Winnipeg, Canada.*
Higgin, R. M. R. (1976). Solar heating for building in Ontario, *Proc. ISES/SESCI Conf., Winnipeg, August* **3,** 212–227.
Hoffmann, E. W. (1975). Four years operation of a solar house, *Proc. SESCI Conf., Ottawa, June* pp. III 27–34.
Hooper, F. C. (1955). The Possibility of Complete Solar Heating of Canadian Buildings, *Eng. J.* November.
Hopkinson, R. G., *et al.* (1966). "Daylighting," Chapter 2, pp. 29–58. Heinemann, London.
Hottel, H. C., and Woertz, B. B. (1942). The Performance of Flat-Plate Solar-Heat Collectors, Solar Energy Conversion Project Publ. No. 3, M.I.T., Cambridge, Massachusetts, Transactions of the ASME, February. American Society of Mechanical Engineering, New York.
Houghton, F. C., Gutberlet, C., and Blackshaw, J. L. (1934). Studies of solar radiation through bare and shaded windows, *Trans. Am. Soc. Heat. Vent. Eng.* **40,** 101–116.
Hutchinson, F. W. (1945). *Heat. Vent.* 96–97.
Hutchinson, F. W. (1946). *Heat. Vent.* 53–57.
Hutchinson, F. W. (1947). *Heat. Vent.* 55–59.
Ishibashi, T. (1975). Yazaki experimental Solar House One, *ISES Meeting, Los Angeles, California, August* Paper 42/2.
Ishibashi, T. (1976a). Personal communication.
Ishibashi, T. (1976b). Yazaki experimental Solar House No. 1, Asahi Kuzuha solar house and hot water driven absorption refrigeration machine, *J. Soc. Heat., Air-Conditioning,* and *Sanit. Engs.,* **50** (4) (Japanese).
Ishibashi, T. (1976c). Kuzuha Solar House, *J. Jpn. Solar Energy Soc.* **2,** No. 2 (in Japanese).
Ishibashi, T. (1978). Ishibashi Solar House, *J. Soc. Heat., Air-Conditioning and Sanit. Eng.,* **52** (10) (Japanese).
Jagadish, B., and Peres, J. R. (1977). Etude de Cheminée Solaire, Tech. Rep., Laboratoire d'Energetique Solaire, C.N.R.S., Odeillo.

Jeldres, R. (1978). Etude d'un mur collecteur d'Energie solaire, Thesis, Ecole Polytechnique, Montreal.

Jeldres, R., Bilgen, E., and Vasseur, P. (1977). Etude Thermique d'un capteur solaire du type. "Mur Trombe", Rapport Tech. No. EP-77-R-52.

Jensen, M. H. (ed.) (1976). *Proc. Solar Energy Food-Fuel Workshop, Univ. of Arizona, Tucson, Arizona.*

Jensen, M. H., and Hodges, C. N. (1976). Residential environmental control utilizing a combined solar collector-greenhouse, *Proc. Solar Energy Food-Fuel Workshop, Univ. of Arizona, Tucson, Arizona.*

Johnson, T. E. *et al.* (1974/1975). Exploring Space Conditioning with Variable Membranes, Dept. of Architecture, M.I.T. Rep. for the Period January 1, 1974 to April 30, 1975. National Science Foundation, Research Applied to National Needs.

Jones, W. P. (1973). "Air Conditioning Engineering," 2nd ed. Arnold, London.

Jordan, R. C., and Liu, B. Y. K. (1977). "Applications of Solar Energy for Heating and Cooling Buildings," p. 206. American Society Heating, Refrigeration, and Air-Conditioning Engineers, New York.

Kasten, F. (1977). Der einfluss der bewölkung auf die kurzund langwelligen strahlungsflusse am boden, *Ann. Meteorol. Neue Folge* **12,** 65–68.

"Kathopanishad" (1950). *Bhawan's J.*

Keller, S. F. (1977). A New Generation of Material for Solar Collector Covers, Kalwall Corp., Manchester, New Hampshire.

Kendall, J. M. (1973). *Proc. Symp. Solar Radiat., Rockville* p. 190. Smithsonian Institution, Washington, D.C.

Kimura, K. (1975). Design and year-round performance of Kimura Solar House, *ISES Meeting, Los Angles, California, August.*

Kimura, K., and Miyazaki, T. (1976). Solar preheating system of factory ventilation, *Trans. AIJ* (October), 395–396 (in Japanese).

Kimura, K., and Tanaka, S. (1975). Solar heating, cooling and domestic hot water supply at Soka Solar House, *ISES Meeting, Los Angeles, California, August* Paper 41/2.

Klein, S. *et al.* (1975). A method of simulation of solar processes and its application, *Solar Energy* **17,** No. 1, 29–38.

Klein, S. A., Beckman, W. A., and Duffie, J. A. (1976). A design procedure for solar heating systems, *Solar Energy* **18,** 113–127.

Koizumi, N. N., Kawada, Z., Murasaki, H., Ito, T., and Matsui, T. (1976). Living performance of heating at Toshiba Solar House, *Tech. Meeting Jpn. Solar Energy Soc., 2nd, Tokyo, December* (in Japanese).

Kolor, V. (1965). Heat transfer in turbulent flow of fluids through smooth and rough tubes, *Int. J. Heat Mass Transfer* **8,** 639–683.

Kondratyev, K. Ya. (1969). "Radiation in the Atmosphere." Academic Press, New York.

Kondratyev, K. Ya, and Fedorova, M. P. (1960). Scattered and global radiation income to inclined surfaces in the presence of snow cover, *Vestn. Leningrad State Univ.* No. 16, 67–73.

Kondratyev, K. Ya., and Fedorova, M. P. (1976). Radiation regime of inclined surfaces, *UNESCO/World Meteorol. Organizat.-Solar Energy Symp., Geneva,* August 30–September 3.

Kondratyev, K. Ya., and Nikolsky, G. A. (1973). *Proc. Symp. Solar Radiat., Rockville* p. 203. Smithsonian Institution, Washington, D.C.

Kreith, F. (1965). "Principles of Heat Transfer." International Textbook Co., Scranton, Pennsylvania.

Kreith, F. (1973), "Principle of Heat Transfer," 3rd ed. Intext Educational Publ., New York.

Kudo, K., Noguchi, T., and Hiyoshi, K. (1976). Outline of solar cooling and heating system of Yorii Telegraph and Telephone Station, *Tech. Meeting Jpn. Solar Energy Soc., 2nd, Dec.*

Lane, G. A. *et al.* (1975). Heat of fusion systems for solar energy storage, Proc. Workshop Solar Energy Storage Subsyst. Heat. Cool. Buildings, Charlottesville, Virginia NSF-RA-N-75-041, pp. 43–55.

Lawand, T. A. (1975), Environmentally designed housing incorporating solar energy, heliotechnique and development, *Proc. Int. Conf. Dhahran, Saudi Arabia,* pp. 211–235, November 2–6 US Bureau Mines Bull. 504.

Lawand, T. A., Alward, R., Saulnier, B., and Brunet, E., (1974). The development and testing of an environmentally designed greenhouse for colder regions, *Proc. ISES (U.S. Section) Meeting, Fort Collins.*

Leva, M., Weintrab, M., Gunmer, M., Pollchik, M., and Strock, H. H. (1951). Fluid flow through packed and fluidized systems, US Bureau Mines Bull., 504.

Liu, R. C., and Carlson, G. E. (1976). Proposed solar greenhouse design, *Proc. Solar Energy Food-Fuel Workshop, Univ. of Arizona, Tucson, Arizona.*

Liu, B. Y. K., and Jordan, R. C. (1960). The inter-relationship and characteristic distribution of direct, diffuse and total solar radiation, *Solar Energy* **4,** 1–19.

Löf, G. O. G., and Hawley, R. W. (1948). Unsteady state heat transfer between air and loose solids, *Ind. Eng. Chem.* **40,** 1061.

Löf, G. O. G., and Tybout, R. A. (1974). The Design and Cost of optimized systems for residential heating and cooling by solar energy, *Solar Energy* **16** (1), 9–18.

Lorriman, D. P. (1975). Mississauga Solar House demonstration project, *Proc. SESCI Conf., Ottawa, June* pp. III 35–40.

McConnell, Beaudet, Piche, and Maille (1976). The use of off-peak electricity for solar heated homes, *Proc. ISES/SESCI Conf., Winnipeg, Canada, August,* **9,** 29–36.

McCormick, P. O. (1976). Performance of non-integral solar collector greenhouses, *Proc. Solar Energy Food-Fuel Workshop, Univ. of Arizona, Tucson, Arizona.*

McCormick, P. O. (1975). Analytical modeling group report, *Proc. Workshop Solar Energy Storage Subsyst. Heat. Cool. Buildings, Virginia, April 16–18* NSF-Sponsored.

Meyer, C. F., and Todd, D. K. (1973). Heat storage walls, *Water Well J.* October, 35–41.

Michel, J. (1973). *Archit. d'Aujourd'hui* **167.**

Michel, J. (1974). Annales de l'Institut Techique du Bâtiment et des Travaux Publics, "La Maison Solaire." Translation No. 623 (1975), "The Solar House."

Murcray, D. G. (1969). Balloon Boyne Measurement of the Solar Constant, Rep. No. AFRCL-6¼-0070, Univ. of Colorado, Denver, Colorado.

Nakahara, N., Miyakawa, Y., and Yamamoto, M. (1975). Experimental study on house cooling and heating with solar energy using flat plate collector, *ISES Meeting, Los Angeles, California, August* Paper 42/11.

Nakajima, Y., and Ohaski, K. (1977). Project and operation results of a solar house with long term heat storage using underground soil (I and II), *Tech. Meeting Jpn. Solar Energy Society, 3rd, Dec.* (Japanese).

Newton, A. B. (1976). Optimizing solar cooling systems, *ASHRAE J.* **18,** No. 11.

Nisen, A. (1969). "L'Eclairement Naturel des Serres." Les Presses Agronom de Gembloux, Paris.

Nishijima, S., *et al.* (1977). Operation results of solar hot water, heating and (cooling) system in SE Solar House, *Tech. Meeting Jpn. Solar Energy Soc., 3rd, Dec.* (Japanese).

Noguchi, N., Hiyoshi, K., and Kudo, K. (1976). Outline of solar heating and cooling system of Yorii Telegraph and Telephone Station, *Tech. Meeting Jpn. Solar Energy Soc., 2nd, Tokyo, December* (in Japanese).

Ohanessian, P. (1976). Numerical Modeling of a Passive Solar Energy House Heating System. M. Eng. Sci. Thesis, Univ. of Melbourne, Australia.

Ohanessian, P., and Charters, W. W. S. (1975). Theoretical Performance of a Natural Solar Energy Collection System for House Heating, Paper 40/15 ISES 75, Los Angeles, California, August.

Olgyay, V. (1963). "Design with Climate." Princeton Univ. Press, Princeton, New Jersey.

Page, J. K. (1961). The estimate of monthly mean values of daily total short wave radiation on vertical and inclined surfaces from sunshine records for lat. 40° N–40° S, *Proc. U.N. Conf. New Sources of Energy, Rome*.

Page, J. K. (1975). Supplementary note on values of D/K on cloudless days for solar altitudes below 30°, *Proc. Conf. U.K. Meteorol. Data and Solar Energy Appl.* pp. 37–39. UK-ISES.

Page, J. K. (1976). The estimation of monthly mean values of daily shortwave irradiation on vertical and inclined surfaces from sunshine records for Lat. 60° N–40° S, Building Science, Univ of Sheffield, Note No. 32, p. 33.

Page, J. K. (1977). Application of building climatology to the problems of housing and buildings for human settlements, WHO Tech. Note-150, Geneva, p. 64.

Page, J. K., and Nall, A. (1976). Solar Energy: A UK Assessment, UK-ISES.

Paltridge, G. W., and Platt, C. M. R. (1976). "Radiative Processes in Meteorology and Climatology," 1st ed. Elsevier, Amsterdam.

Paltridge, G. W., and Proctor, D. (1976). Monthly mean solar radiation statistics for Australia, *Solar Energy* **18,** 235–244.

Pandey, L. P. (1963). "Solar Worship in India." Allahabad.

Parmelee, G. V. (1945), The transmission of solar radiation through flat glass under summer conditions, *Trans. Am. Soc. Heat. Ventil. Eng.* **51,** 317–344.

Parmelee, G. V. (1954). Irradiation of vertical and horizontal surfaces by diffuse solar radiation from cloudless skies, *Heating, Piping, and Air Conditioning* August, pp. 129–135.

Pepper, C. D. (1976). Solar Heat Your Home. Privately published booklet.

Pickering, E. E. (1975). Residential hot water solar energy storage, *Proc. Workshop Solar Energy Storage Subsyst. Heat. Cool. Buildings, Charlottesville, Virginia* NSF-RA-N-75-041, pp. 24–37.

Plamondon, J. A. (1969). *JPL Space Program, Summary* **3,** 162.

Price, D. R., Wilson, G. E., Froehlich, D. O., and Crump, R. W. (1976). Solar heating of greenhouses in the Northeast, *Proc. Solar Energy Food-Fuel Workshop, Univ. of Arizona, Tucson, Arizona* (M. H. Hensen, ed.).

Prigmore, D. R., and Barber, R. E. (1975). Cooling with the sun's heat: Design considerations and test data for a rankine cycle prototype, *Solar Energy* **17,** 3.

Quirouette, R. L. (1975), Solar Heating, The State of the Art, Note 102, Division of Building Research, National Research Council, October.

Rabl, A., and Nielsen, C. E. (1975). Solar ponds for space heating, *Solar Energy* **17,** 1–12.

Ranz, W. E. (1952). Friction and Transfer Coefficient for Single Particles and Packed Beds, *Chem. Eng. Progr.* 48–247.

Robinson, N. (ed.) (1966). "Solar Radiation," p. 346. Elsevier, Amsterdam.

Rodgers, G. G., Page, J. K., and Souster, C. G. (1977). An interactive computer design methology for the design of solar homes, *Proc. NELP/UNESCO Conf., July* Paper 4-2.

Rush, W. F., Wurm, J., and Wright, L. (1975). A description of the solar-MEC field test installation, I.S.E.S. *Int. Solar Energy Congr. Exposit., Los Angeles, California*.

Russell, A. W. (1974). Savings in Home Heating Costs Resulting from a Solar/Oil Hybrid Heating System. Brace Research Institute Publ. Number S.P.4, November.

Sakai, I., Takagi, M., and Terakawa, K. (1975). Solar Space Heating and Cooling with Bi-Heat Source Heat Pump and Hot Water Supply System, *ISES Meeting, Los Angeles, California, August* Paper 45/1.

Sargent, S. L., and Teagan, W. P. (1973). Compression Refrigeration from a Solar-Powered Organic Rankine Cycle Engine, ASME Paper No. 73-WA/Sol-8.

Sasaki, J. R. (1975). Solar Heating Systems for Canadian Buildings, Building Research Note No. 104, Division of Building Research, National Research Council of Canada, December.

Sasaki, J. R. (1976). Recent Canadian activities in solar heating, *Proc. ISES/SESCI Conf., Winnipeg, Canada, August* **1** 106–109.

Sayigh, A. A. M. (1975a), Solar Energy Availability Prediction from Climatological Data, 2nd Course on Solar Energy Conversion. International College of Applied Physics, Naples, Italy, August.

Sayigh, A. A. M. (1975b). Saudi Arabia and its energy resources, *Int. Solar Energy Conf., COMPLES, Dhahran, Saudi Arabia* November.

Sayigh, A. A. M. (1976a). The energy prospects in the Arab world, *Int. Conf. Mech. Eng. Main Emphasis Energy Univ. of Engineering and Technology, Lahore, Pakistan, March.*

Sayigh, A. A. M. (1976b). The world energy situation and the Islamic countries, *Islamic Solidarity Conf. Sci. Technol., Riyadh, Saudi Arabia March.*

Sayigh, A. A. M. (1976c). Summer Night Cooling in Saudi Arabia, Solar Cooling and Heating, A National Forum, Miami Beach, Florida, December 13–15.

Sayigh, A. A. M. (1976d). Passive Cooling of Building, International College of Applied Physics, 3rd Course on Solar Energy Conversion, Catania, September 5–17.

Sayigh, A. A. M. (1976e). The world energy situation and the Islamic countries, *Islamic Solidarity Conf. Sci. Technol., Riyadh, Saudi Arabia, March.*

Sayigh, A. A. M. (1976f). Thermal Energy Storage for Buildings, 3rd course in Solar Energy Conversion, International College of Applied Physics, Catania, September 15–17.

Sayigh, A. A. M. (1977a). Thermal pile solar energy storage, *Izmir Int. Symp.—I on Solar Energy Fundamentals Appl., Izmir, Turkey, August 1–5.*

Sayigh, A. A. M. (ed.) (1977b). "Solar Energy Engineering," 1st ed. Academic Press, New York.

Sayigh, A. A. M. (1977c). The Technology of Flat Plate Collectors, Fourth Course on Solar Energy Conversion, International Centre for Theoretical Physics, Trieste, September 6–24.

Sayigh, A. A. M. (1977d). Estimation of total radiation intensities—A universal formula, *IAGA/AMAP Joint Assembly Conf., August 22–September 3, Seattle, Washington.*

Sayigh, A. A. M., and El-Salam, E.M.A. (1975). Preliminary Design Data for a Solar House in Riyadh, Saudi Arabia, *Int. Solar Energy Conf. COMPLES, UPM, Dhahran, Saudi Arabia, November 2 5.*

Sayigh, A. A. M., and Shaalan, M. R. (1976). Some Experimental Data for Thermal Pile, Solar Cooling and Heating, A National Forum, Miami Beach, Florida, December 13–15.

Scheidegger, A. E. (1957). "The Physics of Flow through Porous Media," 1st ed. Macmillan, New York.

Schlichting, H. (1960). "Boundary Layer Theory," McGraw-Hill, New York.

Selcuk, M. K. (1971). Analysis Design and Performance Evaluation of Controlled-Environment Greenhouses, *Trans. ASHRAE* Paper No. 2172.

Seshadri, T. R. *et al.* (1959). "Climatological and Solar Data for India." Sarita Prakashan, Meerut, India.

Severns, W. H., and Fellows, J. R. (1964). "Air Conditioning and Refrigeration." Wiley, New York.

Sharp/Eidai Experimental Solar House (1976). *TEMPTROL* **3,** No. 8, August.

Short, T. H., Roller, W. L., and Badger, P. C. (1976). A solar pond for heating greenhouses and rural residences—A preliminary report, *Proc. Solar Energy Food-Fuel Workshop Tucson* (M. H. Jensen, ed.).

Shurcliff, W. A. (1976). Solar Heated Buildings: A Brief Survey, 12th ed. 19 Appleton Street, Cambridge, Massachusetts.

Stein, D., and Wexler, B. M. (1974). Solar House Heating, Brace Research Institute, Publ. Number E.P. 5.

Stephenson, D. G. (1967). Tables of Solar Altitudes, Azimuth Intensity and Heat Gain Factors for Latitudes from 43° to 55° North. Technical Paper No. 243, Division of Building Research, National Research Council of Canada.

Swartman, R. K. (1976). Solar Homes: An Old Idea Revisited, *Alternatives* **5,** No. 3–4, 18–21, 29.

Swartman, R. K., Ha, V., and Swaminathan, C. (1974). Comparison of Ammonia-Water and Ammonia-Sodium Thiocynate in a Solar Refrigeration System, COMPLES, 1st Semester.

Swinbank, W. C. (1963). Long Wave Radiation from Clear Skies, *Q.J.R. Meteorol. Soc.* **89,** 339–348.

Szokolay, S. V. (1975). "Solar Energy and Building." The Architectural Press, London and Wiley, New York.

Tabor, H. (1963). Solar ponds—Large area solar collectors for power production, *Solar Energy* **7,** 189–194.

Tan, H. M., and Charters, W. W. S. (1970). An experimental investigation of forced-convective heat transfer for fully-developed turbulent flow in a rectangular duct with asymeteric heating, *Solar Energy* **13,** 121–125.

Tanaka, S., and Suzuki, T. (1970). Experimental study on radiant floor heating by solar energy, *Trans. AIJ* September 175–176 (in Japanese).

Tavassoli, M. (1975). "Architecture in the Hot Zone." Marvi Publ., Tehran, Iran (in Persian).

Telkes, M. (1974). Solar energy storage, *ASHRAE J.* **80B,** 38–44.

Temps, R. C., and Coulson, K. L. (1977). Solar radiation incident upon slopes of different orientations, *Solar Energy* **19,** 179–184.

Thekaekara, M. P., Kruger, R., and Duncan, C. H. (1969). *Appl. Opt.* **8,** 1713.

Thekaekara, M. P. (1975). Survey of quantitative data on the solar energy and its spectral distribution, *COMPLES Conf., Dhahran.*

Thompson, D. E. F. (1976). Evaluation of a Solar Heating Installation in Granton, Ontario, Master's Thesis, Faculty of Engineering, Univ. of Western Ontario.

Thompson, D. E. F., and Swartman, R. K. (1976). Solar retrofit of a home in Granton, Ontario, *Proc. ISES/ESCI Conf., Winnipeg, August* **4,** 92–104.

Tricaud, J. F. (1976). Contribution à l'Estimation des Resources Energétiques solaires, Thesis, Univ. Paris VII.

Trombe, F. (1971). Climatisation des Habitations, Bilan Schématiques des Réalisations, 1956–1972, Annexe B2, au rapport d'activités du Laboratoire, C.N.R.S.

Trombe, F. (1973a). Heating by Solar Radiation, CNRS Internal Rep. B-1-73-100.

Trombe, F. (1973b). Techniques Françaises. Bâtiments, *Trav. Publ. Urban.* **1,** 1–4.

Trombe, F. (1974a). *Tech. Ing.* **3,** C777.

Trombe, F (1974b). Maisons solaires, *Tech. Ing.* **3,** 1–5.

Trombe, F., and Michel, J. (1970). French Patents No. 2.144.066 and 2.189.686. See also U.S. Patent No. 3.832.992.

Trombe, F., and Michel, J. (1975). Night and Day Climatisation, French Patent No. 7532921.

Trombe, F., Robert, J. F., Cabanat, M., and Sesolis, B. (1976). Caracteristiques de Performance des Insolateurs Equipant Les maisons à chauffage Solaire du C.N.R.S., *Conf. Workshop Proc. Passive Solar Heating and Cooling, Albuquerque, New Mexico, May 18–19* LA-6637-C, pp. 201–222.

Tuller, S. E. (1976). The relationship between diffuse, total and extra terrestrial solar radiation, *Solar Energy* **18,** 259–264.

Uchiyama, T. (1976). Solar house, *J. Jpn. Solar Energy Soc.* **2,** No. 3 (in Japanese).

Udagawa, M., and Kimura, K. (1976). Design and Construction of Solar Space Heating and Hot Water Supply System for Multi-Family Housing, *ISES U.S. Section Meeting, Winnipeg, August.*

Unsworth, M. H. (1975a). Variation in the short wave radiation climate of the U.K., *Proc. Conf. U.K. Meteorol. Data Solar Energy Appl., February* UK-ISES.

Unsworth, M. H. (1975b). Long wave radiation at the ground, (ii) Geometry of interception by slopes, solids and obstructed planes, *Q. J. R. Meterol. Soc.* **101,** 25–34.

Unsworth, M. H., and Monteith, J. L. (1972). Aerosol and solar radiation in Britain, *Q. J. R. Meterol Soc.* **98,** 778–797.

Unsworth, M. H., and Monteith, J. L. (1975). Long wave radiation at the ground, Angular distribution of incoming radiation, *Q. J. R. Meterol. Soc.* **101,** 13–29.

Vale, B. R. (1976). "The Autonomous House," Thames and Hudson, London.

Villeneuve, G. O. (1967). "Sommaire Climatique due Quebec," Vol. 1. Ministere des Richesses Naturelles, Quebec, M-24.

Walsh, J. W. T. (1961). "The Science of Daylight," pp. 128–129. McDonald, London.

Walton, J. D., Jr. (1973). Space heating with solar energy at the C.N.R.S. Laboratory, Odeillo, France, *Proc. Solar Heat. Cool. Buildings Workshop, Washington, D.C.* Part I, NSF-RA-N-73-004, pp. 127–139.

Willson, R. C., and Stallkamp, J. A. (1971). Radiometer Comparison Tests, Rep. No. 900–446, Jet Propulsion Laboratory, Pasadena, California.

Yanagimachi, M. (1961). Report on two and half year's experimental living in Yanagimachi Solar House II, *UN Conf. New Sources Energy, Rome, May.*

Yellott, J. I. (1973). Utilization of sun and sky radiation for heating and cooling of buildings, *ASHRAE J.* December.

Yellott, J. (1976). Solar radiation and the atmosphere, *Proc. Conf. Workshop Passive Solar Heat. and Cool. Los Alamos Scientific Laboratory, Los Alamos, New Mexico, May 18–19.*

Zarem, A. M., and Erway, D. D. (1963). Introduction to the Utilization of Solar Energy, Univ. of California, Engineering and Science Extension Series.

Zornig, H. F., Davis M., and Bond, T. E. (1976). Design criteria for greenhouse-residences, *Proc. Solar Energy Food-Fuel Workshop, Univ. of Arizona, Tucson, Arizona.*

Index

I

J

K

L

M

N

O

P

Q

R

S